Also in the Collected Studies Series

JOHN GASCOIGNE
Science, Politics and Universities in Europe, 1600–1800

SILVIO A. BEDINI
Patrons, Artisans and Instruments of Science, 1600–1750

Y. TZVI LANGERMANN
The Jews and the Sciences in the Middle Ages

JERRY STANNARD
Herbs and Herbalism in the Middle Ages and Renaissance

JERRY STANNARD
Pristina Medicamenta: Ancient and Medieval Medical Botany

ANDREW WEAR
Health and Healing in Early Modern England: Studies in Social and Intellectual History

DONALD R. HILL
Studies in Medieval Islamic Technology: From Philo to al-Jazari – from Alexandria to Diyar Bakr

EDWARD S. KENNEDY
Astronomy and Astrology in the Medieval Islamic World

IAN INKSTER
Technology and Industrialisation: Historical Case Studies and International Perspectives

IAN INKSTER
Scientific Culture and Urbanisation in Industrialising Britain

DAVID M. KNIGHT
Science in the Romantic Era

DAVID OLDROYD
Sciences of the Earth: Studies in the History of Mineralogy and Geology

W.H. BROCK
Science for All: Studies in the History of Victorian Science and Education

JOHN BROOKE
Thinking about Matter: Studies in the History of Chemical Philosophy

VARIORUM COLLECTED STUDIES SERIES

The 'Creed of Science' in Victorian England

Professor Roy MacLeod

Roy M. MacLeod

The 'Creed of Science' in Victorian England

Ashgate
VARIORUM

Aldershot · Burlington USA · Singapore · Sydney

Published in the Variorum Collected Studies Series by

Ashgate Publishing Limited
Gower House, Croft Road,
Aldershot, Hampshire GU11 3HR
Great Britain

Ashgate Publishing Company
131 Main Street,
Burlington, Vermont 05401
USA

Ashgate website: http://www.ashgate.com

ISBN 0–86078–669–2

British Library Cataloguing-in-Publication Data
MacLeod, Roy, 1941–
The 'Creed of Science' in Victorian England.
(Variorum Collected Studies Series: CS 598).
1.Science – England – History – 19th Century.
I. Title.
509.4'2'09034

US Library of Congress Cataloging-in-Publication Data
The Library of Congress Catalog Cardnumber has been preassigned as:
97–076867

The paper used in this publication meets the minimum requirements of the American National Standard for Information Sciences – Permanence of Paper for Printed Library Materials, ANSI Z39.48–1984. ∞ ™

Printed by St Edmundsbury Press, Bury St Edmunds, Suffolk

VARIORUM COLLECTED STUDIES SERIES CS598

CONTENTS

This volume contains xxiv + 318 pages

ACKNOWLEDGEMENTS

As with the first volume of my essays, this second owes much to the encouragement of John Smedley and Ruth Peters, to whom I am greatly indebted. Most of these essays were written whilst I held appointments at the University of Sussex, which encouraged a bold look at the 'map of learning', and occasional experiments in education, too. I remain grateful for the comradeship of my colleagues in the History and Social Studies of Science, who were among the pioneers in the new social history of science. For the studies of Victorian periodical publishing, I am grateful to the inspiration of John Maddox, then the Editor (not, as was said of his more autocratic predecessors, the Author) of *Nature*, who commissioned these essays for the centenary of *Nature* in 1969. The silver-bound issue in which they first appeared is now a collector's item. However, their unusual journalistic format made for difficulties in rendering them into conventional book form, so we have word-processed them, with a few important corrections, with the result that they are more readable (if less striking) in appearance.

For the research that led to the several essays on Victorian Cambridge, the Natural Sciences Tripos, and the 'examination revolution', I am grateful for the generous support I received from the (then) Social Sciences Research Council. These essays formed part of a larger whole, from which several other essays were produced. I am pleased to find they have been widely cited, and that the methodology I developed with Dr. Russell Moseley has since been applied to other studies in historical sociology.

The final essay derives from work done not at Sussex, but at Sydney where, since 1982 I have held a chair of History. Over the years, I have been grateful to my Australian colleagues for encouraging me to look at what E.P. Thompson famously called the 'peculiarities of the British', and to ask how their understanding can be enriched by an approach from the periphery. In this case, the story is one which also appealed to my 'co-conspirator', Professor Richard Jarrell, from York University, Ontario; and which both hybridizes and reconciles the language of difference in the history of imperial and colonial science.

For permission to reprint these essays, I am grateful to the (I) the Royal Society of London; (II) the Council of the British Society for the History of Science, London; (III) Sage Publications, Inc., Thousand Oaks, California; (IV)

Elsevier Science, Amsterdam; (V) Frank Cass & Co Ltd., Newbury Park, Essex; (VI) Taylor & Francis, London; (VII) Carfax Publishing Co., Abingdon; (VIII) Nafferton Books, Driffield, Yorkshire; (IX, X, XI) Macmillan Magazines, Ltd., London; (XII) Canadian Science and Technology Historical Association. I am also grateful to Professor W.H. Brock and Dr Russell Moseley for permission to reproduce articles which were originally co-authored with them. Finally, I wish to thank Ms Jill Barnes, who, amidst many competing obligations, helped work the text for final production.

ROY MacLEOD

Sydney
December, 1999

INTRODUCTION

If H.G. Wells were to wake up at the beginning of a new millennium, whether in Pimlico or Parramatta, he would perhaps be surprised by the extent to which science is established, yet contested, in everyday life. While still a student of T.H. Huxley in London, Wells sensed that modern physics and biology would soon ask whether one needed a purposeful Creator to make a world. All that was left of Christmas, it was said, was the pudding. Wells wrote at the end of a century of unparalleled enthusiasm for the promises of science. If Matthew Arnold, as T.S. Eliot put it, had set Culture in the place of Religion, Huxley's students promised to make Science the pacemaker of Culture. Henceforward, it seemed, the principal question before mankind, was to be 'how', rather than 'why'.

Today, at the close of a century that opened with such pride in the triumph of science over belief, we confront a future whose outlines are unclear, its dimensions, uncounted. With the transformational power of information technology, our mental landscapes are now as different from those of Huxley and Wells as is the Earth from Mars. We have limitless faith in ingenuity. Yet, we appear to have lost faith in progress. Science – and its potentially liberating applications – has not produced an end to war, but new weapons of mass destruction. Scientific rationality has not erased the fundamentalism, nationalism and ethnic hatreds that threaten domestic peace and international stability. Poverty, hunger and malnutrition are still insults to the intelligence of institutions. A vocabulary of 'distorted development' and 'underdevelopment' has been coined to describe the life experienced by much of the human race, despite the advances of science and the triumphs of global capitalism.

A century ago, such a prospect was a contradiction in terms. Scientific rationality held within it the seeds of salvation. The applications of science would transcend superstition, overcome poverty, and proclaim, in Tennyson's famous lines, the coming 'Parliament of Man/The Federation of the World'. Future governments were to be led by men of reason, peopled by the parties of progress. Knowledge – scientific knowledge – would be the currency of power. Class would not disappear, perhaps, nor divisions by race. But the usages of science would make these better understood, and their relations, more mutually beneficial. This, in language that scarcely concealed its evangelistic zeal and adventure, was the optimistic message of the 'creed of science', and of its secular manifesto. Its coming would be led, in a missionary metaphor of its own choosing, by a 'church militant' of scientists, a 'movement' of scientific ideas and methods that, by undermining revealed religion, would unite all mankind in a common reading of the Book of Nature.

Since the Enlightenment, natural philosophers had sought to substitute rationality for superstition, demonstrable propositions for traditional authority, self-evident truths for revelation, and happiness for salvation. The tests of the eye and the hand, of observation, experiment, and practice, would bring victory for light over darkness. 'Merchants of light', in Bacon's memorable phrase, would consolidate the revolution, based upon a systematic view of nature – incorporating rationality, objectivity, experimental method, and the mathematization of nature's laws. By the middle of the 19th century, scientific naturalism included three features: a universal and neutral model of science, a distinction between knowledge and belief, and a belief in science as the ultimate yardstick against which to measure civilizations and their place in human history.[1] Science rejected as superstition the practices of non-Western peoples and the teachings of the Church. Science was unique, with a unity that reflected the unity of nature. It was universal, and found universality in both Nature and Reason. This universality was dominated by hierarchies of power which, in the context of the time, were typically white, European, bourgeois and male.

If science was a method, arguably, it was also an end in itself. In 1874, Francis Galton expressed the hope that a scientific priesthood might arise, whose'high duties would have reference to the health and well-being of the nation in its broadest sense...'. Such was the familiar representation, or caricature, of those 'modern prophets' – 'professors of Atheism', according to William Allingham in 1877 – who were carrying all before them. It was, of course, too broad a brush. Few 'prophets' were united in their prophecies, fewer were atheists. For some, including T.H. Huxley, a new language of 'agnosticism' had to be invented. Against the established clerical order, it was common to engage in a kind of inverted snobbery. Adrian Desmond recalls a moment during T.H. Huxley's celebrated visit to America, when, before a New York audience, a preacher refused to call him an atheist. 'These scientists' he put it, 'are nothing of the kind'. On the contrary, 'They have a religion of their own'.[2]

'Without earthquake or tornado', Allingham put it, such 'scientists' (even those who, like Huxley, disliked the term) were creating a 'Creed of the Future'.[3] Those who saw in their very existence a danger to Religion were propelled by a sense of urgency that is impossible to overstate. The sense of an unprecedented crisis – political, moral, religious, intellectual – with potentially disastrous consequences for public order, reverberates through the meetings of the celebrated Metaphysical Society.[4] Materialism was the enemy and, as recent French critics argued, the prelude to Socialism. The dangers of both were real enough to those who chose to find them in John Tyndall's famous Belfast Address to the British Association in 1874, and in his earlier support of Henry Thompson's proposed experiment on the value of prayer.[5] It was Tyndall, possibly more pantheist than materialist, who was widely viewed as the doyen of the scientists' movement;[6] just as he was 'dean' of that celebrated cathedral of science at the Royal Institution, where lecturers celebrated the theory of conservation of energy, the revelations of spectrum analysis, and the marvels of the 'new chemistry'. Yet, Huxley is remembered as 'the Devil's disciple', for his constant interrogrations of tradition.[7] The devil was already within the gates, if one took literally as catechism W. Kingdom Clifford's question, 'What is the domain of Science?' and his answer: 'It is all possible human knowledge which can rightly be used to guide human conduct'.[8] If science was displacing theology, and assuming sovereignty over ethics, organized religion was unnecessary, or worse. As Clifford put it, in a phrase, 'the Priest is always and everywhere the enemy of Mankind'.[9] By implication, the Scientist was Mankind's friend.

Such projections, whether promises or portents, proved wide of the mark. The contest can be constructively seen as a conflict not just (or not even) between science and religion, but between two cultures, two ways of thinking. Indeed, this is how many contemporaries viewed it.[10] Just as the 'scientists' movement' reached its rhetorical peak in the 1870s and 1880s – the high-water mark was, perhaps, Victoria's triumphal Jubilee in 1887[11] – critics rose against the idea of any culture having a monopoly on truth. As Allingham put it, 'The Dogmatism of Science in human affairs is to be as strictly guarded against as the Dogmatism of Theology, and Science has of late unquestionably shown a tendency to overbear and tyrannize in the domain of Human Thought'.[12] In 1881, William Graham's eponymous *The Creed of Science: Religious, Moral and Social,* promised to give readers 'the chief conclusions reached by Modern Science on the central questions of religion, morals and society'.[13] But within its first few pages, it was clear that physicists or naturalists had been pushed off a moral pedestal. It was vital, Graham observed, to distinguish between scientific faith and scientific fact. Science occupied one domain; moral authority, another. On questions of morals and ethics, men (and women) of science, *per se,* had no special authority; outside the world of experiment and observation, if there *were* a creed of science, it, too, was false.

Within a decade, the context changed. In the 'heavy quarterlies', there came about a fashionable vogue for the description of 'scientific fallacies' and shortcomings, while 'antis' of differing persuasions reduced the proconsuls of science to the status of 'unfeeling' brutes.[14] At another level, the seeming arrogance of the 'modern prophets' that had inflamed and outraged the public, was replaced by a more commonsense approach to the marvels of technology that did not threaten to replace religion or the established order. In Britain, compromise fueled consensus. Some leading scientists, such as George Gabriel Stokes, sought to reconcile revealed and natural religion.[15] Others agreed to disagree – to let scientists go their way, and metaphysicians, theirs. The situation was more problematic in France, where government engaged the 'cult' of science, embraced literary naturalism, elevated men of science to high position, and awarded positivism a special prominence in the educational system, and came under attacks from the idealist right that shook the establishment.[16] As an essay in this volume (No. III) describes, French scientists were obliged to regroup, and in some cases, to redefine their position, in an atmosphere in which scientific ideas remained politically charged. At the same time, the promise of natural knowledge as a pathway to indefinite progress was severely tested, by arguments from both biology and thermodynamics.[17] In his much-discussed Romanes Lecture of 1893, Huxley mooted the possibility that evolution, the governing process of the universe, did not, and could not, in itself, produce what High Victorians could confidently call material 'progress'; and counseled a more careful interpretation of nature as applied to politics.[18] The essence of what has been called his 'masterpiece of concealed debate' was an argument that rejected socialism as an evolutionary outcome, and justified a Liberal imperialist foreign policy instead.[19]

In 1894, Huxley marveled at the extent to which the old conflicts of science and religion had subsided.[20] A truce of sorts had been declared in the 'culture wars'. In 1911, *The Nation* looked to what it called the current 'plebification' of science, and recalled nostalgically those 'conspicuous representatives of science' forty years before, 'ready enough to assume the robes which it was thought theology was on the eve of casting off'. It had been, the leader writer concluded, 'in the main...a wholesome and educative influence'.[21] In England at least, scientific naturalism had distanced itself not only from ethics, but from revolution as well. Indeed, by the late 1890s, the 'scientific movement' entered a new dimension, in which debates with religion were deferred, apparently without resolution, whilst attention turned the applications of science to industry and the interests of global commercial and strategic capital. Following the Treaty of Berlin, Britain and the Empire were confronted by new and more powerful international and colonial rivalries. The rapid industrialization of Germany and the United States, swept Britain up in what contemporaries saw as a race among nations, in which the 'survival of the fittest' was to be measured by success in achieving 'national efficiency'. In this, the methods of science were the essential instruments. The rhetorical translation of science and its creeds, from the threatening language of materialism and socialism, to the instrumental language of management had begun.

As instrument and ideology, science proved an active element in European colonialism, and was a central feature in the 'second wave' of European colonial expansion that reached a crest in the 1890s. If, in Britain, the state was content to leave science to private individuals and local enterprise, colonial governments were more engaged in applying scientific knowledge, both directly and through their public officials. Whether in India, Africa, Asia or Australasia, a creed of *science sans doctrine* came into common use, and endorsed policies that assured European dominion over inanimate and animate nature, and conveyed the right to shape and regulate colonial life.

If Britain's dominions, Australia, New Zealand, India, Canada and South Africa, took eagerly to science and its applications – often discounting the endless debate between science and religion, as incapable of producing practical solutions to practical problems – the 'scientific movement' remained largely a metropolitan push. For all its obvious success in raising debate, in proselytizing for evolution, in seizing the high ground of public education, it remained far less successful at the business of government. The debacle of the South African War revealed a Britain that had failed to absorb and apply the first principles of scientific rationality in its machinery of government.[22] A new set of 'ginger groups' was required, and rose to the occasion, as the newly established British Science Guild, with its programme of public science, came to the fore.[23] It was to *Nature,* born in the spirit of the 1860s and driven until his death in 1920, by its founding editor, Norman (later Sir Norman) Lockyer, that the new generation of reformers looked for leadership. The blame for Britain's economic failure was set at the feet of Britain's traditional institutions – parliament and government, education and industry – which had seemingly failed to embrace science either as creed or craft. Reformers chose their targets at will, variously singling out the shortcomings of national education, the rigidities of public examinations, and the dominance of Oxbridge and the Classics in the higher reaches of government. British industry seemed to enjoy a culture that favoured the effortless gentleman amateur over the industrious player.[24] In each, the intellectual interests of science were evident, and evidently manipulated, by a wider culture that accepted and elevated science to a rhetorical pinnacle, and gave it public space, but remained uneasy of its public acceptance.

-II-

Within this broad framework of negotiated naturalism, with its many promises and panaceas, rivalries and recriminations, the essays in this volume were conceived. In their origins, they marked the reopening of historical interest in the relations of science and sensibility, of science and cultural authority in the shaping of late-Victorian England. As such, they attempted to depart from the historical language of 'conflict' that, for over a generation, and in parallel literatures had popularized the virtues of science, martyred by religion; and celebrated the rise of the 'professional' scientist over the cowering amateur.[25] Conflict, like the Kuhnian emphasis on 'revolution', has figured prominently in the vocabulary of historians of science. However, social and cultural historians, perhaps more accustomed to 'normal' science – in the phrase also popularized by Kuhn – looked for, and found, factors disposing more towards conservativism and consilience with the establishment. Rather than outright, 'Big Bang' conflicts, there was more steady-state controversy, as different positions jostled for cultural leadership. Thus an intellectual revolution could be achieved without a political revolution – indeed, even without over-rapid change. Unlike Marx, the object for English naturalists was not to change the world, but to understand it. Acceptance of new ideas was a process that was incremental, politically Peelite – evolutionary, not revolutionary. In this process, emphasis unavoidably turned as much to matters of professionalism, codes, and the curriculum, as to the fine structure of scientific theory or experiment. The new intellectual culture was Broad Church and centre-right, a position not ill-suited to those wishing to secure influence, whether in the cloisters and corridors of Westminster and Whitehall, or in the interlocking directorships of the universities and learned societies. Personal influence would lead to increased state aid to research, and perhaps position in government. By extension, it would bring science to industry, and to everyday life.

In this sense, these essays marked an attempt to bridge the interests of social and cultural historians, and those of an emerging generation of social historians of science. In the phrase of the hour, 'externalist' historians of science were looking at ways to see why people who believed as they did, acted as they did – the history of ideas in action.[26] The rules of engagement were less immediately concerned with the ideological dimensions of science, and its failures to achieve its Enlightenment promises – clear though these were, and elsewhere addressed – than with the sources of dynamism and creativity shown by the 'clerisy' of science, in the language of Carlyle, or of the episcopacy, in the phrase of his critics.[27] This leadership inherited forms which it adapted and replayed, and from which it re-created the social and institutional basis of the modern British 'scientific establishment'. That establishment would have its own cathedrals. Darwin was, after much discussion, buried in Westminster Abbey.[28] Yet, when, in 1903, the time came to honour Herbert Spencer, the surviving members of the X-Club were divided. Lubbock and Hooker supported a proposal for a commemorative plaque in the Abbey, but Huxley opposed; in his view, the Abbey was not a pantheon, but a place of worship.[29] The appeal failed. The establishment of science would choose and celebrate its heroes and martyrs in its own way.

-III-

The period canvassed by these essays begins in the mid-1860s – with England, still in the shadow of the Great Exhibition – and continues through four decades, in which the common context was convulsed by *Essays and Reviews*, the post-Darwinian debate, the second industrial revolution, the coming of electricity, steam printing, national secondary education, and colonial imperialism. The twelve essays begin with a study of the famous X-Club – what I called the 'Albemarle Street Conspiracy', the network sometimes seen by historians as a coterie of friends and allies; sometimes, an elitist caucus for professional reform.[30] In either case, from well before its establishment in 1864, it formed a party of science *avant la lettre* – certainly, a party of principle well before Jules Ferry adopted 'scientisme' as the political ideology of the Third Republic. [31]

This essay has an interesting history. The X-Club (or X-network, as some prefer) had been known for many years, thanks to several marmoreal, double-decker High Victorian 'Life and Letters' written by or for its members. It was not until the early 1970s, however, that its more general significance became widely appreciated. By lifting it from its semi-biographical isolation, and casting it into the language of a 'ginger group' within the emerging 'scientific movement', I suggested that it was an important 'research site', a place of conversation that became a centre of calculation. Its nine members – T.H. Huxley, John Tyndall, J.D. Hooker, Herbert Spencer, George Busk, Edward Frankland, John Lubbock, William Spottiswoode and T.H. Hirst – in many respects shared a common view of the world, and appealed to a Broad Church of support (if not always complete agreement) on history, nature and other subjects amenable to naturalistic explanation.[32] In its original form, this paper caught the crest of a wave of interest in such social formations in British science. Appearing, unexpectedly, at nearly the same time (but from a different perspective) as a paper by a North American colleague,[33] it became the point of departure for at least one American doctoral dissertation, several papers,[34] and at least one major revisionist interpretation of the 'conflict' between science and religion.[35] In the phrase of James Moore, it was 'the most powerful coterie in late-Victorian science.[36] By 1970, its day had come.

The idea of a 'club' of English naturalists was in itself hardly new; and indeed 'clubs' were part of the vernacular of English science since the seventeenth century. The learned societies of London continued the idea, in glittering gatherings, dinners and *conversazione,* for which the Philosophical Club and the Royal Society Club, among others, were justly celebrated. But the historiographical juxtaposition of the 'X-Club', as a sub-political party of reform, within the established social and political context of London, and during the 'Age of Reform' that preoccupied British governments between 1830 and 1870, helped promote a better understanding of the way that science in Britain was actually managed, promoted, encouraged and supported.

The X-network was not, it must be stressed, unique; on the contrary, historians would see it as one in a long line of reforming networks in English science, of which the Cambridge Analytical Society is a well-known, earlier example.[37] It was also not the only voice of reform, which had been a constant theme since Babbage and the 'Declinist' school of the 1830s, and its more recent advocates in the Philosophical Club of the 1840s.[38] Insofar as there were interesting differences, these groups promoted institutional reform, without necessarily sharing a common agenda. With the 'Nine', however, one senses a new empathy, and a larger vision, informed by common assent to the values of scientific naturalism. The friend's remarkable, almost spiritual alliance, saw action on several fronts, the contours of which are suggested in the 'Scientist's Declaration'(No. II) dealing with an incident that involved future members of the Club, was contemporary with its creation.

Throughout this study, the membership network, and not simply the 'club' itself, was the significant reality; because, however fascinating as an entity, the 'Nine' were only one wing of the larger 'scientific movement' in Victorian England.[39] At the same time, its success in bringing about reforms in science, and scientific reforms in society – whilst tempting to assert – is often difficult to prove. Such reforms came about not so much through direct action, as by 'permeation'; such was the case with the movement for the 'endowment of research' and the entry of leading publishers into the field of scientific publishing. However, in the case of the International Scientific Series (No. IV), one does find a spectacular hallmark of the X-network. This influential corpus of 'good books on great ideas' was an enterprising attempt at literary social Darwinism, and used an evolutionary metaphor to link all human knowledge, from music to metaphysics. Published more or less simultaneously in four countries and three languages, mobilizing a potentially vast American and Continental as well as British and Empire audience, it epitomized the way in which the X-network influenced publishing interests, and how those interests in turn intersected and permeated the educational and cultural structure of Britain.

If the International Scientific Series (ISS) launched some of the 'Great Books' of scientific naturalism, it was even so not the most radical literary venture in which the X-network participated. It has been variously argued that at least some of the 'Nine' advocated Darwinism because it legitimated their cultural ambitions.[40] Possibly a more reasonable interpretation would view them as strong individuals, committed, to varying degrees, to an evolutionary understanding of 'Man's Place in Nature'. Whatever their ambitions, their influence was undoubtedly formative – and not only behind the scenes – in the establishment of a popular readership for science – in several quasi-literary journals, but especially in the most famous of British science journals. *Nature,* published by Macmillan from 1869, was not the first,[41] but – for reasons that I explain (Nos. X and XI) – became the best produced, best-sold, most widely-read, and (as survival was a principal objective) one of the most enduring journals in modern science. For at least the next hundred years, *Nature* became both the medium and the message of the 'scientific movement' throughout the English-speaking world

Following the ISS, five essays (Nos. V, VI, VII, VIII, IX) turn to an examination of the scientific movement at work in higher education. Since D.S.L. Cardwell's pioneering work,[42] the social history of science in Britain been greatly enriched by studies not only of institutions and organizations, but also of interests and practices – accounts of which contributed in the 1970s and 1980s to the emerging sociology of scientific knowledge. In the last decade, the changing roles of scientists as 'actors', as trusted agents of enquiry, as agents of political and intellectual change, have become issues central to our understanding of science.[43] Historians have cultivated the 'research sites' of experiment and observation, the expedition and the laboratory, as well as the sites of representation in the library and the museum. Among these, we find 'centres of calculation', which, in Bruno Latour's useful phrase, have helped order the content, rate and direction of scientific thinking, and consolidate its recognition and reward.[44] In this context, I argued that the extension of public education was as significant as the success of experiments or expeditions in shaping the cultural authority of science, and the scientific consciousness of Victorian Britain.

These essays begin with No. V, a summary account of the leading public debates surrounding the extension of scientific and technical education in the late nineteenth century. This essay, originally commissioned for an Irish University Press reprint of the relevant British Parliamentary Papers, attempts to show how the reception of science changed during a turbulent and often traumatic period in Britain's cultural and economic history. The place of Britain as the 'workshop of the world', celebrated in the Great Exhibition of 1851, was thereafter threatened by the rise of industrial Germany and the United States. In response, the support of science, – and more important, the increased diffusion of scientific methods into industry and commerce – were seen as the answer, repeated, with variations in emphasis, to a succession of government enquiries. From giving evidence, members of the X-Club were not spared, although it was not evolutionary theory, nor scientific naturalism that caught the headlines, but the need for practical applications of biology, physics and chemistry, and for the applied sciences underlying technology and engineering.

To historians familiar with the long-term economic performance of the British economy, the prospect of Britain enjoying a science-led economic recovery now seems, at best, wishful thinking. The issue, however, remains perennial, and with the advent of new information technologies, has again achieved prominence in British political and educational circles. It remains to be seen whether the cultural constraints and characteristics discussed so fervently in the 1890s say as much today, a century on, to Europeans who in many ways are more affected by the manipulations of corporate globalization and political harmonization than by national or international competition.

Whether science, however lacking in spirituality, was to be the means of salvation, the place of science in education was secure. It was made so by the diffusion of science into the system of public, national and university examinations. The examination 'revolution' in Victorian England has been called a 'silent revolution'. In the context of these studies, however, it is appealing to see this development far more in contextual and evolutionary terms. This is the subject of the next three papers (VI, VII and VIII). Examinations were the hallmark of Victorian liberalism and free trade, and the extension of opportunity to 'all the talents'. At the same time, they were battlefields for new disciplines, seeking access to 'intellectual space' on the 'mental maps' of learning.

These essays trace the gradual advance of science, its frontier skirmishes, and its ultimate, sovereign settlement within the 'curricular space' formerly occupied by the mathematics and the classics in Cambridge during the second half of the nineteenth century. They discuss – in what today looks like rather heroic prose – the successes of scientific reformers (some of whom, like members of the X-Club and its extended network, actively supported the principle of 'external examiners') in taking a prominent role in the examination process throughout the country. It will be correctly observed that the history of the university on the banks of the Cam is in many ways unique – not least in relation to its sister on the Cherwell – and generalizations about scholarly practices, even within Cambridge, are notoriously difficult to sustain. Nevertheless, we felt secure in the belief that in the negotiation of intellectual boundaries and the arbitrage of ideas, Cambridge was an venue at which to explore questions that were uppermost in the minds of contemporary educators throughout the kingdom.

Having selected Cambridge, Russell Moseley and I then chose to examine the Natural Sciences Tripos, which gained for science an expanding presence within previously closed structures. Since these essays were published, scholars have explored more deeply and widely the causes and consequences of the 'examination revolution', and its significance for British educational thought and curricular development. Twenty years on, however, we still await comparative accounts for Oxford, and for the Scottish and provincial universities.[45] When they come, they will undoubtedly reflect the vocabulary of class, codes and classification, in the language of Basil Bernstein; and cultural reproduction, after Pierre Bourdieu.[46] The issues these essays engage speak to the transformation of an ostensibly static curriculum into one that remains, to this day, expanding in diversity and changing in style.

In retrospect, these essays had a 'meta-purpose'. Outside the more notable personalities and well-known research sites – such as the Cavendish and the Clarendon – the 'idea of the university' has not often benefited from the attentions of historians of science. A possible explanation lies in the observation that historians of science are typically interested more in the origins of scientific ideas than in their dissemination through science education – more in the context of discovery, to paraphrase Karl Popper, than in the context of justification. However, in defining scientific practice, the emergence of new disciplines, and the response of educational institutions to external pressures, the 'examination idea' affords a useful window, revealing a great deal about the ways in which the sciences have assumed, established, and defended their cultural authority.

As ideology, craft and practice, bent on demystifying the world, the 'creed of science', the project of scientific naturalism, the ideology of the scientists' movement, became mission, mandate, and metaphor for knowing all that could be known about Man, Nature and Society. Between 1860 and 1900, it embraced a new set of mentalities and curricular geographies; it became a feature of government and an ally of industry. Its dynamics fired the imagination and fuelled research; its institutions transformed travel and trade, agriculture and commerce, art and architecture, exploration and discovery, landscape painting and literature.[47] The 'creed' did its work on the imperial periphery, too, establishing the cultural pre-eminence and prestige of Western science, and the dominance of its assumptions, throughout the settler colonies (No. XII) and the non-Western world. Science wrote the manifesto of modernism, and if it seemed to reduce God to the role of a benevolent but distant uncle, it would remain for future generations to rediscover the limitations of human beings, and to reawaken interest in metaphysics.

Today, there are many new creeds in science.[48] Knowing the 'mind of God,' in Stephen Hawking's phrase, is not yet reducible to common sense, but within the next century, the 'scientists' movement' will undoubtedly press for greater clarity. If the advance of science has so far brought in its wake, in the phrase of Max Weber, the 'disenchantment of the world', it has been perhaps a small price to pay. But whether science will ever redeem the promises of Bacon, and see natural knowledge used exclusively for the relief of Man's estate – as many mid-Victorians fervently hoped, and expected – raises hostages to a future that no one at the end of the Millennium can safely predict.

Endnotes

1 *See* Michael Adas, *Machines as the Measure of Men. Science, Technology, and Ideologies of Western Dominance* (Ithaca and London: Cornell University Press, 1989).

2 For Galton, *see* Trancis Galton, *Englishmen of Science: Their Nature and Nurture* (London: Macmillan, 1874), 360. Adrian Desmond, *Huxley: Evolution's High Priest* (London: Michael Joseph, 1997), 98.

3 [William Allingham], 'Modern Prophets', *Fraser's Magazine*, XVI (ns), (September 1877), 273–292. *See also* Sydney Ross, '*Scientist*: The Story of a Word', *Annals of Science,* 18 (1962), 65–85.

4 On which the best source remains: Alan W. Brown, *The Metaphysical Society: Victorian Minds in Crisis, 1862–80* (New York: Oxford University Press, 1944). *See also* Robert Young, 'The Impact of Darwin on Conventional Thought', in A. Symondson (ed.), *The Victorian Crisis of Faith* (London: SPCK, 1979), and his influential survey, 'The Historiographic and Ideological Contexts of the Nineteenth-Century Debate on Man's Place in Nature', in Robert M. Young and M. Teich, (eds), *Changing Perspectives in the History of Science: Essays in Honour of Joseph Needham* (London Heineman, 1973), 344–438.

5 Frank Turner, 'Rainfall, Plagues and the Prince of Wales: A Chapter in the Conflict of Religion and Science', *Journal of British Studies,* 13 (1974), 46–65; Christopher Hamlin, 'Providence and Putrefaction: Victorian Sanitarians and the Natural Theology of Health and Disease', in Patrick Brantlinger, (ed.), *Energy and Entropy: Science and Culture in Victorian Britain* (Bloomington: Indiana University Press, 1989), 93–123.

6 *See* Ruth Barton, 'John Tyndall, Pantheist: A Rereading of the Belfast Address', *Osiris,* 3 (second series), (1987), 111–134.

7 Adrian Desmond, *Huxley: The Devil's Disciple* (London: Michael Joseph, 1995).

8 W.K. Clifford, 'The Ethics of Religion', *Fortnightly Review*, 22 (ns), (July 1877), 35–52.

9 Clifford, Ibid. *See also* Clifford's influential, 'On the Aims and Instruments of Scientific Thought', *Macmillan's Magazine*, 26 (1872), 499–512.

10 A good example is Thomas Sinclair, *The Mount: Speech from the English Heights* (London: Trubner, 1878), who coins the adjective 'scientistic' in his enthusiasm to dispose of 'prevailing monsters'.

11 *See* Grant Allen, 'The Progress of Science from 1836 to 1886', *Fortnightly Review*, 41 (1887), 868–884.

12 Allingham, op. cit. 292.

13 William Graham, *The Creed of Science: Religious, Moral and Social* (London: Kegan Paul, 1881).

14 *See*, e.g., Ouida [Louse de la Ramé], 'Some Fallacies of Science', *North American Re-*

view, 142 (1886), 139–152; *also* Richard French, *Antivivisection and Medical Science in Victorian Society* (Princeton: Princeton University Press, 1975). *See also*, Leo Tolstoy, 'The Law of Force and the Law of Love', *Fortnightly Review*, 85 (1909), 461–485 and 689–719.

15 David B. Wilson, 'A Physicist's Alternative to Materialism: The Religious Thought of George Gabriel Stokes', in Brantlinger, (ed.), op. cit., 177–205.

16 *See also* Harry W. Paul, 'The Debate over the Bankrupcy of Science in 1895', *French Historical Studies*, 5 (Spring, 1968), 299–327.

17 Grey Myers, 'Nineteenth-Century Popularizations of Thermodynamics and the Rhetoric of Social Prophecy', in Brantlinger, (ed.), op. cit. 307–339.

18 Thomas H. Huxley, *Evolution and Ethics* (London: Macmillan, 1893). For a compelling contemporary critique of prevailing assumptions, *see* Julia Wedgwood, *Nineteenth Century Teachers and other Essays* (London: Hodder and Stoughton, 1909), 'Ethics and Science', 304–323.

19 Michael S. Helfand, 'T.H. Huxley's "Evolution and Ethics": The Politics of Evolution and the Evolution of Politics', *Victorian Studies*, 20 (2), (Winter 1977), 159–177. *See also* H.C.G. Matthew, *The Liberal Imperialists: The Ideas and Politics of a Post-Gladstonian Elite* (Oxford: Oxford University Press, 1973).

20 *See* David Newsome, *The Victorian World Picture: Perceptions and Introspections in an Age of Change* (London: John Murray, 1997), 213.

21 [Anon], 'Scientists and the Masses', *The Nation*, 92 (4 May 1911), 441.

22 Geoffrey Searle, *The Quest for National Efficiency* (Oxford: Basil Blackwell, 1971); and *Eugenics and Politics in Britian, 1900 – 1914* (Leyden: Norrdhoff International Publishing, 1976).

23 Roy MacLeod, 'Science for Imperial Efficiency: Reflections on the British Science Guild, 1905–1936', *Public Understanding of Science*, 3 (1), (1994), 155–194.

24 One of the more celebrated analyses of this phenomenon is Martin Wiener, *English Culture and the Decline of the Industrial Spirit, 1850–1980* (Cambridge: Cambridge University Press, 1981).

25 Sheridan Gilley, 'The Huxley-Wilberforce Debate: A Reconsideration', in K. Robbins, (ed.), *Religion and Humanism: Studies in Church History* (Oxford: Basil Blackwell, 1981), 325–340.

26 For more recent collections of essays, similarly devoted to a 'contextualist' historiography of Victorian science, but separated by another decade, *see* James Paradis and Thomas Postlewait, (eds), 'Victorian Science and Victorian Values: Literary Perspectives', *Annals of the New York Academy of Sciences*, vol 360 (New York: The New York Academy of Science, 1981) and Bernard Lightman, (ed.), *Victorian Science in Context* (Chicago: University of Chicago Press, 1997).

27 *See* Frank Turner's engaging, 'Victorian Scientific Naturalism and Thomas Carlyle', *Victorian Studies*, 18 (1975), 325–343, and his other essays *in Contesting Cultual Authority: Essays in Victorian Intellectual Life* (Cambridge: Cambridge University Press, 1993).

28 James R. Moore, 'Charles Darwin lies in Westminster Abbey', *Biological Journal of the Linnaean Society*, 17 (1982), 97–113.

29 Hannah Gay, 'No "Heathen's Corner" Here: The Failed Campaign to Memorialize Herbert Spencer in Westminster Abbey', *British Journal for the History of Science*, 31 (1998), 41–54.

30 James R. Moore, 'Deconstructing Darwinism: The Politics of Evolution in the 1860', *Journal of the History of Biology*, 24 (1991) 353–408.

31 Following the defeat of France in the Franco-Prussian war, *scientisme*, denoting the superiority of an 'objective' view of Nature and Reason, became a dominant ideology, popularized by positivists, including Ernest Renan, who spoke about the need 'to scientifically organize humankind'. *See* Ernest Renan, 'Dialogues Philosophiques', in *Oeuvres complètes* (Paris: Calman-Lévy, 1947), vol 1, 599: Renan's influence upon English men of science and letters is often cited, but little

discussed. In the 1930s, neo-positivism brought these ideas back to the intellectual scene. I am indebted to Patrick Petitjean for making this connection.

32 For a recent view of disagreements, as well as agreements, among the 'X', *see* James Strick, 'Darwinism and the Origin of Life: The Role of H.C. Bastian in the British Spontaneous Generation Debates, 1868–1873', *Journal of the History of Biology*, 32 (1999), 51–92.

33 J. Vernon Jensen, 'The X Club: Fraternity of Victorian Scientists', *British Journal for the History of Science*, 5 (1970), 63–72; 'Interrelationships within the Victorian X Club', *Dalhousie Review*, 51 (1972), 539–552; '"The Most Intimate and Trusted Friend I Have": A Note on Ellen Busk, Young T.H. Huxley's Confidante', *Historical Studies*, 17 (1977), 315–322; and his *Thomas Henry Huxley: Communicating for Science* (Newark: University of Delaware Press, 1991).

34 Ruth Barton, 'The X- Club: Science, Religion and Social Change in Victorian England', (Unpublished PhD dissertation, University of Pennsylvania, 1976), from which have arisen (among other papers) 'An Influential Set of Chaps: The X-Club and Royal Society Politics, 1864–85', *Isis*, 23, (1990), 53–81; and 'Huxley, Lubbock and Half a Dozen Others: Professionals and Gentlemen in the Formation of the X-Club, 1861–1864', *Isis*, 89 (1998), 410–444. *See also* Andrew Harrison, 'Scientific Naturalists and the Government of the Royal Society, 1860–1900', (Unpublished PhD dissertation, Open University, 1988).

35 Frank M. Turner, 'Victorian Scientific Naturalism', in *Between Science and Religion: the Reaction to Scientific Naturalism in Late Victorian England* (New Haven: Yale University Press, 1974); and 'The Victorian Conflict between Science and Religion: A Professional Dimension', *Isis*, 69 (1978), 356–376.

36 James R. Moore, 'Theodicy and Society: The Crisis of the Intelligentsia', in Richard J. Helmstadter and Bernard Lightman, (eds), *Victorian Faith in Crisis: Essays on Continuity and Change in Nineteenth Century Religious Belief* (London: Macmillan, 1990), 172.

37 Susan Faye Cannon, *Science in Culture: The Early Victorian Period* (New York: Dawson/ Science History Publications, 1978).

38 Roy MacLeod, 'Whigs and Savants: Reflections on the Reform Movement in the Royal Society, 1830–1848,' in Ian Inkster and J.B. Morrell, (eds), *Metropolis and Province: Science in British Culture, 1780–1850* (London: Hutchinson, 1983), 55–90. *See also* I.R. Morus, 'Correlation and Control: William Robert Grove and the Construction of a New Philosophy of Scientific Reform', *Studies in the History and Philosophy of Science*, 22 (1991), 598–621.

39 The term acquired some currency among contemporaries, and soon had its own historian. *See* A.W. Rücker, 'Professor Tyndall and the Scientific Movement', *Nature*, 36 (923), (7 July 1887), 217–218.

40 *See* Barton, (1998), 415; *see also* Moore, 'Deconstructing Darwinism', op. cit. 375–376.

41 For recent work on its precursors, *see* Ruth Barton, 'Just before *Nature*: The Purposes of Science and the Purposes of Popularization in some English Popular Science Journals of the 1860s', *Annals of Science*, 55 (1998), 1–33.

42 D.S.L. Cardwell, *The Organisation of Science in England: A Retrospect* (London: William Heinemann, 1957; 1972).

43 Much of this early work was led by scholars at Edinburgh University, of whom the most important was Steven Shapin; *see*, for example, his 'The Politics of Observation: Cerebral Anatomy and Social Interests in the Edinburgh Phrenology Disputes', in Roy Wallis (ed), *On the Margins of Science: The Social Construction of Rejected Knowledge*, Sociological Review Monograph 27 (Keele: University of Keele, 1979), 139–178.

44 Bruno Latour, *Science in Action: How to Follow Scientists and Engineers through Society* (Cambridge, Mass: Harvard University Press, 1987), ch 6, 'Centres of Calculation', 215–257.

45 Comparative work is now underway by Christopher Stray. *See* his important *Classics*

Transformed: Schools, Universities and Societies in England, 1830–1960 (Oxford: Clarendon Press, 1998). Others are being encouraged by Jonathan Smith and the members of the Teaching and Learning at Cambridge group, based at Trinity College, Cambridge.

46 Basil Bernstein, *Class, Codes and Control* (London: Routledge and Kegan Paul, 1971-5), 3 vols; and *Class and Pedagogies: Visible and Invisible* (Paris: OECD, 1975); *Pedagogy, Symbolic Control, and Identity: Theory, Research, Critique* (London: Taylor and Francis, 1996); Pierre Bourdieu and Jean-Claude Passeron, *Reproduction in Education, Society, and Culture* (London: Sage, 1977). *See also* Derek Robbins, *The Work of Pierre Bourdieu: Recognising Society* (Bletchley: Open University Press, 1991).

47 For a valuable introduction to a range of influences, *see* U. C. Knoepflmacher and G.B. Tennyson, (eds), *Nature and the Victorian Imagination* (Berkeley: University of California Press, 1977).

48 *See*, for its entire Victorian optimism and contradictions, E.O. Wilson, *Consilience: The Unity of Knowledge* (New York: Knopf, 1998).

PUBLISHER'S NOTE

The articles in this volume, as in all others in the Collected Studies Series, have not been given a new, continuous pagination. In order to avoid confusion, and to facilitate their use where these same studies have been referred to elsewhere, the original pagination has been maintained wherever possible.

Each article has been given a Roman numeral in order of appearance, as listed in the Contents. This number is repeated on each page and quoted in the index entries.

I

THE X-CLUB
A SOCIAL NETWORK OF SCIENCE IN LATE-VICTORIAN ENGLAND

AMONG the more significant features of nineteenth-century British science was the emergence of social and professional networks which helped to shape or influence the course of scientific activity. By and large, the most conspicuous arbiters of scientific values were the learned societies, the British Association and the universities. These formal bodies, however, almost always reflected the attitudes and assumptions of small, informal, often obscure and sometimes anonymous, clusters or networks of individuals. Until recently, most of these groups have been overlooked or neglected (1), but growing interest in the social history of science has stimulated fresh research into the history of scientific societies generally, and into scientific *élites* in particular. The 'X-Club', recalled by the contemporary American historian, John Fiske, as 'the most powerful and influential scientific coterie in England' (2), was one of the most important and instructive of these groups. The following pages will inquire into the origins, the development, and the probable significance of this 'coterie' in the social climate of late-Victorian science.

It is well known that clubs for discussion over dinner and port wine have played an important part in the development of scientific life in Britain (3). Many existing professional or semi-professional associations and scientific bodies began as dining clubs, several during the so-called 'Age of Clubs' of the late eighteenth century. Reflecting a spontaneous desire by men of similar tastes and interests to meet colleagues and friends socially, 'to prove their title to good fellowship' (4), the informal club became the accepted social context for the exchange of new information and ideas, literary, political and scientific.

The importance of these social functions did not disappear, when, with the passage of years, informal fraternities matured into formal societies. But formal institutions did not always provide the environments in which leading men of science could count on both a 'feast of reason and a flow of soul' (5). Moreover, friendships and rivalries within the scientific 'community' rendered large, infrequent and conventional scientific meetings inappropriate for the discussion of confidential matters affecting the conduct and reward of scientific work, and the reform of scientific institutions.

In response to these different needs, several informal, or at least 'unofficial', groups were created. Among them were the Philosophical Club begun in 1897 as a 'ginger group' seeking reform within the Royal Society. At the opposite extreme was the Red Lion Club, founded by Edward Forbes, at the British Association meeting in Birmingham in 1839 as a 'protest against dons and donnishness in science' (6). As one participant recalled, the Red Lions held feasts of 'spartan simplicity and anarchic constitution, with rites of Pantagruelistic aspect, intermingles with extremely unconventional orations, and queer songs . . . by way of counterblast to the official banquets of the British Association . . .' (7). To some distinguished foreign visitors, the sight of grave English professors waving and wagging their coat tails by way of applause, was, at best, an unusual experience. The Red Lions carried on intermittently until the First World War. But this form of gathering tended to degenerate into mere socializing. It lacked the dimension of serious professional discourse that some men of science felt necessary.

A more serious and more specialized attempt to create this dimension was made by a group of chemists about 1865. This group formed the 'B-Club' from Section B (Chemistry) of the British Association in the belief that 'there were other ways of promoting the welfare of chemical science than those contemplated by the more formal society' (8). The 'hives of B's' as they called themselves, met once a month during the London season, and combined specifically chemical gossip with friendly sessions of poetry and word-play. But neither the Red Lions, nor more specialized professional clubs were sufficiently intimate to satisfy the small group of men who became known as the 'Young Guard' of Victorian science.

At the Cheltenham meeting of the British Association in 1856, Joseph (later Sir Joseph) Hooker (1817-1911), Director of Kew Gardens, met young Thomas Huxley (1825-1895) and urged him to help enlist botanists and zoologists to 'discuss some plan that would bring about more unity in our efforts to advance science' (9). 'As I get more and more engrossed at Kew', Hooker continued, 'I feel the want of association with my brother Naturalists, especially of such men as you, Busk, Henfrey and Carpenter.

> We never meet except by pure accident, and seldom then as naturalists, and if we want to introduce a mutual friend, it is only by cut and thrust into one another's business hours. It is the same with our publications . . . Without some recognised place of resort that will fulfill the conditions

> of being a rendezvous for ourselves, an inducement to our friends to take an interest in Natural History, and at the same time a profitable intellectual resort—we shall be always ignorant of one another's whereabouts and writings' (10).

The Red Lions, 'admitting the rag-tag and bobtail of literature and the Arts, together with the dregs of scientific society' could never meet Hooker's conditions. 'We want some place where we should never be disappointed of finding someone or something worth going out for' (11).

Huxley agreed. 'Otherwise', he wrote anxiously to Hooker, 'I wonder if we are ever to meet again in this world' (12). Huxley himself provided the answer. In January 1864, he suggested a regular meeting to unite his closest friends (13). After preliminary discussions, a dinner was held on Thursday, 3 November 1864, at St George's Hotel, Albemarle Street, to launch the new enterprise. On 7 November, Herbert Spencer reported the event to his father:

> In pursuance of a long-suspended intention, a few of the most advanced men of science have united to form a small club to dine together occasionally. It consists of Huxley, Frankland, Tyndall and Hooker, Lubbock, Busk, Hirst and myself. Two more will probably be admitted, but the number will be limited to ten (14).

'There is no knowing', Hirst wrote in his diary, 'into what this club, which counts amongst its members some of the best workers of the day, may grow . . .' (15). Certainly, this November evening brought together a group who would, for nearly a quarter of a century, wield far-reaching influence on the style and conduct of English science.

They were first of all, friends of long standing. Edward (later Sir Edward) Frankland (1825-1899) met John Tyndall (1820-1893) at Queenswood College, Hampshire, about 1847, soon after Tyndall began teaching mathematics and surveying. Both left for Germany in 1848 to study with Bunsen at Marburg and at Berlin, and both received Ph.D. degrees there. Frankland returned in 1850 to take up Playfair's chair at Putney College, and then went to the first chemistry chair at the new Owen's College in Manchester in 1851. Tyndall returned briefly to Queenswood, and then moved to the Royal Institution in 1853. In 1858, frustrated by repressive conditions in Manchester, Frankland moved to London and to a post at St Bartholomew's Hospital. In 1863, with Tyndall's help, he moved to the Royal Institution, where he spent the next six years—'the happiest in my life', he later recalled (16).

Thomas Archer Hirst (1830-1892) met Tyndall in 1846, when both were boys articled to Richard Carter, a land agent and surveyor of Halifax in West

Yorkshire. Hirst followed Tyndall to Marburg in 1849, took his Ph.D., and completed his education with Gauss and Weber at Göttingen, and with Jacob Steiner at Berlin. In 1853, he briefly succeeded Tyndall at Queenswood College, and in 1860 became maths master at University College School in London. In 1865, he was made professor of physics at University College, and in 1867, on the death of De Morgan, succeeded him as professor of pure mathematics (17).

Tyndall and Huxley met at the Ipswich meeting of the British Association in 1851, and Hooker joined their circle in 1857. Spencer met Huxley at the British Association in 1852. Through Huxley he met Tyndall in 1853. Around 1861 Spencer became acquainted with George Busk (1807-1886) and his wife, and soon became a close friend of the family. Busk, a surgeon by profession, had been the first naturalist to encourage Huxley on his return from the *Rattlesnake* Expedition in 1850. Busk was quiet and reserved, and much less well known to the outside world than to the scientific societies in which he participated. Mrs Busk was remembered as being 'scientifically cultivated in a degree rare among ladies'. It was through the Busks that, in 1862, Spencer met Sir John Lubbock (1834-1913). Lubbock had been a friend of Huxley since 1856, when Huxley had encouraged him to stand for the Royal Society, and helped to secure his election (18). Lubbock, in turn, met Hooker in the Botanic Garden at Oxford, during the British Association meeting of 1860 (19).

Initially, therefore, the eight came together from three 'sets' of friends—those of Huxley, Spencer and Tyndall. The eight had much in common. All but Lubbock were of middle-class origin. All were 'of one mind on theological topics' (20), and, in Leonard Huxley's phrase, 'animated by similar ideas of the high function of science in this country'. In 1864, all but Lubbock lived and worked in London, and Lubbock's home at Lamas, in Chislehurst, was not far away. Frankland, Tyndall and Huxley taught at the Royal Institution; Spencer, after 1866, lived near Lancaster Gate. Hirst lived in Marylebone, and Hooker at Kew.

All, except Lubbock and Busk, were approaching middle age and had comparatively recently made their mark in the scientific world. In 1864, Lubbock, the youngest, was thirty; Hirst was 34; Frankland and Huxley were 39, and Spencer and Tyndall were 44. Hooker was 47, and Busk the oldest was 57. With the exception of Hooker and Spencer, the eight were among the 'new men elected to the Royal Society under the revised statutes of 1847. Hooker became an F.R.S. in 1847, Busk in 1850, Huxley in 1851, Tyndall in 1852, Frankland in 1853, Lubbock in 1858 and Hirst in 1861. Hooker had been one of the original members of the Philosophical

Club with Bowman and Grove. Huxley and Tyndall became members in 1855, and Busk, Treasurer in 1863. But they and the rest were, in 1864, still outside the Royal Society establishment.

Of the eight who attended the first dinner, only three represented the physical sciences. It was logical that the next member should come from the physical sciences as well, and the following month William Spottiswoode (1825–1883), mathematician and friend of Tyndall, was elected (21). Spottiswoode was the same age as Huxley and Frankland, and was, like Frankland, elected to the Royal Society in 1853. A tenth member was considered but for some reason never appointed. As Spencer later recalled, no one was found who fulfilled the two requirements—that he should be of adequate mental calibre and that he should be on terms of intimacy with the existing members. 'These meetings after all, were intended to be social gatherings of friends.' Eventually the subject of a tenth member was dropped (22).

At the first dinner it was agreed to meet subsequently on the first Thursday of each month except during July, August and September. Thursday was chosen because it was the day of the Royal Society meeting when all would normally be 'up in town'. Dinner began at six, so as to finish in time for the Royal Society at 8.00 (23). By stages, their dinners moved from Albemarle Street to Almond's Hotel, Clifford Street, and after Spottiswoode's death in 1886, to the Athenaeum, where all remaining eight were members (24).

The Club, according to Spencer, had no rules, except to have none. Each member held in turn the offices of secretary and treasurer, with the duty of collecting accounts, sending notes of meetings, and, according to Frankland, keeping rough notes of all proceedings (25). The question whether formal minutes should be taken was apparently considered, but ultimately abandoned. In November 1885, a resolution ordering the Treasurer to keep regular notes was defeated, when Tyndall pointed out that such a rule would violate the agreement against having rules of any kind (26). Informal notes, however, were kept of several meetings by Thomas Hirst, with assistance from Huxley, Hooker and Tyndall. The first eight of these appear in Frankland's biography, but the notebooks themselves have recently been discovered among the Tyndall papers at the Royal Institution (27).

Soon arose the question of a name. Several alternatives were suggested —the 'Thorough Club' was one; the 'Blastodermic', referring to the part of the ovum where the rudiments of animal organization first appear, was another. In the end, Spencer recalls, Mrs Busk suggested they end the matter by calling themselves the 'X-Club'. All approved. 'Beyond the advantage that it committed us to nothing' (28), as Spencer said, this name could secretly give notice of meetings by use of postcards bearing the notation 'X=6', or whatever date the first Thursday of the month fell upon. In a more romantic vein, Cyril Bibby believes the 'X' stood in triple symbolism—for the undecided name of the Club, for the originally envisaged ten members and for the undetermined tenth who was never chosen (29).

Between 1865 and 1880 the Club enjoyed its most exciting phase—particularly during 1870-1878, when Hooker (P.R.S.), Huxley (Sec. R.S.) and Spottiswoode (Treas. R.S.) all held office in the Royal Society at the same time. During these years the members of the X-Club also began to win their first scientific awards. In the course of the next twenty years, the nine earned five Royal Medals, three Copley Medals, one Rumford Medal, two Darwin Medals, and one Lyell and one Wollaston Medal from the Geological Society. Among them they received eighteen honorary doctorates, one Order of Merit, and one Prussian 'Pour la Mérite'. Two received knighthoods, one became a Privy Councillor and one a Justice of the Peace. Three became Corresponding Members and one a Foreign Associate of the Académie des Sciences (30).

The social importance of the Club equalled its scientific eminence. It was perhaps an 'Albemarle Street Conspiracy', boasting considerable practical experience of research, considerable experience of foreign lands, and certain fixed ideas about the place of science in society. Hooker (1873-1878), Spottiswoode (1878-1883) and Huxley (1883-1885) successively held the Presidency of the Royal Society. Spottiswoode was Treasurer between 1870 and 1878, Huxley became Senior Secretary between 1872 and 1881, and Frankland was Foreign Secretary between 1895 and 1899. Hirst was on the Council between 1864-1866, 1871-1873 and 1880-1882.

Their influence outside the Royal Society was equally pronounced. Five of the nine became Presidents of the British Association (31). Busk took an influential role in the Royal College of Surgeons, successively as member of Council (1865), Examiner (1868), and President (1871). Hirst became President of the London Mathematical Society (1872-1874) and Frankland, President of the Chemical Society (1871-1873).

The Club had not begun with any formal purpose, although there seems to have been a vague intention to discuss scientific and philosophical questions. Spencer recalled 'that much time was spent chiefly in lively talk, of which *badinage* formed a considerable element' (32). In May 1866, for example, the 'X' nicknamed each other, sometimes with greater accuracy than kindness. Thus we find the Xquisite Lubbock, and the Xcellent Spottiswoode, the Xperienced Hooker, the Xalted Huxley, the Xcentric

Tyndall, the Xemplary Busk, the Xpert Frankland, the Xtravagant Hirst, and the Xhaustive Spencer (33). But 'besides personal friendship', recalled Hirst, 'the bond that united us was devotion to science, pure and free, untrammelled by religious dogmas. Amongst ourselves there is perfect outspokenness, and no doubt opportunities will arise when concerted action on our part may be of service' (34). In this cordial atmosphere, controversies on science and religion and the views of Bishop Colenso waged freely between discussions on atomic structure and the merits of Bacon as the originator of inductive method (35).

Similarly, it was appropriate for the nine to rally round and encourage one another professionally, as in 1866 when together they subscribed to 250 copies of Spencer's *System of Philosophy* to save their colleague from financial embarrassment (36). Again, in 1866, Hooker successfully interceded in a dispute between Tyndall and Francis Palgrave at the *Saturday Review*. Tyndall, he assured Palgrave, was 'so stirling and amicable, and his faults are so positively heart-affections, that I never can bear to see him hurt and got the better of without the strongest sympathy and wish to resent' (37).

Inevitably, conversation also turned to social issues of science. It was only natural, for example, that at their first meeting in 1864, when the weekly *Reader* was about to change hands, the importance of a lasting scientific newspaper and journal should concern the Club, and that its members should support the efforts of Tyndall and Huxley which ultimately brought *Nature* into being (38) likewise, it was expected that E. L. Youmans, the American publishing entrepreneur, would visit the 'X' to win its help in arranging American editions of English scientific works.

The brief descriptions we have of the Club's early meetings are tantalizing. In October 1865, the X 'suggested' the ballot for the forthcoming Royal Society Council elections, and Lubbock put forward a scheme to endow a Christie Lectureship at the Royal Institution. In November 1865, the group discussed the advisability of Tyndall's accepting a chair at Oxford, and the unsatisfactory method of nominating officers of the Royal Society. In December Spottiswoode recorded discussions on the strength of the liberal 'Reform Party' at Oxford. Meetings between 1866 and 1868 discussed the relationship between Faraday and Davy, the possibility of speeding publication of the *Philosophical Transactions*, the Duke of Argyll's theory of valleys, and the annual appointment of sectional presidents for the British Association. In June 1867, the 'X' voted to support Lubbock's candidature for Parliament. On other occasions, the Club discussed whether the Royal Society was justified in urging the Board of Trade to inspect ships' compasses (April 1868), and whether Lubbock's name should be added to the

British Association Committee on Scientific Education (January 1869). Other sessions discussed Sir George Airy's suitability for the Presidency of the Royal Society (January 1870), Tyndall's work on 'dirt and disease' (1874), and the creation of a Science Museum (1877). With such events as the impending transfer of the British Museum's scientific section to South Kensington and Huxley's election to the London School Board (November 1870), the nomination of officers for the major learned societies, and the adjudication of pension and medal claims, the dinners must certainly have assumed political importance. At one point (March 1871) Spencer is noted to have 'protested against the transaction of so much business' (39). But it is most unlikely that his protest was heeded.

It was never intended for the Club's discussions to become public, but Spencer, at least, believed they had a significant public effect. 'In course of time the existence of the Club became known in the scientific world, and it was, we heard, spoken of with bated breath—was indeed, I believe, supposed to exercise more power than it did' (40). Huxley recalled overhearing two members of the Athenaeum speaking one day: 'I say, do you know anything about the "X-Club"?' 'Why—I have heard of it.' 'What do they do?' 'Well, they govern scientific affairs, and really, on the whole, they don't do it badly' (41). The self-possessed but unassuming manner in which the X-Club considered major items was certainly disarming. Thus Tyndall wrote to Hirst in late 1876:

> We had our X meeting on Thursday last . . .
> We had a good deal of talk about the disposal of the £4,000 which the Government had placed in the hands of the Royal Society. A good deal of heart burning is likely to flow from this same gift. It is not one into the need of which we have fairly and naturally flowed, so that it will have to be managed instead of healthily assimilated. Spencer, as usual, was laying down the law with an *a priori* definiteness, but was by no means left at peace with his conclusion (42).

The myth of the X-Club's pervasive influence was enhanced by the galaxy of eminent public and scientific 'men of mark' invited to its monthly dinners (43). Such men included W. K. Clifford, the mathematician, Charles Darwin, David Masson (Professor of English at University College and sometime editor of the *Reader*), and Robert Lowe, Chancellor of the Exchequer. Distinguished foreigners included Auguste Laugel, Helmholtz and Cornu from the Continent; and Asa Gray, Louis Aggasiz and E. L. Youmans from America. The X-Club almost naturally became a powerful instrument for wielding scientific opinion in winning political battles. Thus, in 1872-1873, Huxley, Tyndall, Frankland and Busk, working jointly through Lubbock in the House of Commons, rallied to the support of Hooker in his administrative contest with Acton Ayrton, Gladstone's temperamental First Commissioner of Works. In the same year, the good offices of Huxley and Spottiswoode were clearly seen in Hirst's appointment in 1873 to the Directorship of Naval Studies at the Royal Naval College (44), by the new Chancellor and guest of the X-Club, G. J. Goschen.

For a time, the X-Club succeeded in combining scientific discussion with domestic holidays. Every year, in June, the Club held its monthly meeting on a Saturday, and scheduled week-end outings in the country (45). On these occasions wives were invited, and postcard announcements were addressed to 'X's and Yv's' (46). Each year about fifteen assembled for drives or walking expeditions, often to Maidenhead, but also to Windsor Forest, or to Leith Hill, Oxford, where they could listen to Huxley reading Tennyson (47).

After 1874, however, these country meetings became less frequent. 'Several motives'—probably including the deaths, within the space of a few months, of Hooker's first wife, Frances, and Frankland's first wife, Sophie—brought these happy occasions to an end (48). The end of June excursions, however, did not mean the end of all domestic gatherings. On occasions when a member of the Club presided at the British Association—such as at Liverpool in 1870 when Huxley was President, at Dublin in 1878 when Spottiswoode was President, and at York in 1887, when Lubbock presided—the members and their wives took a suite of rooms at the chief hotel and, in Spencer's phrase, 'united their forces' (49). Safety in numbers was perhaps a consideration when, as after Tyndall's presidential address at Belfast in 1874, the gales of controversy blew with special force (50).

In January 1876 the 'X' celebrated its 100th meeting and the Club entered a new phase of development. Its members had by now acceded to positions of power in the scientific community. But while its outward appearance grew more impressive, its internal harmony was repeatedly disturbed. Some arguments were generated by the deaths of members, and the question of whether to elect new ones. In 1883, Spottiswoode, the last to join, was the first to die, from typhoid, at his London home. 'The fact is', Spencer wrote to Youmans, 'we are all beginning to break up in one way or another. Of the members, one is going and of the remaining eight, only two are in good health' (51). In October, 1885, Huxley mourned that only Frankland and Lubbock were in sufficiently good spirits to attend (52). In 1888, Lubbock and Frankland urged the election of two new members and mentioned General Strachey, Michael Foster, John Evans and Francis

Galton. Hooker was neutral; 'I am not anxious for recruits to the "X" but do not want to sit alone at the table' (53), he wrote. Huxley indicated his general support for new elections, but Tyndall strongly opposed them. His decision was not on personal grounds; Galton, had long shared Tyndall's love of mountaineering (54) and both Strachey and Foster were friends. Perhaps Tyndall felt nostalgically that the election of new men would give formal recognition to the swift passage of time, and change the character of the group. 'In the end', Hirst wrote, 'the subject was deferred, although there were feelings that because Tyndall had not recently attended meetings he ought not to count' (55). According to Huxley, it was agreed to limit new members to those whose names contained all the consonants absent from the names of the old members. Since, as the story goes, they had no Slavonic friends, there was an end to it (56). The more one learns about the X-Club, the less apocryphal this explanation seems.

But the heat generated by the question of new members was mild compared to the bitterness the Club experienced in the 1880s over Gladstone's posture in Egypt, and the death of Gordon at Khartoum. The frustrations of public affairs exploded round the dinner table. As Huxley wrote to Professor Bartholomew Price at Oxford in May: 'The X-Club is going to Smithereens, as if a charge of dynamite had been exploded in the middle of it' (57). In September, Tyndall remarked to Hirst that 'the Table Round of the X will, I fear never meet again in its pristine vigour' (58). In December 1885, a month after his release from the duties of President of the Royal Society, Huxley himself railed against 'politics, scandal and the three classes of witnesses—lairs, d——d liars and experts' (59). In 1887, emotions were roused again by the spectacle of Stokes standing for Parliament and by the break-up of the Liberal Party. At the November dinner, it was decided to draw up and sign a kind of 'scientific declaration' in support of the Union, and Tyndall was deputed to write it in strong but moderate language. 'It was thought that nearly every scientist of note would sign it' (60). Many, in fact, did so, and the Club (excepting Lubbock) became a notable outpost of opposition to Gladstone, and support to Derby and Salisbury.

In the absence of new ideas or fresh members personal rifts were predictable. As late as 1885, Huxley and Spencer were good friends, bantering each other in playful prose. Thus, when Spencer missed a dinner, Huxley consoled him:

> We were very sorry to miss you yesterday—were reduced to five; but we contrived to keep our spirits up and positively sat till after ten o clock —all except Lubbock, who had to go to the Linnaean. I don't think that

> anything of a profound character was said—in fact, in your absence, I am afraid we inclined to frivolity.

and Spencer promptly replied:

> And so you sat till 10. Well, really this is too bad. Considering that I am always the one to protest about the early dissolutions that habitually take place, that you should seize the occasion of my absence for making a night of it adding insult to injury. It would really seem from the fact that you deliberately bring before me that I have hitherto been the cause of the prompt breaking up of the party! I shall have to bring the question before the next X, and ask what it is in my behaviour which leads to this obvious anxiety to get away as soon as possible when I am present (61).

Four years later, however, an almost accidental dispute over the philosophical justifications for a land nationalization policy drove Huxley and Spencer into angry and opposite camps. Their private differences, boiling over into the pages of the *Nineteenth Century*, caused grave concern in the 'X'. Mutual friends tried to heal the wound, but their advice seemed only to annoy Spencer even more. Instead of meeting Huxley's attempts at compromise, Spencer withdrew from society and sulked. He even prepared a letter of resignation from the Club, which only Hooker, with the greatest difficulty, persuaded him to withdraw. Tyndall urged Spencer to meet Huxley and sort out their difficulties amicably: 'You may well believe that this newspaper controversy has been a source of mourning to my wife and me. Many a time since it began have I wished to be at your side, or better still, to have you and Huxley face to face. With a little tact and moderation, the differences between you—if a difference exist at all—might have been easily arranged' (62). To Tyndall, Spencer merely replied:

> Doubtless you and others of the Club do not fully understand the state of mind produced in me . . . The effect on me had been such that the thoughts and irritations have been going round in my brain day and night as in a mill, without the possibility of stopping them (63).

It was not until the end of 1893 that cordial relations were restored.

In the meantime separation and advancing age were contributing to the Club's gradual decline. After 1886, Spencer's failing health and his rest cures at Bournemouth led him frequently to miss dinners and to decline his friends (64). In May 1886, only Hooker and Frankland attended (65), and by June, Huxley felt the Club was wasting away. 'We really must make up our minds', he told Lubbock, 'what is to be done for the future' (66). Spencer's declining health reflected the 'Table Round's' declining fortunes (67). In

1889, Hooker rebuked Tyndall for not coming: 'There were Hirst, Frankland and Lubbock. Poor Spencer, looking the ghost of misery, dined apart' (68).

Gradually the group moved from London; Hooker to Sunningdale, Lubbock to High Elms in Kent, Tyndall to Hindhead in the Lake District, Huxley to Eastbourne in 1890, Spencer to Brighton in 1898, and Frankland to Reigate in 1885 (69). Busk's death in 1886 (70) was followed by Hirst's long illness and death in 1892. In early 1892, Hooker wrote to Tyndall:

> I cannot get over the loss of Hirst . . . in respect of the X meetings . . . How I wish you could return to us; except to meet Huxley I care now very little about it. This is blasphemy, but if it is so, so it is (71).

Tyndall wept with grief to Huxley:

> He is gone, and the loss to all of us is great—to me especially. We have been intimate with each other for more than five and forty years and without a moment's chill to our affections. To me his life was almost more or less of a tragedy. I know his intellectual power to be great, but I saw that power perpetually broken by imperfect health (72).

In April, Hooker wrote again to Tyndall, in even gloomier terms:

> I fear the poor X is on its last legs. Frankland is the only dependable attendant. Spencer is always ill, or thinks he is, and that is as bad; Huxley lives too far off and it is always a toss up as to whether Lubbock can and will come . . . (73).

Petty annoyances acquired undue importance, until the Club's difficulties seemed to outweigh its advantages. The last entries in the X-Club Notebook were made on 10 March 1892. To Huxley, Hooker wrote in April 1892, that the 'amount of correspondence for every meeting held, indicates the feeble hold it has on its members; in most cases, of necessity, but in such a case as Lubbock's, mere convenience'. His doubts gathered force with time: 'I am coming to the conclusion that at our age these clubs are an anachronism . . . as I have said all along, they were made for younger men and younger men should settle their future' (74). In May, he exploded:

> I have lost all patience with the X Club. It is an incessant bombardment of summonses without response, or unsatisfactory ones, previous to every date of meeting. You are the only one who has answered the last, and as I greatly fear you may not be able to attend, I am writing to the others to put off the 2nd (*sub voce*) *sine die*. The truth is that, except Frankland,

> we are all crippled by circumstances of health, distress or in Lubbock's case, of other demands.
>
> My idea is that it is best to let it die out unobserved, and say nothing about its decease to anyone (75).

The experience of the meeting of April 1892, seemed to confirm his worst predictions. 'The last X two months ago was a fiasco. Only Frankland came, and he left at 8 p.m.' (76). Tyndall agreed: 'The X has gone to pieces by spontaneous fission' (77).

In the meantime, Britain was fast becoming a different place from that which the young X-Club had known. 'What a state of the world we are living in,' wrote Spencer to Tyndall in January 1893, 'with its socialist anarchism, and all kinds of wild ideas and destructive actions. The prophesies I have been making from time to time ever since 1860 as to the results of giving to men political power without importuning to them equal political burdens, are becoming true far faster than I had anticipated' (78). By March 1893, Hooker wished to concede defeat. Out of six meetings during the past session, Hirst attended four, Huxley two, Hooker five, Spencer three, Lubbock two, Frankland five and Tyndall none. Hooker wrote:

> Frankland and I were alone at the last meeting, and he had to leave early. Neither Lubbock nor Spencer are to be depended upon—nor can you or I well be frequent attendants at stated intervals. If Hirst had survived, I would have stuck by him to the last battle, but I feel that with him gone and you away, the spell is broken. Tyndall's revivication could help matters, and he, poor fellow, is only another fallen leaf from the once goodly tree (79).

Their meeting of March 1893, was to be the X-Club's last.

To Lubbock Spencer wrote in May 1893: 'I fear that now the X is dead there is but little chance of our meeting . . . I wish it were otherwise (80)'. In December 1893, Tyndall died tragically from an overdose of chloral. 'Another of us has gone', Hooker wrote to Huxley, 'what a tragedy it all is, it seems to take a bite out of one's life . . . He was quite the purest, brightest creature I ever knew to be a philosopher' (81).

In 1895, Huxley died, and by the time of his death, the X-Club had already become more a memory than an experience. 'The X-Club', Hooker wrote to Huxley's wife, 'died with him. I have never had the heart to ask for another meeting even to wind up' (82). Frankland died in 1899, and at the turn of the century only Hooker, Lubbock (now Lord Avebury) and Spencer remained. They rarely met or corresponded. Spencer reminded

Hooker in 1901 that 'It is a long time since any news passed between us—a year and a half, I think. Superfluous letter writing is, at your time of life, and even at mine, a thing to be avoided; but still, I should like to have a few lines telling me how you fare in your contest with the inevitable' (83). Hooker replied in kind and thanked Spencer for his evidence of abiding fellow feeling. '. . . the dear old Club is rapidly, with us, I fear, approaching the vanishing point. How curious it seems to me that we who were, I think, considered its oldest members, should be amongst the three survivors' (84).

The Club's continuous history falls clearly into three phases—the early heroic age between 1865–1880; the more mature counsels of the mid-1880s and early 1880s, and the disenchantments, disagreements and the elderly vagaries of the late 1880s and early 1890s. While the harmony of the first two decades welded the group together, it is not surprising that the passing years saw their share of dissent. Spencer remarked in 1904 that its ranks were 'never thinned by desertions or by differences' and that 'during these years nothing has occurred to disturb the harmony of our meetings' (85), but it is clear that this description was more nostalgic than accurate. No ordinary group could contain Tyndall's stubborn Liberalism, Lubbock's breezy self-assurance, Huxley's acid brilliance and Spencer's weighty dogmatism.

Altogether the Club met 240 times (86). Although an average of seven attended each dinner, all nine members appeared together only 27 times (87). Whatever effect such 'poor attendance' may have had on the group's effect as a lobby, it certainly did not diminish its public stature. For the historian, it is a tragedy that the practice of keeping formal records was abandoned (88). Though Hirst's Journals and the Club Notebook are highly suggestive, the absence of minutes has left obscure the Club's role at critical moments. Hints of its influence will certainly continue to grow as new MSS. come to light, and Frankland's private papers, if ever they become available to scholars, will undoubtedly help place the Club in perspective.

Given its background and assumptions, the X-Club could not outlive the deaths of its members, and was as unsuited as the Philosophical Club to the changed social circumstances of science after about 1900. The rise of specialization, and the development of science at the universities, signalled the end of the subtle monopoly of power held by the London scientific societies and the 'Young Guard'. While the Royal Society and the Athenaeum naturally continued to bring to focus the nation's scientific *élite*, influential scientific networks began to revolve more around university departments. Nonetheless, immortalized by the memory of its members, the legend of the X-Club has survived. Perhaps it has served as a model for other informal groups of influential scientists in Britain and abroad. Certainly, its history has a continuing appeal for those who believe that informal *élites* still decide the most important questions of scientific policy. The concept of an 'open conspiracy' was of lasting importance. If the idea of the X-Club did not exist, others would soon have invented it.

Acknowledgements

This research was supported by a grant from the Department of Education and Science. Quotations from the Huxley Papers have been made by kind permission of the Governors of Imperial College. The Tyndall Papers and the Hirst Journals have been quoted by permission of the Royal Institution. I am exceedingly grateful to Dr W. H. Brock, of Leicester University, for important information on Edward Frankland, and for comments on an earlier draft of this essay.

Notes

(1) Notable exceptions include Walter Cannon, 'The Role of the Cambridge Movement in Early Nineteenth Century Science', *Proceedings of the Xth International Congress of the History of Science*, 1962; and l : 'Scientists and Broad Churchmen: an Early Victorian Network', *J. British Studies*, **4**, 65–88 (1964). All work on the Victorian 'scientific aristocracy' must be indebted to Cannon, and to the pioneering accomplishment of Noel Annan's *Leslie Stephen* (London, 1951).

(2) Letter of Fiske to wife, 8 December 1873, quoted in H. H. Harper, *The Personal Letters of John Fiske* (Cedar Rapids, 1939), p. 144, and with modifications in E. F. Fiske, *The Letters of John Fiske* (New York, 1940), p. 283. See 'The Human Side of Great Scientists', *Scientific American*, 15 June 1912, p. 530.

(3) See Sir A. Geikie, *Annals of the Royal Society Club* (London, 1917), p. 1; T. G. Bonney, *Annals of the Philosophical Club of the Royal Society* (London, 1919).

(4) Joseph F. Payne, *History of the College Club of the Royal College of Physicians in London* (London, 1909); T. E. Allibone, 'The Club of the Royal College of Physicians, the Smeatonian Society of Civil Engineers and their Relation to the Royal Society Club', *Notes & Records Roy. Soc. Lond.* **22**, 186–192 (1962).

(5) [W. H. Smyth], *Sketch of the Rise and Progress of the Royal Society Club* (London, 1860), p. 60.

(6) O. J. R. Howarth, *The British Association for the Advancement of Science: A Retrospect: 1831–1931* (London, 1931), p. 91. See also John Timbs, *Clubs and Club Life in London* (London, 1872), p. 258.

(7) Quoted in H. G. Hutchinson, *Life of Sir John Lubbock, Lord Avebury* (London, 1914), p. 63. The quotation derives from Huxley's obituary of Tyndall in the *Nineteenth Century*, **35** (January, 1894), p. 6.

(8) Alexander Scott, Presidential Address to the Chemical Society, *J. Chem. Soc.* **109**, 342–351 (1916). Around the turn of the century the 'B Club' was succeeded by the Chemical Club.

I

(9) *Huxley Papers* (Imperial College), vol. 3, f. 23. Hooker to Huxley, Summer 1856.
(10) *Ibid.*
(11) *Ibid.*
(12) Leonard Huxley, *The Life and Letters of T. H. Huxley* (London, 1900) (hereafter *L. L. Huxley*), vol. 1, p. 256.
(13) *Ibid.*
(14) H. Spencer, *An Autobiography* (London, 1904), vol. 2, p. 115.
(15) *Hirst Journals*, vol. 4, p. 1702. Entry of 6 November 1864.
(16) W.N.W. and S.J.C., *Sketches from the Life of Edward Frankland* (hereafter Frankland) (London, 1902), p. 138.
(17) *Proc. Roy. Soc.* **52**, xii–xviii (1893).
(18) *Lubbock Papers* (British Museum), Add. MSS. 49,638, f. 28. Huxley to Lubbock, 10 December 1856.
(19) *Lubbock Papers*, Add. MSS. 49,670, f. 76. Hooker to Lubbock, 2 July 1907.
(20) Frankland, *op. cit.* p. 51.
(21) Spottiswoode was elected on 8 December. *Hirst Journals*, vol. 4, p. 1709. Entry of 14 December 1864.
(22) Spencer, *op. cit.* vol. 2, p. 115. From Hirst's early minutes it appears that Busk was asked to invite William (later Sir William) Fergusson (1808–1877) to join. Fergusson was then Professor of Surgery at King's College, London, and was afterwards sergeant-surgeon to Queen Victoria and President of the Royal College of Surgeons. His election would have increased the number of medical men in the Club. Apparently, he declined to be elected. Frankland, *op. cit.* pp. 151–152.
(23) *L. L. Huxley*, vol. 1, p. 256.
(24) Naturally the 'X' was not the only club where men of science could meet. But there was no other single venue where these nine could all assemble. Spottiswoode, for instance, did not belong to the Athenaeum, and Spencer did not belong to the Royal Society. (This was, of course, a very sensitive point with Spencer, whose candidacy for the Royal Society in 1874 aroused a small storm in the scientific community. See David Duncan, *The Life & Letters of Herbert Spencer* (London, (1908), p. 169.) Three of the nine, Tyndall, Hooker, and after 1884, Huxley, belonged to 'The Club', the exclusive dining society founded by Sir Joshua Reynolds in 1764, whose members included Kelvin, Rosebery, Salisbury, Gladstone, Grant Duff and William Flower. (See C. J. Cornish, *Sir William Flower* (London, 1904), p. 218.) In May 1890, Hooker had Gladstone to dinner at the Club and in 1892 invited Lubbock to join. See Leonard Huxley, *Life and Letters of Sir Joseph Dalton Hooker* (London, 1918) (hereafter L. L. Hooker), vol. 2, p. 343, and *Lubbock Papers*, Add. MSS. 49,653, f. 39. Hooker to Lubbock, 17 February 1892.
(25) Frankland, *op. cit.* p. 150.
(26) Frankland, *op. cit.* pp. 148–162.
(27) Hirst's *Journals* and Tyndall's diaries, both deposited in the Library of the Royal Institution also contain many passing references to the X Club, but no detailed minutes of Club proceedings.
(28) Spencer, *op. cit.* p. 16.
(29) C. Bibby, *T. H. Huxley: Scientist, Humanist and Educator* (London, 1959), p. 249.
(30) *L. L. Huxley*, *op. cit.* vol. 1, p. 258; Frankland, *op. cit.* p. 51.
(31) Hooker (1868), Huxley (1870), Tyndall (1874), Spottiswoode (1878) and Lubbock

(1881). Under the circumstances, it is curious that of the remaining four, Frankland, was *not* made President of the B.A. Sir Henry Armstrong attributes his 'jilting' [*sic*] to the nation's consistent disregard of chemistry and its applications. See H. E. Armstrong, First Frankland Memorial Oration of the Lancastrian Frankland Society, *J. Soc. Chem. Ind.* **53**, 464 (1934).

(32) Spencer, *op. cit.* vol. 2, p. 116.

(33) *Hirst Journals*, vol. 4, p. 1786, entry 10 May 1866.

(34) *Hirst Journals*, vol. 4, p. 1702, entry 6 November 1864.

(35) *L. L. Hooker*, vol. 2, p. 359. Tyndall held that he saw atoms visually in his mind's eye, but was unable to describe to Huxley what he saw; Huxley quickly replied: 'Ah, now I see myself, in the beginning was the Atom, and the Atom is without form and structure and darkness sits on the face of the Atom.'

(36) Spencer, *op. cit.* vol. 2, p. 136.

(37) *L. L. Hooker*, vol. 2, p. 349.

(38) Spencer, *op. cit.* vol. 2, p. 120. References which follow are cited in the small *X Club Notebook* kept with the Hirst and Tyndall papers at the Royal Institution.

(39) Frankland, *op. cit.* p. 160. In February 1869, Hirst wrote: 'We were more than once called to order by Spencer, for allowing the conversation to become broken up, instead of remaining general.' *X Club Notebook* (R.I), p. 40.

(40) Spencer, *op. cit.* vol. 2, p. 116.

(41) See Huxley, *Nineteenth Century*, **35**, 11 (1894), cf. Bibby, *op. cit.* p. 249.

(42) *Tyndall Papers* (Royal Institution), vol. 2, f. 644, Tyndall to Hirst, 17 December 1876.

(43) Frankland, *op. cit.* p. 162.

(44) *Huxley Papers*, vol. 18, f. 174, Huxley to Hirst, 9 December 1872.

(45) There are contradictory impressions about the date and number of these Saturday meetings. Frankland says (*op. cit.* p. 48) that there were the usual monthly meetings for June; Spencer (*op. cit.* p. 117) insists that they were supplementary.

(46) *L. L. Huxley*, vol. 1, p. 258.

(47) *Ibid.* p. 261.

(48) Neither Frankland nor Spencer, however, directly attributed the Club's change in practice to these events.

(49) Spencer, *op. cit.* vol. 2, p. 219.

(50) *Ibid.* p. 282.

(51) David Duncan, *The Life and Letters of Herbert Spencer* (London, 1908), p. 248.

(52) *Huxley Papers*, vol. 2, f. 276, Huxley to Hooker, 24 October 1895.

(53) *Huxley Papers*, vol. 3, f. 321, Hooker to Huxley, 25 March 1888.

(54) Francis Galton, *Memories of My Life* (London, 1908), p. 191.

(55) Frankland, *op. cit.* p. 161; *Huxley Papers*, vol. 16, f. 268. Frankland to Huxley, 5 December 1886.

(56) *L. L. Huxley*, vol. 1, p. 259.

(57) *Ibid.* vol. 2, p. 119.

(58) *Tyndall Papers*, vol. 16, f. 672.

(59) *L. L. Huxley*, vol. 1, p. 258.

(60) *Huxley Papers*, vol. 16, f. 272, Frankland to Huxley, 15 March 1872.

(61) Duncan, *op. cit.* p. 249, Spencer to Huxley, 7 December 1885.

(62) *Ibid.* p. 332, Tyndall to Spencer, 25 November 1889. For the controversy see Duncan, *op. cit.* pp. 331–335.

(63) *Ibid.* p. 333, Spencer to Tyndall, 9 December 1889.
(64) *Ibid.* p. 327, Spencer to Huxley, 10 March 1886.
(65) *Huxley Papers*, vol. 8, f. 246, Huxley to Tyndall, 7 May 1886.
(66) *Lubbock Papers*, Add. MSS. 49,646, f. 52, Huxley to Lubbock, 2 June 1886.
(67) *Lubbock Papers*, Add. MSS. 49,648, f. 28, Tyndall to Lubbock, 25 October 1888.
(68) *Tyndall Papers*, vol. 3, f. 2767, Hooker to Tyndall, 3 May 1889.
(69) Frankland, *op. cit.* p. 162.
(70) On 15 October, Tyndall wrote to Busk's sister: 'I never knew a man possessed of a more genuine love for science in its best and broadest sense . . . Had he been less noble than he was he might have been more talked about by the public.' *Tyndall Papers*, vol. 1, f. 187.
(71) *Tyndall Papers*, vol. 8, f. 2790. Hooker to Tyndall, 6 March 1892.
(72) *Tyndall Papers*, vol. 9, f. 3153, Tyndall to Huxley, 18 February 1892.
(73) *Tyndall Papers*, vol. 8, f. 2791, Hooker to Tyndall, 30 April 1892.
(74) *Huxley Papers*, vol. 3, f. 389, Hooker to Huxley, 24 April 1892.
(75) *Huxley Papers*, vol. 3, f. 391, Hooker to Huxley, 30 May 1892.
(76) *Tyndall Papers*, vol. 8, f. 2792, Hooker to Tyndall, 20 June 1892.
(77) *Huxley Papers*, vol. 8, f. 442, Tyndall to Huxley, 27 December 1892.
(78) *Tyndall Papers*, vol. 3, f. 1165, Spencer to Tyndall, 30 January 1893.
(79) *Huxley Papers*, vol. 3, f. 400, Hooker to Huxley, 23 March 1893.
(80) Duncan, *op. cit.* p. 327, Spencer to Lubbock, 18 May 1893.
(81) *L. L. Hooker*, vol. 2, p. 349.
(82) *Huxley Papers*, vol. 3, f. 341, Hooker to Mrs. Huxley, 24 October 1897.
(83) Duncan, *op. cit.* p. 456, Spencer to Hooker, 16 March 1901.
(84) *Ibid.* p. 470, Hooker to Spencer, *ca*, June 1903. Spencer died in 1903, Hooker in 1911, and Avebury in 1913.
(85) Spencer, *op. cit.* p. 118.
(86) Frankland, *op. cit.* p. 162. Frankland says he himself attended 186 times; Spencer attended 173 times; Huxley, 171; Hirst, 170; Hooker, 169; Spottiswoode, 160; Tyndall, 147; Busk, 143; and Lubbock, 131.
(87) Spencer, *op. cit.* p. 118.
(88) Frankland, *op. cit.* p. 160.

Corrigenda

p. 306, line 6: 1897 should read 1847
p. 307, line 33: Queenswood should read Queenwood
p. 307, last line: West Yorkshire should read West Riding of Yorkshire
p. 308, line 3: Queenswood should read Queenwood
p. 308, line 6: death read resignation
p. 310, line 15: 'la' should read 'le'
p. 312, line 36: Laugel read Laugier
p. 313, line 14: Oxford should read Surrey
p. 316, line 5: Lake District read Sussex

II

THE SCIENTISTS' DECLARATION: REFLEXIONS ON SCIENCE AND BELIEF IN THE WAKE OF *ESSAYS AND REVIEWS*, 1864-5

W. H. Brock and R. M. Macleod

During the decades following the publication of Darwin's *Origin of species* in 1859, religious belief in England and in particular the Church of England experienced some of the most intense criticism in its history. The early 1860s saw the appearance of Lyell's *Evidence of the antiquity of man* (1863), Tylor's research on the early history of mankind (1863), Renan's *Vie de Jésus* (1863), Pius IX's encyclical, *Quanta cura*, and the accompanying *Syllabus errarum*, John Henry Newman's *Apologia* (1864), and Swinburne's notorious *Atalanta in Calydon* (1865); it was in this period also that Arthur Stanley was appointed Dean of Westminster, and that Bills were introduced in Parliament to amend or repeal the 'Test Acts' as they affected universities. They were the years that witnessed Lyell present the case for geology at the British Association at Bath (1864), the first meeting of the X-Club (1864), and the award of the Royal Society's Copley Medal to Charles Darwin. These were the years in which, as Owen Chadwick has put it, 'the controversy between "science" and "religion" took fire'.[1]

To be sure, only part of this criticism arose directly from the new work of science. From within the Established Church discussion focused on theological issues as central as everlasting punishment, redemption, the nature of the Trinity, and the social role of the Church. From without, the role of the Church and its schools was debated in the universities and in Parliament. The rigours of clerical ritualism, given added force by the political manoeuvres of the papacy, were exciting dissatisfaction throughout Europe. Finally, successive conquests of biblical criticism, proceeding from German example, created fresh rifts which became indelibly associated with the use of new 'scientific' methods of scholarship.

Of course, the confrontations of science and religion were made more complicated, at both social and intellectual levels, by a conflation of issues concerning the sufficiency of natural theology, the doctrine of revealed truth, the belief in biblical literalism, and the unfettered search for new knowledge. The last issue, in particular, has much to do with the difficult relationship we now perceive between the self-defining, self-generating ethos of scientific inquiry, and broader political and philosophical concepts of intellectual freedom. We can hardly neglect to ask whether we may see in the 'warfare' between science and religion an important reflexion of a much more general concern about the realities of political and intellectual repression, a concern which may have given less immediate alarm in England of the 1860s than in mid-century Italy, France, and Germany, but which was certainly no less evident. To what extent was 'freedom of expression' in matters of science seen as a 'test case' for freedom of political and religious expression generally? And, if there was an association of this kind, how did these issues, assimilated in the rhetoric of Victorian liberalism, affect the rhetoric of the debate about 'Man's place in Nature'?

These questions, and others like them, have provided the overall framework for recent inquiries. Such questions, which in our own case derive from work on the mid-Victorian scientific intelligentsia,[2] require a detailed study of the structures and mentalities of different elements in the world of science, where, not surprisingly, political purposes, religious beliefs, and scientific attitudes are often difficult to separate.

A striking example of the importance and difficulty of relating discussion about the nature of knowledge and the requirements of belief to the general context of political and theological debate, is provided by the publication in 1865 of the *Declaration of students of the natural and physical sciences*. This Declaration, generated in a desperate attempt to maintain, in the face of new knowledge, 'a harmonious alliance between Physical Science and Revealed Religion', embodied the assumptions and claims that were to inform debates on scientific naturalism for at least the next quarter of a century.

1. *The Declaration*

In early 1864 the embers of *Essays and reviews* provided the spark. In February, in an historic decision considered by some as the 'final triumph of the secular over the ecclesiastical jurisdiction' in England,[3] the Privy Council voted on an appeal to quash the synodical sentence of the Dean of Arches against Rowland Williams and H. B. Wilson. Colenso's appeal to the Privy Council was still in progress.[4] Consequently, by the spring of 1864 the Church was in an uproar. In despair at the legal vindication of the Essayists, 10 906 clergymen (nearly one half of the clergy of England and Ireland) united in protest against them by way of a 'Declaration' to the Archbishop of Canterbury.[5] This 'Oxford Declaration' declared 'from belief' that the Church 'maintains without reserve or qualification the Inspiration and Divine Authority of the whole Canonical Scriptures, as not only containing but being the Word of God'.[6] These clergy were, in limited terms, successful, and *Essays and reviews* was formally condemned by both Houses of Convocation in June 1864. But the victory was short-lived. Lord Chancellor Westbury ridiculed Convocation, and its guiding spirit, Bishop Wilberforce, in his famous verdict that the synodical judgement was simply 'a well lubricated set of words—a sentence so oily and saponaceous that no one can grasp it'.[7] If 'Soapy Sam' Wilberforce could never forget this sobriquet, nor would the world forget Disraeli's remark at the Oxford Diocesan Conference in November 1864, when he announced his famous preference for the side of the Angels.[8]

On Saturday, 16 April 1864, a group of London chemists, styling themselves 'Students of the natural and physical sciences', decided to follow the example of the Oxford Declarationists. On Thursday, 21 April, Canon Christopher Wordsworth presented their hurriedly circulated memorial, bearing 28 signatures, to the Lower House of the Convocation of Canterbury.[9]

> We, the undersigned Students of the Natural Sciences, desire to express our sincere regret, that researches into scientific truth are perverted by some in our own times into occasion for casting doubt upon the Truth and Authenticity of the Holy Scriptures. We conceive that it is impossible for the Word of God, as written in the book of nature, and God's Word written in Holy Scripture, to contradict one another, however much they may appear to differ. We are not forgetful that Physical Science is not complete, but is only in a condition of progress, and that at present our finite reason enables us only to see as through a glass darkly; and we confidently believe, that a time will come when the two records will be seen to agree in every particular. We cannot but deplore that Natural Science should be looked upon with suspicion by many who do not make a study of it, merely on account of the unadvised manner in which some are placing it in opposition to Holy Writ. We believe that it is the duty of every Scientific Student to investigate nature simply for the purpose of elucidating truth, and that if he finds that some of his results appear to be in contradiction to the Written Word, or rather to his own *interpretations* of it, which may be erroneous, he should not presumptously affirm that his own conclusions must be right, and the statements of Scripture wrong; rather, leave the two side by side till it shall please God to allow us to see the manner in which they may be reconciled; and instead of insisting upon the seeming differences between Science and the Scriptures, it would be as well to rest in faith upon the points in which they agree.
>
> We therefore pray, that the Bishops and Clergy in Convocation assembled, and of the Church of England, will do all in their power to maintain a harmonious alliance between Physical Science and Revealed Religion.[10]

The final paragraph, aimed directly at Convocation, was deleted by the memorialists on finding that it offended many potential signatories. In May the revised document, retitled a 'Declaration' was circulated. Its intention was to draw attention to the nature of the conventional 'Test' of belief in the Thirty-nine Articles to which all members of the Church of England and graduates of Oxford and Cambridge Universities were required to subscribe; and to state explicitly a 'Fortieth Article' of religious belief to which all Christian men of science should be asked to subscribe.

Once a signature was successfully obtained, the certificate was mounted in a book kept for the purpose. Some signatures were written into this book directly; a few others were signed in by proxy. The volume was eventually deposited in the Bodleian Library, Oxford, and a printed list of signatures, in alphabetical order, published.[11]

The text of the Declaration indicates the continuing strength of the tradition of a theology of nature in the 'common context' of Victorian intellectual life.[12] It held to the importance of interpreting scripture in the light of new scientific findings, and also to the belief that God, and by implication God's Word, is implicit in the order of nature and could be elucidated through sound empirical investigation. Such investigation would presumably produce a view of the world in which 'truth' in scripture and in nature would be identical. A charitable reading would also suggest that the Declaration, extremely moderate in tone, argued for a compromise, or an armistice, between the 'two sides'.

On the other hand, the Declaration did set specific constraints upon the interpretation of scripture which the less charitable could extend to interpretations of natural phenomena, and in this sense it appeared highly dangerous. Moreover, the simplistic language of the Declaration, if it deliberately embraced a large cross-section of opinion, specifically lacked a cutting edge. Perhaps it could give some men of science an opportunity to 'justify' themselves, but these advantages would appeal to few. As the *Saturday review* commented:

> The declaration itself is a good illustration of the double faces which such documents are apt to assume. They are drawn so as to seem very trivial and unmeaning to those who are asked to sign them, and who scan their wording closely; but they are timed so as to seem important and full of meaning to the mass of careless readers, who only take into consideration the particular juncture selected for signing them. This scientific declaration, construed quite strictly, lays down simply that the Supreme Being has not told a falsehood . . . But coming at this particular moment, no one would doubt that its practical meaning was to give a general endorsement to the traditional interpretation of the book of Genesis, and to express a conviction that, however irreconcilable the statements upon scientific matters contained in that book might seem to be with the conclusion of science, the time would surely come when the two would be found to agree with perfect accuracy. Of course, this is a belief which a great many people hold, and for which there is a great deal to be said; but it is not a self-evident proposition, nor is it an integral portion of the Christian Creed. It might be proved that the cosmogonical parts of Genesis were a corrupt interpolation; or that they were written by the author of the book, but without Divine warrant; or that the Divine warrant was extended only to the moral and spiritual inferences drawn from them, and not to the scientific accuracy of the statements made; and in any of these cases the Nicene Creed and the Apostles' Creed would be wholly unaffected.[13]

2. *The protagonists*

Despite this sceptical response, the Declaration proceeded. The principal movers in its production were the first five signatories: Alexander William Gillman (1843–?), Herbert McLeod (1841–1923), John

Stenhouse (1809–90), Charles Edward Groves (1841–1920), and David Howard (1839–1916); and the last signatory, the chemist and paint manufacturer, Capel Henry Berger (1839–68).[14] These six were far from being well-known religious controversialists—indeed, they are disturbingly obscure. We know little of their background, and excepting a few fragmentary records, virtually nothing of their dealings with other scientists. In so far as Berger deliberately kept his signature to the end, we may suspect that he was the instigator of the manifesto—he was a member of the same group of London Plymouth Brethren as Philip Gosse (who also signed) and it seems at least possible that he found the problem of reconciling scientific advances with biblical fundamentalism particularly difficult. Indeed, the fact that many of the orally agreed signatories are in Berger's handwriting more than confirms that he was the leader of the project.[15]

Gillman—whose grandfather lived with Coleridge (and treated him with laudanum) and whose father was the chairman of the Prudential Assurance Company—was a brewery chemist and former student of the Royal College of Chemistry.[16] Howard, 'a zealous churchman' who gave liberally to the building funds of London churches, was a grandson of Luke Howard. An occasional student at the Royal College of Chemistry between 1858 and 1860, he entered the family pharmaceutical business and had a distinguished career as an industrialist.[17] Stenhouse, a staunch Presbyterian and the most distinguished of the protagonists, was a pupil of Liebig and famous for his work on narcotics. Then aged 54, his shaky signature reminds us that a stroke paralysed him in 1856 and that Charles Groves, his private assistant and collaborator, in fact executed all his research. Groves himself was a High Anglican, and later became the editor of the Chemical Society's *Journal* from 1885 to 1889. W. A. Tilden, who was briefly employed by Stenhouse, recorded later that:

> both were of earnest religious convictions, and the 'Doctor' [Stenhouse] who could only sit and watch operations, was much disposed to talk during work, which was not unfrequently interrupted by controversy on subjects connected with their respective religious views, Stenhouse being a Presbyterian and a great admirer of the famous preacher Charles Spurgeon, while Groves as a high Churchman stood up for the episcopal establishment.[18]

Groves was a former student at the Royal College of Chemistry, where his fellow Anglican, McLeod, was Hofmann's assistant. Herbert McLeod, described as 'deeply religious',[19] later became a distinguished F.R.S., best known for his magnificent editorial work on the *Royal Society catalogue of scientific papers* (1867 to 1925).

From what we know of these six men it is not easy to formulate a consistent set of behavioural explanations. The Declaration reveals a sense of fear, both of science and of biblical criticism; it also reveals, beneath a mask of apparent reasonableness, a serious confusion of objectives. It makes no distinction between scientific fact and hypothesis, or between experiment, discovery, and verification. Theologically it fails to distinguish between literal authenticity and literal belief. Finally, it signally fails to define the process of 'elucidating truth', and sets no critical standards for men of science, apart from the doubtful goal of bland compromise. At a time when fresh accomplishments in scientific research and Tennysonian visions of progress were vaulting men to new thresholds of understanding, a dampening appeal to the limitations of 'our finite reason' appears distinctly otiose. Indeed, one can only wonder at the means by which the protagonists themselves proposed to reconcile the claims of their science and their theology.

From the foregoing, however, two points do stand out. First, the protagonists seemed deliberately to avoid sectarian appeals presumably in order to capture the largest possible number of scientific signatories. In this they succeeded. The signatories were drawn from all branches of organized Christianity; some even drew attention to themselves as Roman Catholics or Baptists, although these denominational labels were suppressed in the printed Declaration.[20]

Secondly, the whole exercise seems to have been something of a Royal College of Chemistry 'plot'. It seems improbable that the document had any official blessing from the College, and it was certainly not mentioned in the College *Minutes*. But McLeod seems to have been mainly responsible for the postal appeals, while personal signatures were collected at McLeod's laboratory at the Royal College of Chemistry, and at the private laboratory of Stenhouse in Rodney Street, Pentonville. Moreover, no fewer than 39 so-called 'students' of the Royal College of Chemistry signed the document,[21] a fact that did not pass unnoticed by those who complained sarcastically of the lack of experience among the signatories,[22] who (as they themselves confessed privately[23]) were largely unknown. McLeod, in particular, was thought to be over-reaching himself by acting without the approval of either A. W. Hofmann or the sanction of the Department of Science and Art.[24]

3. *The reaction*

According to Augustus De Morgan[25] (who, as professor of mathematics at the 'Godless institution in Gower Street', took a savage delight in the Declaration), papers for signature were sent to Fellows and members of all scientific societies. One wonders who financed the appeal, for there must have been a considerable postal expense. By our calculation, in 1865 there were 6480 members of scientific societies of London, including 673 British Fellows of the Royal Society. According to Leone Levi's analysis of the 'progress of learned societies' reported to the British Association in 1868, there were roughly 40 000 people contributing to, or interested in, science in 1867. We may estimate, therefore, that potentially 40 000 people could have signed the Declaration had they all been approached; but realistically the field was nearer 5000.[26] It was not the first 'Declaration' on 'science and religion' to be circulated among scientific men since the publication of *Essays and reviews*. Indeed, in February 1861 William Spottiswoode and John Lubbock memorialized men of science in support of Frederick Temple and his colleagues, and obtained the signatures of Darwin, Lyell, Horner, George Bentham, Thomas Graham, George Airy, and George Busk.[27] By no means all the 'scientific party' signed, however—exceptions included J. D. Hooker and J. F. Herschel—and many refused on the grounds that such statements might either pledge men 'to partisanships in matters in their nature excessively recondite',[28] or 'create a fission in the "body politic" of scientific men' which would 'countenance a belief amongst outsiders that our religious differences influenced our scientific views'.[29] The danger, in Hooker's view, of aligning 'the young protagonists in science', whose opinions were 'of no weight in religious matters' against 'men of older standing and opposite tendencies (who have nevertheless the confidence of the public)',[30] was sufficient reason to avoid such declarations.

If sympathetic scientists could resist declarations in aid of science, so clergymen could resist declarations defending the prerogative of the Church. Thus Bishops Tait and Thirlwall rejected the Oxford Declaration as a 'melancholy Declaration' to which signatures were obtained 'by a kind of moral torture' and 'in a way quite unworthy of the character of those who put it forth'.[31] Half of all the Anglican clergy stood aloof and when it was presented at Lambeth, only four of the 28 bishops lent it their support.

In the event, the London chemists could have profitably taken into account the experience of these earlier memorials. But the 'scientific Declarationists' were not deterred, and proceeded to muster 717 signatures[32] by the time the document was published in the middle of May 1865.

In the meantime, the first critical public reactions against the Declaration began to appear. Charles Daubeny of Oxford, an Anglican chemist, geologist, and botanist, writing in *The Times*[33] in July 1864, spoke out for those who might have rejected the petition without wanting to put their objections to paper. He found the document of 'doubtful expediency and likely to lead to much misconception'. Since lawyers and other professional laymen were unlikely ever to circulate similar declarations, it implied that scientists 'are peculiarly liable to the charge of infidelity'. This he denied, admitting, however, that 'persons who take a perverse pleasure in opposing received doctrines have forged into weapons wherewith to assail Christianity' certain recent scientific ideas. But such perverse people were not men of science; indeed, to judge from

Essays and reviews, the critics, with the exception of Powell,[34] were found among the clergy themselves.

Some opponents of *Essays and reviews* had used Colenso's reputation as a writer of arithmetic textbooks to suggest he was a 'scientist-heretic', but Daubeny had no sympathy for this argument; Colenso was, in his view,

> an Anglican Bishop, of some mathematical, or at least arithmetical reputation . . . who, so far from being addicted to the study of nature, betrayed how little his pursuits in early life had taken that direction, by confessing that the first doubts which came across his mind as to the reality of an Universal deluge had been suggested to him in Africa by a native convert.[35]

Nevertheless, by whatever naïve means Colenso had come by his doubts, Daubeny clearly shared Colenso's feelings about Divine inspiration. Hence he could not sign a Declaration that appeared to emanate doctrinally from 'the narrowest' scriptural inspiration:

> I am not prepared to declare that conclusions honestly arrived at in the course of scientific investigations, as for instance those relating to the age of the world, the antiquity of the human race, or the prevalence of a deluge over parts of the globe not at the time inhabited by man, ought, as a matter of Christian duty, to be suppressed and ignored.[36]

The whole appeal, thought Daubeny, was the work of 'over-zealous' individuals who would misguidedly 'entangle their scientific brethren within the meshes of a Declaration which, however speciously worded, would have the effect of checking the progress of truth, by inculcating as a duty the suppression of any facts or deductions resulting from inquiries into Nature, which might not harmonize with Scriptural statements on subjects beyond the domain and scope of Revelation'.[37]

There, with this rather verbose comment, the matter could have ended if the promoters had been wiser men. But, through August and September 1864, they continued to collect signatures. Among their early conquests was Sir David Brewster, a devout evangelical Presbyterian and the most eminent signatory captured by the Declarationists. (His degrees and orders occupy thirteen printed lines.) Brewster had waxed indignant in defence of science and Galileo in his *Martyrs of science* (1841). Twenty years on, the issue was now the 'reconciliation' of science and religion.

By September 1864 Berger had collected 210 signatures, including those of the physicists Joule and Smee, the astronomers Glaisher and Main, the geologists Bowerbank[38] and Sedgwick, and the chemists Anderson and Richardson. Typical of the appeals sent out was that received by the mathematician Thomas Archer Hirst, then just beginning his professional career in London:

> Deal, 7 September 1864.
>
> Sir,
>
> I beg to call your attention to the accompanying Declaration, now in course of signature among scientific men, which it is proposed to issue when

a sufficient number of names have been obtained. I trust that you will approve of the spirit of the Document and the terms in which it is expressed, and solicit the favour of your signature to be appended to the Memorial.

I am Sir, your obedient servant
Herbert McLeod.[39]

Unfortunately Hirst destroyed his first reply 'in a prudent moment' and simply sent a curt refusal that the Declaration's 'spirit by no means meets with my entire approbation'.

If Hirst spoke for a section of the younger generation and Daubeny spoke for a section of the more established, of a far greater intellectual and national stature was the scientist and philosopher Sir John Herschel, whose capture by the Declarationists would have surpassed even the glory of Brewster. Faraday had not signed.[40] Would Herschel join them?

Herschel's swift and decisive reaction, which he (and Berger) allowed the *Athenaeum* to publish on 17 September (reprinted in *The Times* three days later) was uncompromisingly direct:

> I consider this movement simply mischievous, having a direct tendency (by putting forward a new shibboleth, a new verbal test of religious partisanship) to add a fresh element of discord to the already discordant relations of the Christian world. I do not deny that care and caution *are* apparent on the face of the document I am called on to subscribe. But no nicety of wording, no artifice of human language, will suffice to discriminate the hundredth part of the shades of meaning in which the most worldwise differences of thought in such subjects may be involved.[41]

In essence Herschel found the declaration troublesome, quite apart from the truth or falsehood of its proposition—a problem which he had previously discussed in his *Discourse on the study of natural philosophy* (1830).

To Berger himself, Herschel replied that even to ask for his signature was 'an infringement of that social forbearance which guards the freedom of religious opinion in this country with especial sanctity' (this was only 35 years after Catholic Emancipation!); but this opinion was certainly *not* to be 'construed into a profession of Atheism or infidelity'. De Morgan thought this a 'stinging answer', and confided to his fellow-mathematician Sir William Rowan Hamilton that 'the result will be a warning not to apply to science to make declarations which are—under very distorted phrases—intended to support ecclesiastics in (exhi)biting their sinuosity'.[42] The *Saturday review* took a similar stance:

> The unfortunate mania which possesses so many people for inviting their neighbours to purge themselves from heresy will seriously aggravate a difference of opinion, which is a very quarrel as it stands; and if Sir John Herschel shall have done anything to check this fashion, he will have rendered good service to Christianity.[43]

Herschel's views were shared by the politician, traveller, linguist, hymnologist, and former editor of the *Westminster review*, Sir John Bowring

(1792–1872), who was approached by Stenhouse. Bowring argued that there were not simply 'two truths', the religious and the scientific, for two truths could not contradict one another anyway; now was the time 'to emancipate ourselves from the tyranny of all dogmatizing creeds'.[44] The cause of truth was better served by latitudinarianism in inquiry.

The issue was ventilated again during the British Association meeting at Bath, between 14 and 23 September 1864, when fresh attempts were made to canvas signatories. Attention focused on the Presidential Address of Sir Charles Lyell, in which he appeared to shift his ground from plutonism to hydrothermalism and to admit fossils (*Eozöon Canadense*) in Azoic, or pre-Cambrian, rocks.[45] The *Saturday review* saw Lyell's address as cautious and frank, and was satisfied that:

> While shibboleths and counter-shibboleths are being industriously circulated to fetter the investigations of scientific inquirers, we may trace, in the modest tone which prevailed at the meeting of the Association, a much more trustworthy safeguard against undue presumption than any which the most amply signed declaration could possibly supply.[46]

Thomas Hirst, who was to become the Association's General Secretary in 1866, was

> pleased to notice the applause with which every protest against fettering science by religious dogmas was received. Colenso has done good work, the first fruits of which are already evident in the increased courage which scientific men have acquired to expressing [sic] frankly their convictions.[47]

The Bath meeting did witness some feeling for the opposite views. The editor of *The Bath Chronicle*, which reprinted the letters by Herschel and Bowring that had appeared in *The Times* nine days earlier, saw a conspiracy at work among a 'dangerous clique' which was

> assuming a prominent place in the Association, and may make it as deservedly unpopular as it is now deservedly popular. From circumstances which transpired at the Bath Congress we think there can be no doubt that a certain section are endeavouring to use the influence of this grand Scientific League in furtherance of heretical teachings, as a prop to the Scepticism which has of late years met with disciples even in the ranks of duly authorised Christian Ministers. When men, however accomplished they may be, begin to act upon their limited reason against Revelation, and oppose their crude deductions against the inspired truth of Scripture, we feel that they trespass beyond the bounds of lawful research, and are led by intellectual conceit into rash and delusive imaginings. Heaven forbid that we should do aught to limit intellectual enquiry—what we deprecate is the puffed up pretension that strives by hasty generalisations in the name of Science to over turn faith in the Book of Books.[48]

Others, however, saw the matter in a different light. In an article entitled 'Sir John Herschel and the New Test', the *Saturday review* joined in condemning the 'inquisitors', 'very active of late', requiring 'oaths of abduration' among men of science. It was clear, however, that scientists and theologists were divided and that blame attached to both sides: some men of science

> seem to value their studies, as an Orangeman values his religion, chiefly for the opportunity it gives them of making their natural enemies uncomfortable. It is impossible to read some recent speculations upon delicate scientific questions without seeing that the author has his old antagonists, the parsons, in his mind's eye all the time, and is experiencing the same kind of glee as a small boy feels when he is tying a tin kettle to a dog's tail.[49]

Certainly no one was more violent in his opposition to the Declaration than De Morgan who, as a contributor to the *Athenaeum*,[50] gave wide publicity to his view that the Declaration would do 'irreparable damage'. Like Herschel and Bowring, and Daubeny, De Morgan regarded the Declaration as a 'vote of censure on free inquiry'; but even more, as he wrote in October 1864, as a theological plot to prevent theological inquiry itself.[51] 'This Declaration sins in nothing except the assumption that theology is perfect, and the discordance between theological assumption and scientific theory are to be left side by side until it shall please God to allow us to see the manner in which they may be reconciled.'[52] 'Is this the way which it has hitherto pleased God that we shall be allowed to see? Not at all: apparent contradictions have never been reconciled without an examination of both sides.'[53]

He then printed in the *Athenaeum* a clever Broad Church paraphrase of the Declaration for 'Students of theology and nature', to which he added:

> We, the undersigned Students of Theology and of Nature, desire to express our sincere regret, that common notions of religious truth are perverted by some in our own times into occasion for casting reproach upon the advocates of demonstrated or highly probably scientific theories. We conceive that it is impossible for the Word of God, as correctly read in the Book of Nature and the Word of God as truly interpreted out of Holy Scripture, to contradict one another, however much they may appear to differ . . . [54]

Perhaps Herschel and Bowring would feel more disposed to sign this, De Morgan speculated. Perhaps the geologist Adam Sedgwick would sign, for De Morgan felt certain the Declarationists had only the signature of an obscure 'Adam Sedgwick, M.D.' (He was wrong!)[55]

Privately, De Morgan had penned an even more delicious translation of the 'illogically worded' declaration for the benefit of his fellow logician, Hamilton:

> Colenso and others are inquiring into the distinction between historical record and revealed doctrine. This won't do; our system is so shaky that it will not stand substantial repair. Why! we dare not knock in a nail, for fear the rotten beam should crumble. We must have a declaration that two truths cannot disagree. Some must sign for love, and more for fear; we must say in conversation that none but an atheist would refuse to sign.[56]

Efforts to collect further signatures continued during October, and De Morgan instanced their failures: Richard Owen, Faraday, Whewell, Airy, Lyell, Murchison, Tyndall, and, notably, Sabine.[57]

The Declaration was in the foreground of events surrounding the first meeting, on the 3 November 1864, of the X-Club, whose nine members dedicated themselves to a 'devotion to science, pure and free, untrammelled by religious dogmas'[58]—a sentiment underlying the recommendation of George Busk, one of the X-Club, that the Royal Society should award its Copley Medal to Charles Darwin.[59]

At the end of the month, on 30 November 1864, the issue found an echo in Sabine's presidential address to the Royal Society, and particularly in the citation accompanying the award of the Copley Medal to Charles Darwin for his 'Important researches into geology, zoology and botanical physiology'. After deliberately excluding Darwin's *Origin of species* from the grounds of his award in his spoken address, Sabine, challenged by Huxley and Hooker, softened his published address with a weak defence that 'every real *bona fide* inquiry into the truths of nature is in itself essentially legitimate'.[60]

In the meantime De Morgan dropped his anonymity, for he had at last been solicited by the Declarationists—like a 'grocer's supplication for tea and sugar patronage'[61]—and blasted forth against a hostile Jericho. To Herschel's reasons for rejection—intrusion and tolerance of different views—he added, in a letter to the *Athenaeum* of 19 November 1864, his own speciality of logic:

> Even if I did regret [the divorce of religion and science] I should be ashamed to put my name to bad chemistry made to do duty for good reasoning. The declaration is an awkward attempt to saturate sophism with truism; but the sophism is largely left in excess.

The promoters were 'lineal successors of Tony—Fire—the faggot' whose 'unholy fire must be trodden out' or at least confined to theology. The whole exercise was in bad taste and liable, as Daubeny had said, to produce intolerant disputes and partisanship within the scientific community.[62] Sarcastically, he offered his signature, provided it was asterisked, and his letter was also published. Perhaps De Morgan's ferocity was not unconnected with his interest in Spiritualism, which gave his protest for freedom of inquiry[63] a special significance.

In Ireland, Hamilton, fortified by gossip from De Morgan, followed the issue of what he labelled 'the Fortieth Article of Religion'[64] with great interest. On learning of Herschel's reply, he drolly speculated, in the *Athenaeum* of 31 December 1864, about who had drawn up such a declaration, 'At least we are not in *this* case asked to sign for "the love of God" ' (a reference to the 39 Articles). Hamilton obviously waited with keen anticipation the pleasure of saying, No!

> I am happy to say, that I have not been hitherto asked to *sign* the document; perhaps an *Astronomer* is regarded as scarcely a *Student* of the *Natural Sciences*.[65]
> If I were a Professor of *Mathematics* I might have a still better chance of escaping.[66]

But when the invitation came in December 1864, he declined simply on grounds that the Declaration lacked authority.

At the end of 1864, when there was still no sign of publication, De Morgan speculated that the 'young' promoters were experiencing difficulties from the withdrawal of names, the poverty of the list, and the 'manifest absurdity of a scientific protest which has no scientific authority'.[67] The manuscript list, however, reveals few withdrawals,[68] and at last, in May 1865, the Declaration was published. The *Chemical news* reported that 'for the small charge of twopence, anyone can possess this Directory of Orthodoxy'.[69]

4. *The signatories*

(i) THE ROYAL SOCIETY AND THE DECLARATION

If this were all the *Chemical news* was prepared to say about it, despite the presence of chemists on the list, De Morgan probed more deeply in the *Athenaeum*.[70] In particular, he examined the list for Fellows of the Royal Society, 'as the oldest, largest and highest' of learned societies. His analysis, which we shall extend, is of considerable interest. Indeed, of all the possible scientific groups represented among the signatories perhaps none occupied a position more central than the Royal Society. Even though Fellows constituted less than ten per cent of the total list, their views would have undoubtedly counted for a great deal in society generally. For many years the religious attitudes of Fellows of the Royal Society have attracted the attention of historians of science[71] and religious apologists.[72] It therefore seems appropriate to digress briefly in order to consider the religious profile of the Society as revealed by the Declaration. Was there, for example, any connexion between the Declarationists and the 'amateurs' whom Babbage and Granville described in 1830?[73] Or did the 'scientists' revolt' of the 1830s and 1840s bring into the Society more tolerance of religious diversity?

It would be extremely interesting to chart a religious tableau for the Fellowship in this period, similar in intention to those now being constructed for the early Fellows in the middle of the seventeenth century. For the moment, unfortunately, an elaborate statistical analysis lies beyond our resources, and we have deliberately restricted ourselves to a few descriptive comments.

On St Andrew's Day 1864, there were 605 Fellows. Of this number 65 (or roughly ten per cent) had signed the Declaration.[74] Now, as De Morgan fairly observed, one had to 'abate one out of ten as inaccessible, living abroad, travelling, etc., and [therefore] we must see that eight have refused where one has consented. Refused, remember; for "yes" or "no" has been put to everyone'.[75] He then used the Babbage-Granville characteristic of 'the F.R.S. who Fairly Represents Science', namely the criterion of publication in the *Philosophical transactions*. (He recognized that many scientists published elsewhere, but claimed that this did not seriously affect the argument.)[76] Of the 605 Fellows alive in 1864, 166 had published in the *Philosophical transactions* (or 27 per cent); of these, only 18 were signatories.[77] De Morgan therefore asserted that the Declarationists captured only about one out of nine Fellows who published scientific work. If one disregards De Morgan's severely exclusive criterion of scientific status and accepts instead the authority of the *Dictionary of national biography*, 48 of the 65 may be considered 'scientific'. Of the 65 Fellows who signed the Declaration, the majority (44) were elected before the reforms of 1847–8; of these, 15 were 'amateurs'. Of the 21 elected in or after 1848, only one (General Sir Henry Rawlinson, an Assyriologist) could be considered an 'amateur'. Eleven of the 65 were clergymen of the Church of England, eight of whom were elected before 1848. Although there may have been a predisposing tendency to have signed the Declaration 'attacking' science among those elected before 1848 (whose average age in 1865 was 66) as against those elected after the reforms (whose average age was 55), there is no justification for asserting that scientific or 'amateur' status *per se* necessarily reflected attitudes towards the Declaration. Nor did age seem material.[78] The Declaration was, quite simply, a list of scientists with a point of view.

Perhaps only a dozen of the Fellows of the Royal Society listed in Table 1 are particularly well known to historians of science (Brewster, Challis, Calvert, Gilbert, Glaisher, Gosse, Harley, Joule, Moseley, Reade, Sedgwick, and Stenhouse). Of these, Brewster, Sedgwick, and Joule were undoubtedly the promoters' most distinguished signatories. Brewster's daughter later recalled:

> . . . with what kind of proud satisfaction he put into my hands a list of scientific men of high standing [sic] who had avowed their faith in Scripture. He referred to this as a token that there was no natural tendency in science to shake the faith of men in the Word of God, and no justice in regarding men of real science as more inclined to infidelity than others.[79]

Sedgwick signed, considering the Declaration 'a kind of peace offering by a body of men who were honestly searching after truth and ready to abide by it wheresoever found'.[101] As for Joule, John Tyndall was distinctly disparaging; Joule, he wrote to Debus, had 'put his name to that drivelling declaration which they are now sending round regarding Science and Religion. It appears to me that he has never yet raised himself into the real region of philosophy to which his experiments point'.[102]

One would have liked to see a pattern emerging among the Fellows

TABLE I

SIGNATORY FELLOWS*

Name	*Dates*	*Date of election to Royal Society*
Rev. Henry Hervey Baber	(1775–1869)	1816
J. Hutton Balfour	(1808–1884)	1856
T. Graham Balfour	(1813–1891)	1858
James Bateman	(1811–1897)	1838
Samuel Husband Beckles[80]	(1814–1890)	1859
*Thomas Bell	(1792–1880)	1828
Rev. Joseph Bosworth	(1789–1880)	1829
*James Scott Bowerbank	(1797–1877)	1842
*David Brewster	(1781–1868)	1815
*Charles Brooke	(1804–1879)	1847
Alexander Bryson	(1802–1869)	1854
Major T. Seymour Butt[81]	(1805–1890)	1836
*Rev. James Challis	(1803–1882)	1848
General Francis Rawdon Chesney	(1789–1872)	1834
Rev. Henry Christmas (Noel Fearn, pseud.)	(1811–1868)	1842
Rev. William Clark	(1788–1869)	1836
Earl of Enniskillen[82]	(1807–1886)	1829
*Frederick Crace Calvert	(1819–1873)	1859
Edward William Cooke	(1811–1880)	1863
Mathew Curling Friend[83]	(1792–1871)	1820
*Joseph Henry Gilbert	(1817–1901)	1860
*James Glaisher	(1809–1903)	1849
*Philip Henry Gosse	(1810–1888)	1856
Rr. Admiral William Hutcheon Hall	(?1797–1878)	1847
Rt. Hon. Robert Adam Christopher Nisbet-Hamilton[84]	(1804–1877)	1833
John Scandrett Harford	(1785–1866)	1823
*Rev. Robert Harley	(1828–1910)	1863
James Heygate[85]	(1823–1872)	1843
*John Higginbottom[86]	(1788–1876)	1852
John Hogg	(1800–1869)	1839
*Samuel Elliott Hoskins	(1799–1883)	1843
Charles Hood[87]	(1805–1889)	1843
Robert Hudson[88]	(1801–1883)	1834
*Robert Hunt	(1807–1887)	1854
Cuthbert William Johnson	(1799–1878)	1842
Percival Norton Johnson[89]	(1793–1866)	1846
Charles Handfield Jones	(1819–1890)	1850
*Thomas Rymer Jones	(1810–1880)	1844
*Thomas Wharton Jones[90]	(1808–1891)	1840
*James Prescott Joule	(1818–1889)	1850
George Lowe[91]	(1788–1868)	1834
*John Denis Macdonald	(1826–1880)	1859
Gilbert Wakefield Mackmurdon[92]	(1799–1869)	1839
Rev. Robert Main	(1808–1878)	1860
Sir John Maxwell, Bart.[93]	(1791–1866)	1829

Name	*Dates*	*Date of election to Royal Society*
*Rev. Henry Moseley	(1801–1872)	1839
Sir Edwin Pearson[94]	(1802–1883)	1833
Gen. Sir Henry Creswicke Rawlinson	(1810–1895)	1850
Rev. Joseph Bancroft Reade	(1801–1870)	1838
Sir John Richardson	(1787–1865)	1825
Henry D. Rogers[95]	(1808–1866)	1858
Rev. Adam Sedgwick	(1785–1873)	1821
William Sharp	(1805–1896)	1840
Alfred Smee	(1818–1877)	1823
Samuel Reynolds Solly[96]	(1781–1866)	1823
John Spencer Stanhope[97]	(1787–1873)	1816
*John Stenhouse	(1804–1880)	1848
Alexander John Sutherland[98]	(1811–1867)	1846
*Col. William Henry Sykes	(1790–1872)	1843
Alexander Tweedie	(1794–1884)	1838
Nathaniel Bagshaw Ward	(1791–1868)	1852
John Waterhouse[99]	(1806–1879)	1834
James Whatman[100]	(1813–1887)	1840
Rev. John Wilson	(1804–1875)	1845
James Yates	(1789–1871)	1839

* An asterisk indicates the Fellow published in *Philosophical transactions*. All appear in the *D.N.B.* except those annotated.

who signed, which would enable us to draw conclusions concerning the possible influence of age, status, and the coincidence of scientific and religious attitudes generally. The list in Table 1 is, however, so small that it is extremely difficult to draw such substantive conclusions. One may reasonably speculate that the list includes a smaller proportion of mathematicians, physicists, engineers, and physicians than the proportion of these disciplines within the Royal Society at the time, and proportionally more astronomers, chemists, surgeons, biologists, and geologists than would be found in a distribution by discipline of Fellows in the 1860s.[103] For a wider view, however, we must turn from the Royal Society to the other scientific societies. In doing so we may add a few details to the picture that De Morgan described.

(ii) OTHER SIGNATORIES

These comments about the age and fields of interest of the signatories apply only to Fellows of the Royal Society. What of the other 652? Of that number, probably about 500 bore some claim to be considered 'men of science'. Among the better known of these (21 of whom were subsequently elected to the Royal Society, see Table 2) were: Thomas Anderson, J. Lloyd Bullock (a pharmaceutical manufacturer and a founder of the Royal College of Chemistry), Edward Divers, Alexander

Stewart Herschel (Herschel's son!), Edward J. Mills (then another factotum of Stenhouse), Frederick Penny, J. H. Pepper (of Pepper's Ghost), W. H. Perkin, and Thomas Richardson.

TABLE 2

SIGNATORIES WHO BECAME FELLOWS AFTER 1865

Name	*Dates*	*Date of election to Royal Society*
James Risdon Bennett	(1809–1891)	1875
Rev. Miles Joseph Berkeley	(1803–1889)	1879
John Jeremiah Bigsby	(1792–1881)	1869
William Carruthers[104]	(1830–1922)	1871
George Dickie	(1812–1882)	1881
Edmund Divers[105]	(1837–1912)	1885
Sir Walter Elliot	(1803–1887)	1878
Charles Edward Groves[106]	(1841–1920)	1883
Alexander S. Herschel	(1836–1907)	1884
John Eliot Howard	(1807–1883)	1874
John Whitaker Hulke	(1830–1895)	1867
Edward Hull[107]	(1829–1917)	1867
Sir George Johnson	(1818–1896)	1872
Herbert McLeod[108]	(1841–1923)	1881
Edmund J. Mills[109]	(1840–1921)	1874
Sir William Henry Perkin	(1838–1907)	1866
Thomas Richardson	(1816–1867)	1866
William Chandler Roberts	(1843–1902)	1875
George James Snelius	(1837–1906)	1887
George James Symons	(1838–1900)	1878
John Wood	(1825–1891)	1871

Note: All appear in the *D.N.B.* except those annotated.

Of interest are a few unsolicited comments with which some signatories embroidered their signatures. The naval surgeon, J. D. (later Sir John) MacDonald, F.R.S., signed 'with perfect concurrence, well knowing that "a little knowledge is a dangerous thing" and is the basis of so much presumption amongst us poor weak human beings'. The Revd W. Lockwood, a Fellow of the Linnean Society (no. 359), offered his signature with the honest observation that 'I can hardly class myself with scientific men'. Vice-Admiral William Ramsay (no. 579) agreed wich the spirit, not the wording, of the document: 'I believe that the scriptures are given to show us the way to salvation, and not to give instruction either directly or indirectly in the natural sciences.' By signature 647, it is obvious that the promoters were finding great difficulty in obtaining further scientific signatures. The end of the list was dominated by solicitors and clergymen, and even a sprinkling of mustered foreigners. No women were among the signatories. Most of the men were upwards of fifty years of age.

De Morgan's greatest scorn was reserved for the totally unqualified signatories. In the letter 'B' (taking his cue from 'Berger') he calculated eight names possessing no degrees:

> one is an 'acting paleontologist' [to the Geological Society of Ireland], one is F.K.Q.C.P., which we do not know how to elongate, unless it be into 'Fellow of the Kew Queer Collection of Plants',[110] [another] is described as 'Missionary to China',[111] and one only as 'J.P.'.[112]

Indeed, nearly 100 names bear no degree or qualifications; others were obviously of school teachers, including some from Harrow, Berger's old school.[113]

De Morgan was perhaps intolerant of Irish medical qualifications, but his other criticisms seem justified. No wonder the *Saturday review* was so scathing:

> A considerable number of the signatories are not of sufficient importance to have made it worth their while to inform the world whether they thought science and religion reconcileable or not. No doubt, however, it was an inviting opportunity of placing the fact that they were men of science beyond the reach of any captious doubt.[114]

The largest single group among the other signatories was, as might be expected, the medical profession, with 251 signatures (35 per cent), followed by the Fellows of the Geological Society with 116 signatures (18 per cent), the Chemical Society with 78 signatures (12 per cent), the Linnean Society with 73 signatures (11 per cent), and the Astronomical Society with 64 signatures (9·8 per cent). In proportion to the membership of these learned societies, the Declarationists represented about 13 per cent of the Geological Society, 23 per cent of the Chemical Society, 16 per cent of the Linnean Society, and 14 per cent of the Astronomical Society. (The medical signatories represented only 2 per cent of the something like ten thousand people who could be described as 'medical men'.) As far as the geologists are concerned, none of the signatories was on the Council of the Geological Society in 1864–5 (although some were from time to time officers of the society). While Sedgwick signed, other influential geologists (notably Murchison, Lyell, and W. L. Hamilton, the President of the Society in 1864–5, did not). One suspects that those in the Geological, and other societies, who did sign, represented a significant 'backbench', possibly provincial, element.

More detailed studies of these fairly significant minorities could be revealing. How did systematic religious views (and, by implication, systematic belief systems generally) inform the actual work of men of science? More specifically, how did these individuals see the connexion between their science and their beliefs. Such questions function at both cognitive and institutional levels. It is well known that epistemological assumptions and social values momentarily governing scientific fields or disciplines have worked either to encourage or resist intellectual or institutional transformations.[115] Did the values of the Declarationists operate to oppose experimentation in some fields or to intensify study in others?[116]

It is also clear that certain institutions are particularly prominent among the signatories to the Declaration. For example, we have already noted that all the protagonists were close to the Royal College of Chemistry; that thirty-nine 'students' of the college signed (one of whom, G. J. Snelius, became F.R.S.); and that others had been taught at the college or taken its chemistry classes while studying at the Royal College of Mines. It is, however, extremely difficult to proceed from these facts to larger historiographical generalizations. Although it would be important to be able to demonstrate epistemological justifications for the chemists' support of the Declaration (and to compare their views on religion with, say, the defence of vitalism against the development of animal chemistry[117]), such possibilities are at present mere speculation. We may also speculate that the chemists who signed did so, not only through belief but through social pressure as well, owing to the strong 'old-boy network' which had come to exist around the Royal College of Chemistry by 1865. This network was directly inspired by the College's first director, A. W. Hofmann. Although Hofmann did not sign the document, given his close friendship with Herbert McLeod (who went to Germany with him for a time during 1865), it seems unlikely that he would not know about it. It is also conceivable that the loss of Hofmann to Germany in 1865 made the staff and students of the College, both past and present, fear for its future. Consciously, or unconsciously, they may have wanted to stress the 'orthodoxy' of chemistry, and of their beloved institution, through the Declaration.[118]

5. *Conclusions*

By 1866 the dust had begun to settle. Although the Declaration would be cited as late as 1880 as evidence of the fundamental unity of scientists,[119] its force was spent. Several of its supporters, including Berger and Brewster, soon died, Brewster being mourned by one clergyman as a great man 'quite contented to make the Bible [his] all'.[120] The Declaration today is nearly forgotten, remembered only as a rear-guard defence of scriptural belief. At the time its chief effects were to be seen not only in the beginning of the X-Club, but also in the foundation of the 'Victoria Institute or Philosophical Society of Great Britain'. The mandate of the Victoria Institute was

> to investigate fully and impartially the most important questions of Philosophy and Science, but more especially those that bear upon the great

> truths revealed in the Holy Scripture, with the view of defending these truths against the opposition of science, falsely so-called.[121]

Its prolegomena, composed in 1865, by James Reddie, a civil servant given to writing pseudo-scientific tracts, recounted events since *Essays and reviews* and described the need for a 'defence movement' against the 'undue influence of scientific coteries'. Reddie appealed for a 'Science of Science . . . a proper correlation of all the various sciences into one grand and consistent Philosophy, which will be the interpretation of the nature of those things as ordained by the one true God'. By December 1866 200 members had been enrolled, among whom at least 11 were F.R.S.s (Sir H. Barkly, J. F. Bateman, T. K. Callard, The Archbishop of Canterbury, Revd J. Challis, Revd W. H. Dallinger, Sir J. Fayrer, F. Bisset Hawkins, Revd T. R. Robinson, Rt Hon Lord Teignmouth, and Canon H. B. Tristam) and at least 30 others had some claim to scientific knowledge.[122]

In the last analysis, what did the Declaration demonstrate? In Berger's view the Declarationists were attempting to

> combat the prevailing opinion that the study of science necessarily tends to infidelity; to show that the deepest researches and a belief in the most wonderful discoveries of our day, are not incompatible with a belief in the Bible; that men of science, and those who study nature for business or pleasure may be and are true Christians.[123]

Many Victorian men of science would have found no difficulty in subscribing to this position. If there were difficulties in supporting the Declaration, they arose more from the fear of sterile controversy and discord among men of science, and from an aversion to the imposition of theological tests, than from an active objection to the Declarationists' claims and intentions. These fears underlay the promotion of science teaching at Rugby by J. M. Wilson, who encouraged studies which would 'help free our successors from the same partition of soul, the same divided allegiance, from which the present generation suffers'.

> At present there is secret, if not avoided, hostility between religion and science, or at any rate a distrustful toleration; nothing but active cooperation will permanently reconcile them. To endeavour not to see the results and tendencies of modern science is folly in the highest degree. The study and knowledge of the seen is sure to react on the study of the unseen; and he will entertain these studies in perfect harmony, and he only, in whom the scientific and religious ideas are allowed to grow up, not in antagonism, but fearlessly and freely, side by side, cooperating in the formation of a reverent, active, and independent mind, and well-balanced judgement.[124]

Huxley, Tyndall, and Spencer would have agreed that a study of science induced lofty ideas, but would have denied that these ideas necessarily supported religious belief. However, a large number of working scientists in the 1860s and 1870s probably did agree with Wilson; certainly the close connexion between Public Schools and the Church implied that most scientific sympathizers on the staffs of such schools would have supported him[125] and were not disposed to join Huxley's metaphysical crusade of scientists clad in shining armour, battling against dark forces of theological obscurantism.[126]

In the wake of the Declaration debate, and with the gradual fragmentation of the common context provided for so long by natural theology, one sees a gradual differentiation of attitudes among men of science. There were those who pronounced themselves to be agnostic or atheist and therefore anti-religious; there were those who remained religious (or uncritical) and who refused to join issue; and there were those who sought to reconcile their activities with their beliefs, either by attempting a reconciliation through (a) a new understanding of biblical literalism, through (b) faith, or through (c) a complete separation of science and religious belief in distinct intellectual compartments. Thus there were some men of science, including George Romanes and John Hall Gladstone, and teachers like James Wilson, who tried, by discovering ambiguities in scripture to stress that Christianity, tested in the white heat of scientific study, would emerge tempered and stronger for the experience. John Gladstone, using the current vocabulary of militancy, wrote:

> The store houses of natural science have often been ransacked for weapons against the old book; the defenders of the faith have sometimes shrieked with alarm, and the assailants have sung their paean in anticipation of victory. Earthworks which form no part of the original fortress have been easily carried, but the citadel itself has remained unshaken and the very vigour of these repeated attacks has proved how impregnable are its valuable walls.[127]

Second, there were careful attempts, such as W. H. Freemantle's *Gospel of secular life* (1882), to urge that religion in fact embraced all forms of inquiry and that no conflict was necessary. More often, however, it appears that the third, and preferred, solution to the problem was to 'reconcile' by 'separation'—to create, in effect, two separate systems of values, two needs, which in A. J. Balfour's influential expression 'stand . . . upon perfect equality'.[128]

By the turn of the century, the practice of holding public meetings of scientists, called together to give their reasoned views on the reconciliation in their own lives, of the demands of science and the rewards of religion, was clearly acceptable. But by then the social and intellectual dominion of science had become firmly placed, so firmly indeed, that it would itself become the target of those who pressed the claims of 'higher truth'.[129]

Whatever the Declaration demonstrated about the attitudes of *scientists*, it was of course widely interpreted in terms of its relevance to *science*. Chadwick instances two forms of opposition between science and religion: (1) opposition arising from the teachings of science as opposed to the literal teachings of the Bible, and (2) opposition arising from the fact that scientific explanations require a system of uniform naturalistic and fundamentally materialistic notions of causation, independent of metaphysical explanation.[130] To these we must add a third possibility: the opposition of free inquiry to the unquestioning acceptance of prescribed dogma. Ironically, the Declaration sowed the seeds of its own destruction, in effect reporting a vote widely interpreted as 'nine to one for unfettered publication of thought'. Yet at the same time it succeeded in demonstrating that the learned scientific community would not parody the theologians and raise its voice against its 'Colenso-Huxleys' or 'Wilson-Tyndalls'. It would not institute club-law to intimidate those who thought differently. How the remorseless critic De Morgan rejoiced that the list of signatures was to be interred by Berger in the Bodleian Library, a relic of Victorian bigotry, for some remote historian to disclose:

> The Declaration has proved something: we know that the followers of natural knowledge did not raise their voices against inquiry; we know that they *would not*. Science was asked to say Don't tell me! and she answered, I like to be told.[131]

This separation of free inquiry from a supposed slavish obedience to literalism—a problem loaded with explosive overtones, at a time when the Church of England was itself anxiously studying the implications of the *Syllabus errarum*[132] and when the Papacy's political role in Italy seemed defined in terms of ecclesiastical resistance to political liberalism—was an important factor in the Declaration debate. In Britain the long struggle to abolish religious tests, beginning in 1864 and lasting until 1871, is revealed in the pages of Hansard as of fundamental political importance. The sense of mortal conflict with the enemies of intellectual freedom, mirrored in the work of Draper and White, was echoed in the lives of embattled agnostics. The devices of 'reconciliation' undoubtedly helped 'save the phenomena', but the consequences of this contemporary separation included a scientific creed separated not merely from religious orthodoxy but from all value systems not specifically functional to scientific endeavour. This creed gave a contextual foundation to the claims of positivist 'objective' science, which rejected the intellectual mediations that an earlier generation found essential. The debate surrounding the Declaration of 1865 crystallized this rejection and provoked questions which escaped easy answer.

NOTES

[1] Owen Chadwick, *The Victorian Church* (London, 1970), Part II, p. 3.

[2] R. M. MacLeod, 'A Victorian Scientific Network: the X-Club', *Notes and records of the Royal Society of London*, xxiv (1969), 305–22; J. V. Jensen, 'The X-Club: fraternity of Victorian scientist', *The British journal for the history of science*, v (1970–1), 63–72; W. H. Brock, 'Prologue to heurism', in History of Education Society, *The changing curriculum* (London, 1971), pp. 71–85;

W. H. Brock and R. M. MacLeod (eds.), *The life and work of Thomas Archer Hirst, F.R.S. (1830–1892)*, in press.

3 P. B. Hinchcliff, *John William Colenso, Bishop of Natal* (London, 1963), p. 39.

4 Colenso's appeal began in June 1864 and lasted until December 1864, with judgement given on 20 March 1865. The appeal swung on the legal question of the jurisdiction of the Natal Bishopric and not on Colenso's theology. He was found legally still Bishop of Natal. His return to South Africa in 1865 subsequently provoked a schism when the Canterbury Convocation resolved to appoint a new Bishop of Natal in 1867.

5 Chadwick, op. cit. (1), p. 84.

6 Described in R. P. Flindall, *The Church of England, 1815–1948. A documentary history* (London, 1972), p. 179, and A. O. J. Cockshut, *Religious controversies of the nineteenth century* (London, 1966), pp. 198–201.

7 Hansard, *Parliamentary debates*, 3rd ser. clxxvi, col. 1546 (15 July 1864), also quoted in Standish Meacham, *Lord Bishop, the life of Samuel Wilberforce* (Cambridge, Mass., 1970), p. 250.

8 W. F. Monypenny and G. E. Buckle, *The life of Benjamin Disraeli* (revised edn., 2 vols., London, 1929), ii. 108.

9 *The chronicle of Convocation, being a record of the proceedings of the Convocation of Canterbury* (1861–64), Lower House, 21 April 1864, p. 1577. No discussion is recorded and no signatures are listed. According to Berger (see note 10, below), this memorial was prepared by Gillman and McLeod, but Wordsworth (later Bishop of Lincoln) attributed it to Stenhouse.

10 *Chronicle*, op. cit. (9), p. 1577. *The Declaration of students of the natural and physical sciences* (London, n.d. [1865]). The only description of the Declaration we know is that of E. G. W. Bill, 'The Declaration of students of the natural and physical sciences, 1865', *Bodleian Library record*, v (1954–6), 262–7. In 1872 the Revd A. H. Berger, a former student of Corpus Christi College, Cambridge, and a son of Capel Berger, presented to the Cambridge University Library a bound collection of replies to the Declaration, together with his father's recollections of the episode. This volume is today kept in the University Archives, Cam. Add.5989 (hereafter 'Berger, *Documents*'). We are indebted to J. D. Burchfield and D. B. Wilson for having brought this volume to our attention. Interestingly, the Declaration has recently been given fresh prominence. See C. A. Russell, 'The end of an era?', Unit 16 of *Science and belief: from Copernicus to Darwin* (London: The Open University, 1973), pp. 66–7.

11 Bodleian MS. Add.C.102, which also contains the printed version, 13×9 inches, xx+67 leaves, presented by Capel Berger. The original signatures occupy 64 pages of this handsome green morocco-bound volume. The printed pamphlet with its alphabetically arranged names was placed in the front and labelled by hand, 'Key to the Volume of Original Signatures'. Inked in against each name was a number referring to the numerical order of signatures in the volume.

12 R. M. Young, 'Natural theology, Victorian periodicals and the fragmentation of the common context' (in press). Cf. also R. M. Young, 'The impact of Darwin on conventional thought', in A. J. Symondson (ed.) *The Victorian crisis of faith* (London, 1970), pp. 13–15.

13 *Saturday review*, 24 September 1864, p. 396.

14 Although Berger's is the last printed signature (no. 716), the manuscript closes with a late-comer (no 717), Robert Oxland, professor of chemistry at Toland Medical College, San Francisco. His signature obviously arrived after publication. Berger attended Harrow, became an active colour chemist, but died of poisoning at the age of 29; see T. B. Berger, *A century and a half of the House of Berger* (London, 1910), p. 62, and *Journal of the Chemical Society*, xxii (1868), p. iv.

15 Chadwick, op. cit. (1), p. 7, gives McLeod the honours. Berger's role was first established by Bill, op. cit. (10), and is confirmed in Berger, *Documents*, op. cit. (10).

16 A. W. Gillman, *Searches into the history of the Gillman or Gilman Family . . .* (London, 1895), pp. 194–5, with portrait. He entered the R.C.C. at the age of 13, and from 15 to 19 he was Hofmann's private assistant; see A. W. Hofmann, *Theories of modern chemistry* (London, 1865), p. ix.

17 Obituary (by J. M. Thomson), in *Journal of the Chemical Society*, cxi (1917), 342–7; *Essex review*, xxvi (1917), 29. Although Berger (Berger, *Documents*, op. cit. [10], p. i) cited Howard as a protagonist, there is no evidence that he played a major role in the proceedings.

18 W. Tilden, obituary of Groves, in *Journal of the Chemical Society*, cxvii (1920), 464–6. Groves left £10 000 to the Royal Institution for the Groves Endowment Fund for Promotion of Scientific Research. Whatever their subsequent fame (or, in the case of Gillman and Berger, obscurity), only Stenhouse was well known in 1864; hence Hamilton's sarcasm in addressing his reply to 'Rev. Charles Grove [sic], or C. E. Grove, Esq', and De Morgan's rhetorical 'fancy a person, whose very name is unknown to scientific men, and is not to be found in the London [Post Office] Directory, at the address given for it, calling upon the Herschels and Hamiltons to sign a Declaration!'; see De Morgan, *Athenaeum*, 31 December 1864, pp. 894–5. Of course, Groves was using Stenhouse's address as a postbox.

[19] H. F. Morley, obituary of McLeod, in *Journal of the Chemical Society*, cxxv (1924), 990–2. At the R.C.C. McLeod helped to discover the aniline dye, Magenta; he later became professor of experimental science (afterwards chemistry) at the Royal Indian Engineering College, Cooper's Hill. He also devised the McLeod gauge for measuring gases at low pressure. See *Proceedings of the Royal Society of London*, Secn. B, xcv (1923–4), 487.

[20] E.g. William Murray, M.D., lecturer in physiology at Durham, added 'Baptist' (no. 78); Robert D. Mordue, M.R.C.S., L.S.A., of Newcastle, added 'Roman Catholic' (no. 79).

[21] De Morgan, in *Athenaeum*, 1 July 1865, p. 19, gave 38, and was so followed by Chadwick, op. cit. (1), p. 7.

[22] De Morgan, *Athenaeum*, 1 July 1865, p. 19.

[23] McLeod saw this as an advantage, a pledge by the younger generation of men of science. See Berger, *Documents*, op. cit. (10), p. 81. The substitution of the term 'Students' for Brewster's recommendation, 'Cultivators', was probably an attempt to forestall criticism.

[24] Ibid., pp. 39–40, an anonymous letter to *The Times*, 22 July 1864, attributed by Berger to Hofmann's colleague, John Percy.

[25] De Morgan, *Athenaeum*, 1 July 1865, p. 19.

[26] L. Levi, *Report of the British Association for the Advancement of Science, 1868* (London, 1869), pp. 169–73, 196–7. Levi analysed the membership figures of 120 learned societies in the United Kingdom and found an aggregate membership of 60 000, which reduced to 45 000 if an allowance was made for overlapping membership. We have further reduced this figure by excluding membership of the Royal Geographical Society, the Antiquarian Society, and various printing societies; i.e. the figure of 40 000 refers solely to membership of natural and physical science societies. For membership figures for 1860, see R. M. MacLeod and E. K. Andrews, *Selected science expenditure and manpower statistics, 1850–1914* (typescript, Science Policy Research Unit, University of Sussex, 1968, 1970). Our 'realistic' figure of 5000 is derived from Levi's data on metropolitan societies.

[27] Horace G. Hutchinson, *Life of Sir John Lubbock* (London, 1914), i. 57–8. A copy of the memorial and John Lubbock's reply can be found in Royal Society MSS., HS, 13.12.

[28] Royal Society MSS., HS, 13.12, Herschel to Spottiswoode or Lubbock (c. February 1861).

[29] J. D. Hooker to J. Lubbock, 29 February 1861, quoted in Leonard Huxley, *Life and letters of Sir J. D. Hooker* (London, 1918), ii. p. 54.

[30] Ibid.

[31] A. P. Stanley, quoted in Cockshut, op. cit. (6), p. 198, and referring to a report in the *Guardian*, 27 April 1864. Cf. 'The new test', *Saturday review*, 5 March 1864, pp. 274–5.

[32] The printed version has 716 signatures. See note 14.

[33] *The Times*, 21 July 1864, p. 9, reprinted in C. P. Daubeny, *Miscellanies* (Oxford and London, 1867), vol. ii., separately paginated section IV, p. 130.

[34] As to his friend Powell, who had died in 1860, Daubeny had already delivered a defence in an anonymous pamphlet, *A few words of apology for the late Professor Baden Powell's essay 'On the study of the evidences of Christianity' contained in the volume entitled 'Essays and reviews'. By a lay graduate*, (Oxford, [?1860]), reprinted in *Miscellanies*, op. cit. (33), ii. section IV, pp. 25–40. For Powell's position, see D. M. Knight, 'Professor Baden Powell and the inductive philosophy', *Durham University journal*, xxix (1967–8), 81–7.

[35] Daubeny, *Miscellanies*, op. cit. (33), ii. section IV, p. 131.

[36] Ibid., p. 132.

[37] Ibid., i. Preface, p. xviii.

[38] Conceivably Bowerbank misunderstood the Declaration's purpose, for he also subscribed to Colenso's Defence Fund, at least, according to De Morgan, in *Athenaeum*, 7 January 1865, p. 22.

[39] *The journals of T. A. Hirst* (Royal Institution), 11 September 1864. See Brock and MacLeod, *Life of Hirst*, op. cit. (2). Hirst's reply is not in Berger, *Documents*, op. cit. (10).

[40] De Morgan, *Athenaeum*, 29 October 1864, p. 566, and 31 December 1864, p. 894. Faraday claimed his views were already public knowledge; see Berger, *Documents*, op. cit. (10), pp. 36, 99.

[41] *Athenaeum*, 17 September 1864, p. 375, and *The Times*, 20 September 1864. De Morgan believed that the promoters tried Herschel because 'over and above his really religious character, they thought him a timid man. They were much mistaken. Herschel is of a nervous and diffident temperament; but perfectly decided in his course of action'; see Robert Percival Graves, *Life of Sir William Rowan Hamilton* (3 vols., Dublin, 1882–9), iii. 620.

[42] Graves, op. cit. (41), p. 618. Cf. *Spectator*, 24 September 1864, p. 1088: 'Some check must certainly be put on the declaration nuisance now growing up. It is especially hard, moreover, to force a declaration on the students of science. There might be more point in extorting the students of religious newspapers like the *British Banner* or *Record*'.

[43] *Saturday review*, 24 September 1864, p. 387.

[44] *Athenaeum*, 17 September 1864, p. 375.

[45] See Charles F. O'Brien, 'Eozöon Canadense, The Dawn Animal of Canada', *Isis*, lxi (1970), 206–23; and idem, *Sir William Dawson. A life in science and religion* (Philadelphia: American Philosophical Society, 1971).

[46] 'The British Association', *Saturday review*, 24 September 1864, p. 383.

[47] *Journals of Hirst*, op. cit. (39), 25 September 1864, f. 1688. Clearly too, Herschel and Bowring's efforts had been effective; Hirst copied their 'excellent' letters into his journal.

[48] *The Bath chronicle*, 19 September 1864, p. 3 (letters of Herschel and Bowring); Editorial, 'A word about ourselves', 29 September, p. 5.

[49] *Saturday review*, 24 September 1864, p. 386.

[50] Identified from a marked file copy in Bill, op. cit. (10), p. 265.

[51] A 'plot' because 'in the case of clergymen, all signs of clerical character (i.e. Revd, etc.) are erased'; see *Athenaeum*, 29 October 1864, p. 566. The printed list does not entirely support this slur. See below.

[52] *Saturday review*, 24 September 1864, commented on this phrase that it was not part of Christian creed even if it was believed by many people. Quoted in Bill, op. cit. (10), pp. 264–5.

[53] *Athenaeum*, 8 October 1864, pp. 463–4.

[54] Ibid., p. 464.

[55] De Morgan withdrew; see ibid., 29 October 1864, p. 566. The degree should have read 'M.A.' not 'M.D.'.

[56] Graves, op. cit. (41), iii. 620 (letter of 2 October 1864).

[57] *Athenaeum*, 29 October 1864, p. 566. Although Berger, *Documents*, op. cit. (10), is clearly only a selection of letters, and those who signed did not necessarily enclose letters, it is of interest that 39 of the 55 replies are negative. These include Babbage, Boole, Huxley, Playfair, Sharpey, and Stokes.

[58] *Journals of Hirst*, op. cit. (39), 6 November 1864, f. 1702.

[59] *Royal Society Council minutes*, iii (1864), p. 219 (3 November 1864).

[60] *Proceedings of the Royal Society of London*, xiii (1863–4), 508. The Copley incident—what Hooker referred to as a 'small breeze at the RS'—is alluded to in F. Darwin (ed.), *The life and letters of Charles Darwin* (London, 1887), iii. 27–9, and L. Huxley (ed.), *Life and letters of Thomas Henry Huxley* (London, 1900), i. 254–5; F. Darwin and A. C. Seward (eds.), *More letters of Charles Darwin* (London, 1903), ii. 255; Huxley, *Life of Hooker*, op. cit. (29), ii. 75–6; Mrs K. M. Lyell (ed.), *Life, letters and journals of Sir Charles Lyell Bart.* (London, 1881), ii. 383.

[61] *Athenaeum*, 19 November 1864, p. 672. It is possible that De Morgan knew Berger through their mutual interest in the decimal and metric systems. (Bowring initiated the British decimal system with the florin in 1849.)

[62] E.g. within scientific societies. It will be recalled that De Morgan was to resign from University College, London, over the College's refusal to appoint the Unitarian Revd James Martineau to the chair of mental philosophy and logic in 1866.

[63] Sophia De Morgan, *From matter to spirit* (London, 1863). Just before his death De Morgan was also much interested in a Free Christian Union, or multi-denominational brotherhood; however, it proved too sectarian. See Sophia Elizabeth De Morgan, *Memoir of Augustus De Morgan* (London, 1882), p. 365.

[64] *Athenaeum*, 31 December 1864, p. 894. Hamilton's biographer, R. P. Graves, thought the Declaration ill-judged, though not ill-meant. See Graves, op. cit. (41), iii. 191.

[65] The astronomers Robert Main, F.R.S., and James Challis, F.R.S., had already signed, and mathematicians like Hirst and Boole had been asked. However, De Morgan felt that 'mathematicians and astronomers are not looked upon as the *elite* of orthodoxy'; see Graves, op. cit. (41), iii. 620. James Challis had, in fact, written *Creation in plan and progress* (Cambridge, 1861), in answer to *Essays and reviews*.

[66] Graves, op. cit. (41), iii. 619.

[67] *Athenaeum*, 31 December 1864, p. 894.

[68] E.g. William Sulkeld, a Royal College of Chemistry student, between nos. 52 and 53.

[69] *Chemical news*, xi (1865), 298. But see Crookes's editorial 'Science, politics and religion', *Quarterly journal of science*, ii (1865), 187, where he estimated that a third of the signatories sincerely believed that the Declaration would help the cause of science and religion, one third signed for the sake of their names appearing, and one third for fear of being branded infidels.

[70] *Athenaeum*, 1 July 1865, pp. 19–20.

[71] Cf. R. S. Westfall, *Science and religion in seventeenth century England* (New Haven, 1958); J. E. McGuire and P. M. Rattansi, 'Newton and the "Pipes of Pan"', *Notes and records of the Royal Society of London*, xxi (1966), 108–43; P. M. Rattansi, 'The social interpretation of science in the seventeenth century', in P. Mathias (ed.), *Science and Society 1600–1900* (Cambridge, 1972,) pp. 1–32.

[72] Arthur H. Tabrum, *Religious beliefs of scientists, including one hundred hitherto unpublished letters on science and religion from eminent men of science* (London: North London Christian Evidence

League, 1910). A second, enlarged edition appeared in 1913. Both editions, which were designed to answer the propaganda of the Rationalist Press Association, carried an Introduction by Revd C. L. Drawbridge who later published, on behalf of the Christian Evidence Society, *The religion of scientists, being recent opinions expressed by two hundred fellows of the Royal Society on the subject of religion and theology* (London, 1932). See also K. A. Kneller, S.J., *Christianity and the leaders of modern science* (London, St Louis, Freiburg, 1911). This first appeared in German in *Stimmer aus Marie-Loach, Katholische Monatsschrift*, Heft 84 u. 85 (Freiburg, 1903). We should like to thank Father J. L. Russell for informing us of these inquiries. A recent example of this genre is Frederick E. Trinklein, *The God of science: personal interviews with thirty-eight leading American and European scientists on the nature of truth, the existence of God and the role of the church* (Grand Rapids, Michigan: Eerdmans, 1971).

73 C. Babbage, *Reflections on the decline of science in England and on some of its causes* (London, 1830); [A. B. Granville], *Science without a head, or the Royal Society dissected* (London, 1830). By 1860 the percentage of peers to ordinary fellows was 4.6 per cent, and the ratio of scientific to non-scientific fellows was 52.6 per cent to 47.4 per cent; see Sir Henry Lyons, *The Royal Society, 1660–1940* (Cambridge, 1944), p. 341.

74 De Morgan gives 62 of 600 Fellows; see the *Athenaeum*, 1 July 1865, p. 19. Our figure of 605 British Fellows is taken from the 1864 Anniversary Report, *Proceedings of the Royal Society of London*, xiii (1863–4), 520, where the total Fellowship is given as 655, of whom 50 were foreign members.

75 *Athenaeum*, 1 July 1865, p. 19.

76 This assumption is unsubstantiated and would repay close scrutiny. Inspection of the *Royal Society catalogue of scientific papers* suggests that the assumption was unwarranted; e.g. John Hogg published nearly 50 papers in the *Philosophical magazine* and many minor natural history journals.

77 De Morgan, op. cit. (75), has 19, but misread the name of Sir John Maxwell (created 1820) for the youthful James Clerk Maxwell: Chadwick, op. cit. (1), p. 8, follows this erroneous reading. Those publishing in the *Philosophical transactions* were Bell, Bowerbank, Brewster, Brooke, Challis, Calvert, Gilbert, Glaisher, Gosse, Harley, Higginbottom, Hoskins, Hunt (who was missed by De Morgan), T. W. Jones, Joule, Macdonald, Moseley, Stenhouse, and Col. Sykes. De Morgan's figure is correct if Hunt is substituted for Maxwell.

78 The overall median age of the signatory Fellows in 1865 was 62; on average they were within fifteen years of their deaths.

79 Mrs Margaret Gordon, *The home life of Sir David Brewster* (London, 1869), pp. 323–4, quoted in Bill, op. cit. (10), p. 266. Brewster was nearly chosen to write a Bridgewater Treatise; see W. H. Brock, in *Notes and records of the Royal Society of London*, xxi (1966), 168–9.

80 Barrister and geologist who found the oldest mammalian remains from rocks of Purbeck group. See F. Boase, *Modern English biography* [hereafter cited as Boase] (6 vols., London, 1892; reprinted, 1965).

81 Officer in Bengal Engineers; wrote miscellaneous scientific papers in India (*Boase*).

82 William Willoughby Cole, third Earl, politician, and fossil collector. His large collection of ichthyic palaeontology was given to the British Museum (Natural History). See *Boase*, and *Proceedings of the Royal Society of London*, xli (1866), pp. ix–xi.

83 Anglican naval officer and port officer at George Town, Tasmania. Returned to England in 1852. Invented the 'Peolrus' for measuring local magnetism in iron ships; see *Australian dictionary of biography*, vol. i.

84 Politician and Privy Councillor. His original surname was altered to Christopher in 1828, and to Christopher Nisbet-Hamilton in 1855; see *Boase*, under 'Nisbet-Hamilton', and Burke's *Landed gentry* (6th edn., London, 1882), p. 1170. His name is strangely absent from the *Record of the Royal Society* (London, 1940).

85 Physician and Vice-President, British Medical Association (*Boase*).

86 Nottingham surgeon and temperance advocate, who wrote on tritons, tadpoles, and frogs (*Boase*).

87 Chemist and ironmaster (*Boase*).

88 Zoologist and Vice-President of Zoological Society (*Boase*).

89 Platinum chemist of famous firm of Johnson and Matthey. See *Boase*; D. MacDonald, *History of platinum* (London, 1960); *Proceedings of the Royal Society of London*, xvi (1867–8), pp. xxiii–v.

90 Ophthalmic surgeon and writer on natural theology; e.g. *The wisdom of the Almighty displayed in the sense of vision* (London, 1851), and *Evolution of the human race from apes, a doctrine unsanctioned by science* (London, 1876). He contributed many papers to the *Proceedings of the Royal Society of London* (*Boase*).

91 Coal gas engineer (*Boase*).

92 Surgeon at St Thomas's Hospital, London (*Boase*).

93 Eighth baronet, politician and economist (*Boase*).

94 Lieutenant of the Yeoman of the Guard, 1836–42 (*Boase*).

95 American geologist, professor of natural history at University of Glasgow, 1857–66; see *Boase*, and *Proceedings of the Royal Society of London*, xvi (1868), p. xxxv.

96 Lawyer. See J. A. Venn, *Alumni Cantabrigiensis* (Cambridge, 1953), Part II, vol. v.

97 Topographer and antiquarian who married the daughter of the agriculturalist, Thomas Coke; published *Topography illustrative of the battle of Plateae* (London, 1817; revised as *Olympia* [London, 1824]). See S. Pickering, F.R.S., *Memoirs of Anna Maria Wilhelmina Pickering, together with extracts from the Journals of . . . J. S. Stanhope* (2 vols., London, 1902–3).

98 Physician to St Luke's Hospital, London, interested in insanity (*Boase*).

99 Halifax astronomer. See *Boase*, and *Proceedings of the Royal Society of London*, xxix (1879), pp. xxvii–xxviii.

100 Politician, Captain in the West Kent militia 1838–54 (*Boase*).

101 Berger, *Documents*, op. cit. (10), p. 86 (Sedgwick to Groves, 13 February 1865). Both Brewster and Sedgwick were invited to write to *The Times* in support of the Declarationists' 'findings', but declined to do so on the grounds of age, infirmity and fear of controversy. The Declaration issue is not discussed in J. W. Clarke and T. M. Hughes, *Life and letters of Adam Sedgwick* (2 vols., Cambridge, 1890).

102 Tyndall papers (Royal Institution, London), typescript, f.2398 (n.d. but c. 16 September 1864).

103 Lyons, in *The Royal Society*, op. cit. (73), calculated that of the 630 Fellows in 1860, 330 could be considered scientific; and that of that number approximately 8 per cent were chemists, 10 per cent geologists, and 36 per cent in medicine. Of the 48 Fellows who could be described as 'scientific' 15 per cent were interested in chemistry, 9 per cent in geology, and 32 per cent in medicine.

104 Botanist and agriculturalist, edited *Annals of natural history*; devout Presbyterian. See *Nature*, cix (1922), 787–8, and *Who was who, 1916–1928*.

105 Chemist, career spent partly in Japan. See *Journal of the Chemical Society*, ciii (1913), 746.

106 See text and note 18.

107 Irish geologist and sometime Secretary of the Victoria Institute. See his *Reminiscences of a strenuous life* (London, 1910).

108 See text and note 19.

109 Philosophically minded physical chemist; professor of technical chemistry at Anderson College, Glasgow. See *Journal of the Chemical Society* (1921), 2130–1.

110 *Athenaeum*, 1 July 1865, p. 19. The signatory concerned was J. T. Banks, professor of physics at Dublin University who was an Hon. Fellow of King's and Queen's College of Physicians.

111 S. P. Barchett.

112 Alexander Beattie.

113 E.g. J. F. Mariller and Gustav Masson. For Berger's Harrow connexions, see Bill, op. cit. (10), pp. 262, 266.

114 *Saturday review*, 24 September 1864, quoted in Bill, op. cit. (10), p. 266.

115 Work extending T. S. Kuhn, *The structure of scientific revolutions* (New York, 1962), is currently underway in the study of factors affecting the emergence of new disciplines. See G. Lemaine *et al., Perspectives in the study of Scientific Disciplines* (The Hague: Mouton, in press).

116 Thus, the vigorous debate concerning the ethics of, and need for, vivisection may have hindered biological research. On this subject, see R. D. French, *Antivivisection and medical science in Victorian Society* (Princeton, 1974). On the other hand, interest in psychic phenomena may have promoted research in physics during the same period. For one example, see W. H. Brock, 'A scientist among the spirits: William Crookes and the radiometer', in W. H. Brock, M. Chapple, and M. A. Hewson, *Studies in physics* (London, 1972), pp. 43–60.

117 E. Benton, 'Vitalism in nineteenth-century scientific thought: a typology and reassessment', *Studies in the history and philosophy of science*, v (1974), 17–48.

118 We should like to thank Dr Gerrylynn K. Roberts for raising these speculations concerning Hofmann and the Royal College of Chemistry. There is no evidence, however, to suggest the existence of an 'established' religion in the College. Any 'uncertainty' concerning the College's future was resolved by the appointment of Frankland (an agnostic) to succeed Hofmann, and by the move to South Kensington in 1872.

119 Chadwick, op. cit. (1), p. 8.

120 J. C. Geikie, *Michael Faraday and Sir David Brewster, philosophers and christians* (London, 1868), p. 13.

121 *Transactions of the Victoria Institute or Philosophical Society of Great Britain*, i (1867), Preface. Eleven signatories of the Declaration were among the founder members of the Victoria Institute.

122 D. Wertheimer, 'The Victoria Institute of Great Britain' (University of Toronto M.A. thesis, 1972).

123 Berger, *Documents*, op. cit. (10), p. 52 (Berger to Lord Belper, 6 September 1864).

[124] J. M. Wilson, 'On teaching natural science in schools', in F. W. Farrar (ed.), *Essays on a liberal education* (London, 1867), p. 290.

[125] A. J. Meadows and W. H. Brock, 'Topics fit for gentlemen. Science teaching in Victorian Public Schools', in B. Simon (ed.), *The Victorian public school* (Dublin, 1975).

[126] Quoted in Chadwick, op. cit. (1), p. 12; T. H. Huxley, 'The Origin of Species', *Westminster review*, April 1860, p. 596 (reprinted in *Darwinian essays* [London, 1899], p. 52). Nor were they interested in doubting the efficacy of prayer; see F. Galton, *English men of science* (London, 1874), and Galton, 'On the causes which operate to create scientific men', *Fortnightly review*, old ser. xii (1872), 345–51.

[127] John Hall Gladstone, 'Points of supposed collision between the scriptures and natural science', *Christian evidence lectures* (London, 1880), p. 164. For interesting letters on reconciliation, see Tabrum, op. cit. (72).

[128] A. J. Balfour, *A defence of philosophic doubt* (London, 1879), pp. 302–3. Cf. W. E. Gladstone's remark: 'Let the scientific men stick to their science, and leave philosophy and religion to poets, philosophers and theologians'; quoted in Alan W. Brown, *The Metaphysical Society: Victorian minds in crisis, 1869–1880* (New York, 1947), p. 106. See also Henry Cotterill, *On the relation between science and religion through the principles of unity and causation* (London, 1880).

[129] Samuel Butler, *Erewhon* (London, 1872), and Frances Power Cobbe, 'The scientific spirit of the age', *Contemporary review*, liv (1888), 126–39. For the obstacles and reactions to science at the end of the nineteenth century, see R. M. MacLeod, 'The support of Victorian science: the endowment of research movement in Great Britain, 1868–1900', *Minerva*, iv (1971), 219–26.

[130] Chadwick, op. cit (1), pp. 1–23.

[131] *Athenaeum*, 1 July 1865, p. 20.

[132] For the Encyclical, the Syllabus, and their aftermath, see E. E. Y. Hales, *Pio Nono* (London, 1954), pp. 255–90. Cf. J. W. Draper, *History of the conflict between religion and science* (London, 1875), chapter 12.

III

The 'Bankruptcy of Science' Debate: The Creed of Science and its Critics, 1885-1900

Historically, science has promoted an image of scientific rationality as the norm of truth, making that which is possible, necessary; that which is necessary, unavoidable; and that which is unavoidable, 'efficient'. The criterion of 'efficiency', as operational model and thermodynamic metaphor, has been the dominating theme in the popularisation of science, often pushing aside the crucial questions: 'efficient' relative to what and to whose interest? Historians are reminded time and again that science has been efficient only in relation to certain governing assumptions about the purposes for which knowledge is employed.

Public criticism of this self-justification has, in the West, assumed major proportions.[1] Yet, remarkably, science is still often taught in ways that assume canonical world-view - fundamentally optimistic if left unchallenged; redemptive, if left unfettered; and purposive, if left in the right hands.[2] As a science educator, I have often been asked by students why, after two generations of protest, such an unself-critical view has persisted.

Few historians have looked closely at the pivotal intellectual controversies of the late nineteenth century which surrounded the 'creed of science'. These debates repeatedly crossed national and cultural frontiers, often drawing strength from influences, attitudes, and mentalities within the natural sciences themselves.[3] This essay cannot do more than outline some directions for future

---

1 For recent academic analysis of science in its social relations, see esp. Jerry Ravetz, *Scientific Knowledge and its Social Problems* (London: Penguin Books, 1973); recent issues of *Science, Technology & Human Values* and *Social Studies of Science*, and the critical bibliographies and essay reviews in Paul Durbin (ed.), *A Guide to the Culture of Science, Medicine and Technology* (New York: Free Press, 1980).

2 John Ziman, *Teaching and Learning about Science and Society* (Cambridge: Cambridge University Press, 1980).

3 Cf. J.L. Heilbron, 'Fin-de-Siècle Physics', in C.G. Bernhard, E. Crawford, and P. Sörbom (eds), *Nobel Symposium*, 52, *Science, Technology and Society in the Time of Alfred Nobel* (Oxford: Pergamon Press, 1982), 51-73, shown to me in manuscript by the

reflection. But these I offer to plead the importance of understanding some past sentiments and conceptual processes that are related to current attempts to hold 'experts' and specialists accountable to the wider purposes of social improvement. Looking back a century, one uncovers a debate that illustrates our current efforts.

Between 1880 and 1895, perhaps 40 years after the natural sciences had begun to enter the secondary school and university curricula of most European countries, an unusual alliance of clerical, academic, and philosophical interests subjected science to bitter attack. Science, it was claimed, was 'bankrupt'. This challenge was not exclusively a romantic reaction against industrialism, not merely an Arnoldian defence of traditional educational values, although such themes were present in the attack. The arguments of the critics of the 1890s were more deeply challenging, asserting not only that science had failed to produce material well-being, but also that its analytical assumptions were themselves self-destructive; if allowed to continue unfettered, the pursuit of science, it was argued, could destroy the ethical framework upon which European civilisation rested.

The same decade that saw Europe in the opening of vast, new, and unexplored vistas - in radioactivity, atomic theory, physiology, and psychology - saw also a peak of public disenchantment with science, with 'scientific' methods and with the buoyant optimism associated with the scientific Enlightenment. By the turn of the century, a new generation of social and historical critics - from Henry Adams to Croce, Durkheim and Weber - were arguing that the basic character of human experience, underwritten by the advance of knowledge, was no longer limitless progress but the *limitation* of progress, the confinement of freedom, the alienation of individuals, and the bureaucratisation of institutions.[4] Critics saw progress, choice, and change as constrained by thermodynamical irreversibilities in the physical universe, and by the boundaries of conscious experience.[5] Such a view led not only to a

---

author; Brigitte Schroeder-Gudehus, 'Characteristiques des relations scientifiques internationale, 1870-1914', *Journal of World History*, 10, (1966), 161-177; Russell McCormmach, 'On Academic Scientists in Wilhelmian Germany', *Daedalus*, 103, (3) (Summer 1974), 157-171.

4 Max Weber, *Critical Studies in the Logic of the Cultural Sciences* translated by Edward Shils and Henry Finch as *Max Weber on the Methodology of the Social Sciences* (New York: The Free Press, 1949); Max Weber, 'The Social Psychology of the World Religions', translated in H.H. Gerth and C. Wright Mills, *From Max Weber: Essays in Sociology* (New York: Glencoe 1949); Anthony Giddens (ed.), *Emile Durkheim: Selected Writings* (Cambridge: Cambridge University Press, 1972), 239-249, 250-268.

5 Stephen G. Brush, *The Temperature of History* (New York: Burt Franklin, 1977), esp. Chapters 5 ('The Heat Death'), 6 ('Realism and Neo-romanticism'), and 8 ('The End').

profound ontological pessimism, but also to an expedient *realpolitik* that claimed not unlimited power to *know*, but asserted the power to direct and control that which is known.

In these two decades, a spectrum of public fears and anxieties was brought into focus which two world wars, the collapse of scientific internationalism, and the demystification of 'objectivity' have done nothing to dispel. With this period, we also associate a convergence of convictions and despairs - in literature, through the rejection of realism and naturalism, and the rediscovery of romance; in art, through the aestheticism typified by Oscar Wilde and art nouveau, both powerful reactions against the philistinism of the Victorian bourgeoisie. In politics, syndicalism and the violence of class struggle attempted a search for justification through the application of scientific laws to social systems. Yet in this period, the Victorian 'scientific movement' also encountered its first sustained opposition.[6] H. Stuart Hughes describes the *fin-de-siècle* period as a 'Revolt against Positivism';[7] Buckley, *in The Victorian Temper*, saw it as a 'Revolt from Reason'.[8] It was a time which, within intellectual circles, embraced both vibrant intellectual development and highly sophisticated anti-intellectualism. Confidence in science as a means of knowing was replaced by a perception of science as means of doing. Science was invested with new values - of instrumentality, commodity, and utility - and its moral value seen as an 'object lesson' in citizenship and efficiency.

'When we have ascertained, by means of Science, the methods of Nature's operations, we shall be able to take her place and perform them for ourselves ...' Winwoode Reade's optimistic claim in 1872 for the future of science coincided with a highwater mark for European belief in the progressive, ultimately benevolent character of natural knowledge.[9] From a working faith in scientific methods, based on an evolutionary perspective, would grow a better understanding of the world, as well as freedom from the tyranny of religion, obscurantism, and faction.[10] In no field were these advantages clearer than in

---

6 Cf. J.T. Merz, *A History of European Thought in the 19th Century* (London: Blackwood, 1904-1912), four volumes; Vol. 4, see esp. Chapters 8 and 9.

7 H. Stuart Hughes, *Consciousness and Society: The Reorientation of European Social Thought, 1890-1930* (New York: Alfred A. Knopf, 1958), 33-66.

8 Jerome Buckley, *The Victorian Temper: A Study in Literary Culture* (London: Frank Cass, 1962).

9 Winwoode Reade, *The Martyrdom of Man* (London: Trubner, 1872), 513. Cf. George Gore, *The Scientific Basis of National Progress* (London: Williams and Norgate, 1882; Cass Reprint, 1970).

10 Frank M. Turner, *Between Science and Religion* (New Haven and London: Yale University Press, 1974).

the 'new' biology. Within the space of a generation, the theory and practice of Darwinian biology, and current advances in experimental physiology, biology and medicine had become the 'new idolatry', a 'new creed' prophesying the destruction not only of revealed religion, but of common humanity as well.[11] And in the 1880s, the synthesis summarised in William Graham's best selling book, *The Creed of Science: Religious, Moral and Civil* (1884), came under attack.[12]

We can trace the sources of the 'bankruptcy' debate to the theological and philosophical opposition of the 1870s against the interpreters of Darwinism, intensified by the claims of naturalism in art, literature, and philosophy.[13] In contemporary usage, this attitude defied precise definition; at different times, it could, and did, embrace elements of such overlapping but articulate concepts as positivism, realism, naturalism, and rationalism. All of these were subjective in place, discipline, and cultural tradition. All could be conflated by advocates into a philosophical position that saw science (or scientific method) as the ultimate court of reference. Those who apotheosised science, who basked in the intellectual luminosity of Littré and Taine, or who subscribed to the literary rationalism of Zola, held in common the fundamental dogma (or 'creed') that, through 'scientific method', reason could understand and know everything.[14] Everything true could be discerned by reason and catalogued by science. From the 1860s, an elite - in France, Paul Bert, Marcelin Berthelot, and Paul Painlevé; in England, John Tyndall, Herbert Spencer, W.K. Clifford, and T.H. Huxley; in Germany, Rudolf Virchow and Ernst Haeckel - had led the vanguard of the scientific movement, eclipsing the 'romantic naturalism' of the 1830s, and producing a new agnostic ideology, a 'scientific naturalism' that, while not necessarily irreconcilable with religion, accepted in principle the possibility of reducing the physical world to material causes.[15] What were the 'articles' of this widely acceptable 'creed of science'? *First*, scientific

---

11 Gertrude Himmelfarb, *Darwin and the Darwinian Revolution* (London: Chatto and Windus, 1959).

12 William Graham, *The Creed of Science: Religious, Moral and Civil* (London: Kegan Paul, 1881). Cf. J.M. Robertson, *A History of Free-thought in the Nineteenth Century* (London: Watts, 1929), 330.

13 For a recent treatment, with disturbing resonances, see Jeremy Rifkin, *Entropy: A New World View* (New York: Viking, 1981); Roy MacLeod, 'Resources of Science in Victorian England: The Endowment of Science Movement, 1868-1900', in Peter Mathias, (ed.), *Science and Society* (Cambridge: Cambridge University Press, 1972), 111-166.

14 Cf. Alexander Welsh, 'Theories of Science and Romance, 1870-1920',*Victorian Studies,* XVII (2), (1973), 135-154.

15 See Turner, op. cit. note 10, 8-23.

naturalism endorsed, through orderly observation, the prospect of universal verification. Scientific knowledge was (or would become) the norm of truth. *Second*, scientific naturalism stressed unity and continuity, not discontinuity or 'randomness'. As Herschel saw the motions of distant stars under universal gravitational control, so the spectroscope revealed harmonies between our sun and those distant stars. Faraday and Oersted revealed that electricity and magnetism were reducible to a unity of forces. Joule and Carnot showed that energies of different forms were continuously conserved. Lyell argued persuasively that natural forms evolved through the application of uniform laws. *Third*, scientific naturalism substituted a dynamic view of the universe for a static one. 'Change' was an irrefutable fact of nature. As Robert Routledge told his popular readership in 1880, with all the fervour of an evangelist, 'We are no longer permitted to look for stability and eternal endurance in any region of the visible world'.[16] *Fourth*, natural knowledge depended upon a simplified Cartesian dualism, a separation of thought and an extension of mind and matter, of subject and object. With this duality, distinctions could be made between the 'testable' and the interpretive, between 'fact' and the 'value' of fact. *Finally*, natural facts could be discussed and regularities described within a Newtonian world of causal relationships, discoverable by direct observation, instruments, or experiment, and generally describable through mathematics. None of these processes of investigation was believed to change the independent world in any way.

The rhetoric of this creed was, with varying usage, based upon the assumption that the universe was ultimately material, mechanical, and uniform. The two great symbolic syntheses - the evolution of species and the conservation of energy - were to be understood through a secular catechism of unification, simplification, correlation, and universality.[17]

In compiling his list of heroes in the 1840s, Carlisle had neglected men of science, but Renan made no such omission two decades later:

> With the saints, the heroes, the great men of all ages, we may fearlessly compare our men of scientific minds, given solely to the research of truth, indifferent to fortune, often proud of the poverty, smiling at the honours they are offered, as careless of flattery as of obloquy, sure of

16 Robert Routledge, *A Popular History of Science* (London: Routledge, 1881), 666.

17 Grant Allen, 'The Progress of Science from 1836 to 1886', *Fortnightly Review*, 41, (1887), 868-884.

the worth of what they are doing, and happy because they possess the truth.[18]

The literary articles of this creed were to be preached by a new, Fifth Estate - what Francis Galton called a 'scientific priesthood'.[19] Their calling had the sanction of natural law. Their creed ultimately derived its strength from acknowledgement of a Supreme power usually described in contemporary political metaphor, such as Huxley's 'Divine Government', Whewell's 'Governor of the Universe', or Herschel's 'Great First Agent'.[20] If, as Wordsworth put it, 'Duty preserved the Stars from Wrong', then Nature's God was the supreme legislator, the man of science his civil judge, and the laboratory his consistory court.

The 'creed' brought into focus a long-standing controversy between the advocates of the 'Divine mind in Nature' and those who preferred a causal theory of matter operating according to the rules of chance. The issue was seen by the scientific movement as one of public instruction. One critic reflected that 'such distrust or dislike of science ... as is to be found among us is due to circumstances which are rapidly disappearing, to misunderstanding and imperfect training ... a tendency in the expounders of scientific doctrine to make too sure of things, to put forward as known fact that which is not yet known fact, but only conjecture'.[21] By the 1880s, however, this distrust was far from disappearing. Such polemical social reformers as Frances Power Cobbe criticised men of science for their 'scientific arrogance';[22] others accused them of being 'in league with the material spirit of the age', forming 'a kind of advanced Radical party among men of learning'.[23] The concepts of unity and uniformity appeared to bring not an appreciation of the 'divine' in nature, but the threatening prospect of an encompassing, atheistic materialism - a prospect that not only made scientists suspect, but also brought the pursuit of science into grave disrepute.

---

18 Ernest Renan, 'The Nobility of Science', translated in *Scientific American*, 49, (17 May 1879), 310.

19 Francis Galton, *English Men of Science: Their Nature and Nurture* (London: Macmillan, 1874; reprinted Cass, 1970), 260.

20 Herbert Dingle, 'The Scientific Outlook in 1851 and 1951', *Proceedings of the Royal Institution of Great Britain*, 35, (1955), 73.

21 W.K. Clifford, *Lectures and Essays* (London, 1879), 243; 'On the Scientific Basis of Morals', *Contemporary Review*, 26, (1875), 650-660, 661-669.

22 Frances Power Cobbe, *The Scientific Spirit of the Age* (London, 1888), 12-13 and 34; appearing originally in *Contemporary Review*, 54, (1888), 126-139.

23 James Clerk Maxwell, *Introductory Lecture on Experimental Physics* [25 October 1871] (London, 1871), 17.

Nothing revealed this change in attitudes quite so clearly as the reaction to John Tyndall's Presidential Address to the British Association at Belfast in 1874.[24] The Belfast Address, with its intimations of materialism, broke the uneasy philosophical truce that had followed the first waves of Darwinian debate. In *Fraser's Magazine*, William Allingham expressed the fear that 'without earthquake or tornado', the professors of 'atheism' and 'scientism' would create a class of modern prophets and a new 'Creed of the Future' in which traditional religion, being merely speculative, would have no practical bearing on human affairs.[25] Other English critics saw 'scientism' (used sometimes interchangeably with 'positivism') as a shallow continental arrogance. The students of Huxley and Tyndall were, it was argued, led astray by this new vanity, in rebelling against the satisfactions of natural theology and divine mystery. Symbolically, one essayist complained that these modern prophets had reduced the symphony of the heavens to a mere lecture on acoustics.[26]

By the early 1880s, intellectual skirmishing on the issue had become, in the 'heavy quarterlies', the order of the day. Gladstone's plea - 'Let the scientific men stick to their science and leave philosophy and religion to the poets, philosophers, and theologians' - was uttered in vain.[27] When the Cambridge historian, J.R. Seeley, observed in 1883 that the 'progress of

---

24 The intellectual crisis of the 1870s is illustrated well by the history of the Metaphysical Society, which began in 1869 in an attempt to 'unite all shades of religious opinion against materialism'. See Alan Brown, *The Metaphysical Society: Victorian Minds in Crisis, 1869-1880* (New York: Columbia University Press, 1947), 21 *et passim.* The impact of reductionism, atomism, and scientific materialism was darkly illustrated by Tennyson, first in his poem 'Lucretius' and later, more optimistically, in 'The Higher Pantheism'. For reactions to Tyndall ('who more than once had severely tried the patience not merely of the public but of a large number of his scientific brethren by the rashness with which he had intruded his speculations into regions far beyond those who are properly the province of the Professor of Natural Science'), see [Henry Wace], 'Scientific Lectures - Their Use and Abuse', *Quarterly Review*, CXLV, (1878), 35-61, esp. 38-39. See also A.S. Eve and C.H. Creasey, *The Life of John Tyndall* (London: Macmillan, 1945), Chapter 15. The Belfast Address was published separately by Longmans, with an explanation by Tyndall in September 1874, and was reprinted with modifications in Tyndall's *Fragments of Science* (London: Longmans, 1879), 6th edition, Vol. II, 193. For an intermediate position on the 'errors of modern physics', see Tyndall's old enemy, P.G. Tait, *Lectures on Some Recent Advances in Physical Science* (London: Macmillan, 1876), 1-26.

25 [William Allingham], 'Modern Prophets', *Fraser's Magazine* , New Series, XVI, (1877), 274-275.

26 Ibid., 291.

27 Quoted in Alan Brown, op. cit. note 24, 106.

science ... might be held to be the principal factor in civilisation, yet ... it by no means varies with civil well-being',[28] the public was excusably concerned. Instead (in G.K. Chesterton's phrase), the 'break-up of the compromise' between science and religion gathered momentum. This controversy paralleled growing conflict between what Chesterton called the 'two great positive enthusiasms' within Western Europe - Christianity and the French Revolution, and their respective political adherents. In the world of science

> there had [been] that easy automatic habit, of assuming science to be an oiled and smooth-running machine, that habit of treating things as obviously unquestionable, when, indeed, they were obviously questionable.[29]

For a century, science had been identified with individualism, with the rejection of tyranny and obscurantism. But the Liberalism of this creed was now cast against the rocks of the methods and assumptions of the new professionalism emerging in European science. E.R. Lankester devoted his address to the Biology Section of the British Association at Southport in 1883 to denying the arguments of those who

> mock at the botanist as a pedant, and the zoologist as a monomaniac; they execrate the physiologist as a monster of cruelty, and brand the geologist as a blasphemer; chemistry is held responsible for the abomination of aniline dyes, and the pollution of rivers, and physics for the dirt and misery of great factory towns.[30]

Possibly in no field was the 'creed of science' more vituperated than in the 'scientific medicine' of the late nineteenth century.[31] Medicine and public health were areas understandably sensitive to critics of a naturalism that seemed predicated on a reductionist materialism. 'Medical materialism', William James wrote, implied science without compassion. Religious visions were product, not of revelation, but of biochemistry. Sickness had become a

---

28 J.R. Seeley, *The Expansion of England* (London: Macmillan, 1883), 5.

29 G.K. Chesterton, *The Victorian Age in Literature* (London: Williams and Norgate, n.d.), 204.

30 E.R. Lankester, *The Advancement of Science: Occasional Essays and Addresses* (London: Macmillan, 1890), 104; from 'Biology and the State', Address to the British Association, Southport, 1883.

31 Sir Samuel Sprigge, 'The Poisoning of the Future', *New Review*, 9, (1893), 445-455; Walter Raleigh, 'The Human Bacillus', *New Review*, 15, (1896), 552-556.

matter of microbes, not of man. The 'humane' movements - including the anti-vivisection and anti-vaccination movements - all witnessed sharp conflicts with authority in this period.[32]

The criticism was further reflected in the literary culture of the day. In *Dr Jekyll and Mr Hyde* (1886), for example, Robert Louis Stevenson dramatised the fantastic prospect of a materialistic and murderous alter-ego lurking behind every respectable medic. In the same year, the novelist Ouida attacked Sir Lyon Playfair's British Association address as self-interested special pleading, and his eulogy of scientific technology as an uncritical, dehumanised vision of the future.[33] How would science-based technology, she argued, set men free from fear or hunger if they only thereby achieved the enslavement of a superior factory? Reflecting sentiments that would surely find resonance among the unemployed in Europe today, Ouida asked what workers set free from drudgery by science would actually gain if, for example, 30,000 shoemakers became instead merely telegraphed clerks or bookkeepers?[34]

Ouida's attack broadened into a denial that new technology was inevitably a Demeter, bestowing blessings with every new invention. Moreover, the approval so eagerly offered by some scientists to the improvement of instruments of war, coupled with the 'curious willingness of the pursuers of science to join hands with tyranny', was 'the darkest menace of the world's future'. She speculated bitterly that

> It is significant of the political immorality and readiness to tyrannise over others in the pursuit of their aims which characterise the scientific classes, that they are willing to admire and support any government which is willing in return to endow their scholarships and erect their laboratories. They are inclined to surrender all political liberty, if by doing so they can obtain a ruler who will build them a number of new colleges, with every new instrument ready to their hands for animal torture and physiological or chemical experiment.[35]

---

32 R.K. French, *Antivivisection and Medical Science in Victorian Society* (Princeton: Princeton University Press, 1975).

33 Ouida (Louise de la Ramée), 'Some Fallacies of Scientists', *North American Review*, 142, (1886), 143-145 and 151-152, and *The New Priesthood* (London: E.W. Allen, 1893), a polemical argument against vivisection. In 1888 Ouida dramatically turned against the new scientific medicine, and condemned what she considered its cruelty, impersonalism, and irreligion. Monica Stirling, *The Fine and the Wicked: The Life and Times of Ouida* (London: Victor Gollanz, 1957), 174.

34 Stirling, ibid., 143-144.

35 Ouida, 'Some Fallacies of Science', op. cit. note 33, 151.

The world would gain nothing, she proclaimed, 'by delivering itself ... from the bondage of the various churches and their priesthoods, if, in their stead, it puts its neck under the yoke of a despotism, more intellectual perhaps, but as bigoted, as arrogant, and as cruel'.[36]

Throughout the 1880s, a number of influential writers continued to denounce the 'arrogance' of scientists who identified man with the 'Kingdom of Nature' and thereby reduced his laws and institutions to the status of mere animal organisation.[37] Social and intellectual resistance to physico-chemical reductionism was illustrated by larger, less specific and less intellectualised references to the 'degrading' tendencies of science, particularly those associated with research in Germany where the spirit of *Wissenschaft* had carried all before it. One writer in the *Journal of Science* even argued that 'the life that is wholly given over to such pursuits (i.e., research) is a misspent one, and is, as an example, positively injurious to Society'.[38] Strindberg, on the other hand, in 'The Father', mocked science as a refuge for psychological failures.[39]

In a similar vein, John Ruskin, claiming that science should not be pursued as an end in itself, but should be subordinated to ethics, opposed the new physical laboratory built at Oxford in 1884.[40] Where science clashed with ethics, science should give way. Ruskin's protest was symptomatic of a broad wave of disillusionment, often unspecific and frequently misdirected, but preoccupied with the alleged menace of science. Werner Siemens felt sufficiently anxious to warn the Association of German Naturalists and Physicians (meeting in Berlin in 1886) against the prophets of doom who had questioned whether man is really better and happier for all the achievements of science and art.[41] To Ruskin, and other critics, the members of this new 'priesthood' had one thing in common: an apparent indifference to ethical or moral values that ought to be common to every man.[42] It was as if the creed of

---

36 Ibid., 152.

37 Robert Ward, 'Scientific Arrogance',*Quarterly Journal of Science*, 3rd series, III, (1881), 249-258.

38 Cited in Frank Fernseed, 'The Future Martyrdom of Science',*Quarterly Journal of Science*, 3rd series, III, (1881), 202.

39 Johan August Strindberg, 'The Father', in *Plays* (Boston, 1912).

40 W.G. Collingwood, *John Ruskin: His Life and Work* (London, 1874), Vol. II, 90 and 223.

41 Cf. *Nature*, 34, (1886), 605-606; *Science*, 8, (15 October 1886), 336-337.

42 Cf. R. Proctor, 'The Dignity of Science', *Knowledge* (1 January 1886), 93-95.

science had been forced upon a subject people with the severity of a Viking raid.[43]

The mood of apprehension crossed to technological America. W.D. Hay's prophetic *300 Years Hence* (1881), for example, foresaw the use of natural selection to justify the expansion of the Anglo-Saxon race. And when Frank Stockton's *The Great War Syndicate* (1889) portrayed scientists attempting to avert war by taking over from the vexatious impulses of politicians, the whole scientific estate came under suspicion.

Meanwhile, the contemporary crises in political liberalism, and the pressures of imperial expansion, economic uncertainty, and international and economic rivalry, all gave additional piquancy to this growing mood of apprehension. Within France and Britain, agricultural and industrial recessions had prompted the growth of workers' movements that owed little to established political parties. In England, for example, the Social Democratic Federation, with an overtly Marxist complexion, attracted fear and attention far out of proportion to its size because its message spread rapidly among the semi-employed casual labouring classes.

A strong connection was drawn by one contemporary observer between the spread of socialism and the loss of faith among the 'masses of working men throughout the length and breadth of the European continent', and the 'large centres of [British] industry'.[44] In his view, the materialistic tendencies of modern science contributed to this decline of faith and to the receptiveness of the working class to the false religion of socialism (as Stephen Jacyna has put it, the *Origin of Species* had thus prepared the way for *Das Kapital*[45]); 'nothing could save society from the present danger but the restitution of genuine religious belief'. Other critics attempted to make religion intellectually respectable again. One endorsed the theory of moral progress enshrined in Mrs Humphry Ward's novel *Robert Elsmere* (1888); in the United States, Josiah Cooke took up the cudgels for science in defending the morality of the 'rational tradition'.[46]

---

43 Grant Allen, 'The Progress of Science from 1836 to 1886', *Fortnightly Review*, 41, (1887), 873.

44 Moritz Kaufman, 'Socialism and Atheism', *Contemporary Review*, 47, (1885), 823-824, 839, quoted in L.S. Jacyna, 'Science and Social Order in the Thought of A.J. Balfour', *Isis*, 71, (1980), 21-22. I am grateful to Dr Jacyna for his insight into A.J. Balfour as both politician and philosopher.

45 Ibid., (Jacyna), 11-34.

46 Josiah Cooke, *The Credentials of Science, the Warrant of Faith* (New York: Robert Carter, 1888).

But, in secularising modern thought, the 'creed of science' offered little room for compromise. The effects of this demoralisation on the elite were no less moving than the alleged effect on the workingmen. 'Something needs to be done for the educated similar to what John Wesley did for the poor', Benjamin Jowett wrote to Frances Nightingale.[47]

Amidst this moral and political uncertainty, questioning spread. Darwinism was not the principal target, but it symbolised scientific naturalism, and the *hubris* of the 'priesthood'. Educators and scientists revealed their own anxieties. T.H. Huxley's 1893 *Romanes Lecture* on 'Evolution and Ethics', for example, disclosed the uncertainty surrounding the relationship between evolutionary processes and moral improvement.[48] The following August, in his British Association Presidential Address, Lord Salisbury appealed to his audience to look beyond empirical material causes in searching for rules to improve mankind.[49] He believed that evolution implied not ethical development, but perhaps the reverse. Huxley denied that naturalism was necessarily materialistic, but even he could not escape a denial of the 'divine mind in nature'. By Autumn 1894, the quarterlies were full of discussions of his agnostic pessimism.[50]

H.G. Wells' *The Time Machine* was published the following year (1895). A scientific romance, Wells' first novel was full of brooding fear. But, by the time it appeared, a wide chasm had truly opened between the scientific movement and its critics. In January 1895, Ferdinand Brunetière, editor of *Revue des Deux Mondes*, published a violent attack against the positivism and *scientisme* long fashionable in France.[51] In his article, Brunetière revealed his own conversion experience following an audience with Pope Leo XIII. A few

---

47 Quoted in Mary Douglas, *Purity and Danger* (London: Routledge and Kegan Paul, 1966), 14.

48 T.H. Huxley, *Evolution and Ethics* (London: Macmillan, 1893); Cf. Michael Helfand, 'T.H. Huxley's Evolution and Ethics: the Politics of Evolution and the Evolution of Politics', *Victorian Studies* , XX (2), (1977), 159-177.

49 Cf. Herbert Spencer, 'Lord Salisbury on Evolution', *Nineteenth Century*, 38, (1895), 740-757.

50 W. Barry, 'Professor Huxley's Creed', *Quarterly Review*, 180, (1895), 160-188; F. Muller, 'Why I am not an Agnostic', *Nineteenth Century*, 36, (1894), 890-895; J. Ward, *Naturalism and Agnosticism*, Gifford Lectures, 1896-1898 (London and Edinburgh: A. and C. Black, 1899), two volumes.

51 Ferdinand Brunetière (1849-1906), professor at the École Normal Supérieure; editor, *Revue des Deux Mondes* from 1893; noted for his *Histoire et Littérature* (1884-1887) and *Le Roman Naturaliste* (1883). Cf. Harry Paul, 'The Debate Over the Bankruptcy of Science in 1895', *French Historical Studies*, 5, (Spring 1968), 303. On the critic, see Blaze de Bury, 'Ferdinand Brunetière', *Fortnightly Review*, Old Series, 64, (1895), 497-511; G. Constant, *Ferdinand Brunetière* (London: 1906).

months later, A.J. Balfour published his *Foundations of Belief*, which assailed the epistemological assumptions of naturalism,[52] and accused science of relegating ethical sentiments to the status of artefacts contrived by the human species to improve its chances of survival. The message was not new, but it was bitterly debated, not least because it came at an important political juncture.

In England, mid-century liberalism was in retreat. Gladstone's hopes of Home Rule had vanished and his party, in tatters since 1887, had barely survived in government through March 1894. The Conservatives' return to power in June 1895, found the liberal cause without a philosophical mandate and attacks against the scientific movement undermined the long-assumed relationship between scientific inquiry and political individualism. Balfour's message made a substantial impression as a 'scientific' plank in the platform of a new philosophical Conservativism.[53] In France, following the war of 1870-1871, the anti-clerical establishment had founded the central ideology of the Third Republic on the cult of reason and science, but by the 1890s, their role was insecure and under repeated attack by the Catholic 'anti-establishment'. The Republic depended upon science but who would vote for *scientisme*?

Smouldering political resentment from within the French literary right now broke upon the 'creed of science'.[54] In the early 1890s, French literary circles embraced the strenuous humanism of Tolstoy, welcoming at an abstract level his testimony that human redemption lay not through science or reason, but through love, effort, and will.[55] In 1889, Paul Bourget's novel *Le Disciple* had vilified a 'modern prophet' (in this case, a deterministic psychologist, an immoral seducer and murderer, who rationalised his evil ways with Spinozian

---

52 Cf. Wilfrid Ward, 'The Foundations of Belief', *Quarterly Review*, 180, (1895), 488-520; James Martineau, 'Foundations of Belief', *Nineteenth Century*, 37, (1895), 557-566; Philip Waggett and R.E. Prothero, 'Through Scientific Doubt to Faith', *Quarterly Review*, 183, (1896), 285-309.

53 Cf. Eighth Duke of Argyll, *The Philosophy of Belief*, review by W.B. Carpenter in *Quarterly Review*, 184, (1896), 214-245.

54 For the French debate, see Paul, op. cit. note 51. I am indebted to Professor Paul, on whose work the following paragraphs draw.

55 Paul, op. cit. note 51, 301. Paul gives evidence suggesting Tolstoy initiated the 'bankruptcy' movement. While suggestive, it seems clear from the following that the term used was in wider circulation, certainly in England. Nevertheless, for an expression of Tolstoy's reaction against 'scientism', see his 'The Law of Force and the Law of Love', *Fortnightly Review*, 91, (1909), 460-485 and 689-720.

precision).[56] The theme became commonplace. Ouida devoted her novella *Toxin* (1889) to the study of a villainous physiologist who murders, with microbes, his rival in love.[57] And in 1895, François Curel's play ‘La Nouvelle Idole’ (the title refers to Pasteurian medical science) featured a physician who inoculates a patient with cancer cells in hopes of curing her, only to learn that she believed she was cured by a visit to Lourdes. The hero wrestles with this paradox, and with the problem of whether he, believing neither in God nor in immortal life, ought to risk his own life for posterity.[58]

This sentiment was not by any means universal. In England, Oscar Wilde, in *The Picture of Dorian Gray* (1891), reviled the nineteenth century as having ‘gone bankrupt through an over expenditure of sympathy’ and suggested ‘science to put us straight’.[59] The advantage of the emotions, Wilde observed, is that they lead us astray; the advantage of science is that it is unemotional. Nonetheless, Wilde found the ‘bankruptcy’ theme turned against him. Symbolically, the novelist George Gissing, a former student at Owen's College, Manchester, had one of his characters describe science as the ‘remorseless enemy of man’.[60]

In this context, Brunetière's clarion was heard sympathetically throughout Europe.[61] His case was disarmingly simple. The promise of science proposed to solve the supreme mysteries of the universe; but, in fact, science had failed to reveal man's nature, his laws of conduct, or his destiny. Therefore, the

---

56 After 1892, Paul Bourget (1852-1935), noted novelist and Academician, turned from the motif of realism and cultivated a fashionable mysticism.

57 ‘The time is at hand’, Ouida wrote, ‘when there will be no priests and no kings, but those of science, and beneath their feet the nation must grovel in terror and writhe in death’, quoted in Stirling, op. cit., 188.

58 ‘La Nouvelle Idole’ (published in Paris, 1895) was first performed in Paris in 1894. To its protagonist, science was ‘la nouvelle idole qui opprime le monde, d'accepter la sanglante offrandes …’, 31.

59 Oscar Wilde, *The Picture of Dorian Gray* (London: Ward, Lock, 1891), 60. The use of this economic metaphor recalls the commercial language applied to scientific discourse since the 17th century. George Basalla has argued that metaphors used by active researchers in the late 19th century drew upon exploration, adventure, popular amusement, and (in England) upon crime detection and puzzle solving. If this is so, the bankruptcy debate reveals evidence of a dramatic clash in cultural communications as well as in forms of discourse. George Basalla (University of Delaware), ‘Science, Scientists and Metaphor’ (unpublished paper 1981).

60 See George Gissing, *Born in Exile* (Melbourne: E.A. Petherick, 1892); and the *Private Papers of Henry Ryecroft* (London, 1903).

61 F. Brunetière, ‘Après une visite au Vatican’, *Revue des Deux Mondes*, CXXVII, (1895), 97-118.

influence of the sciences on history and philosophy was misguided.[62] Scientific methods, in their mechanical accuracy and precision, could not replace religion; scientific reasoning could not provide a morality. In common terms, it had been oversold; its managers were bankrupt. The following year, he repeated the attack in an Introduction to a second translation of A.J. Balfour's *Foundations of Belief*.

Brunetière's critique united idealism, socialism, and catholicism, in a vengeful trinity against materialism, individualism, and mechanism. In any such nebulous war of ideas, of course, it is as difficult to sustain an alliance as to identify the precise target. Language was ill-defined; terms, ambiguous. But, clearly, the scientific priesthood - no matter how it chose to present its creed - was under attack.

Replies to this challenge of 'bankruptcy' came sharp and swift from leading *savants*. Some argued that, in one sense, there was no case to answer. As biologist Charles Richet put it, science had promised nothing except the freedom to inquire and therefore pessimism was not jusitifiable.[63] Notwithstanding the catastrophic prophecies of *fin-de-siècle* novelists, new knowledge did not itself destroy mankind. The irreversible laws of entropy and evolution did not lead necessarily to the death of nature, or to physical deterioration. And what *should* science do for moral development? Surely, to replace misunderstood ideas of charismatic national government and of a traditionally obscurantist theology, with an ideal of 'human solidarity', with Durkheim's 'social physics of morals'. Marcelin Berthelot (a chemist who had served as a Minister of Education and was then Minister of Foreign Affairs) was even more extreme in his defence and blasted any attempt to replace science with mysticism.[64]

The debate in France raged throughout 1896, polarising leaders of the anti-clerical establishment against the Catholic reaction and leaving its traces for generations. Berthelot saw the conflict as serious enough to threaten the stability of the Republic. Nevertheless, simplistic sentiments among littérateurs, and the intellectual egotism of the *savants*, served to sustain a violent antipathy between Church and State. In fact, the party of science had justified itself as a *parti de l'ordre*. It responded, in France, by acquiring a less mechanical, more organic image, and replacing its metaphor of mechanical

62 See Jacques Nanteuil, *Ferdinand Brunetière* (Paris: Blond et Gay, 1933), esp. 89-98.

63 'La Science a-t-elle fait banqueroute?', *Revue Scientifique*, III, (12 January 1895), 33-39. Cited in Paul, op. cit. note 51, 306.

64 Marcelin Berthelot, 'La Science et la Morale', *Revue de Paris* (1 February 1895), 449-469.

positivism with a new empiricism welded to a Durkheimian ideology of social cohesion.

In England, the debate took a different course. Critics of naturalism were more guarded, less given to the ostentatious *cause célèbre*. In any case, there were no clear political barricades to be defended for the sake of science, as in France. Still, in the hands of certain conservative thinkers, the ideological and ethical implications of naturalism were recast. Balfour (and, later, Oliver Lodge) argued that scientific methods could indeed supply a new morality. However, this was not the morality of old-fashioned Liberalism. Instead, science supported the morality of social institutions and the authority of the community as a whole.[65] Lodge used the models favoured by Cambridge physicists to supply a metaphor of unity, government, and providential design which it was man's duty to explore. On this reckoning, the 'ether' provided the primary medium of government, through which Divine Purpose was effected in the physical world.

This redefinition of the alleged function of science gave rise in England to a new set of metaphors. Extending Lodge's argument, Karl Pearson saw science not as a personal adventure, but as a moral discipline; not as an exploratory voyage, but as a route to 'good citizenship'. For Pearson, despite his advanced political views, 'brainworkers' would inherit the earth and would do so by discipline and authority.[66] For Balfour, scientific morality simply repudiated the possibility of unlimited improvement, and the dream of millenarian Baconians had become and idle fantasy.

In the previous decade, scientific inquiry in England and France had been identified with the freedom of the individual. In the late 1890s, the State began to replace the individual as moral guardian of the nation's interests. The scientific spirit that had accompanied liberalism in its support of naturalism in the first years of the century, now found itself identified equally with the forces of Conservatism. This was a function partly of environment, of people, and of ideas. As the generation of individualist 'radicals' disappeared (Tyndall died in 1893, Huxley in 1895), they were replaced by more pragmatic statesmen of science who, in altered social environments, dominated by economic competition and the threat of war, saw different social meanings in science. Natural knowledge was now a resource of social control, an instrument to strengthen empire, to improve national or corporate performance in the race among nations. In England, this new philosophy was incorporated in the

65 Jacyna, op. cit. note 44, 26-29.

66 Donald MacKenzie, 'Karl Pearson and the Professional Middle Class', *Annals of Science*, 36, (1979), 125-145.

popular campaign for national efficiency and in the educational campaign for scientific instruction as a route to good citizenship.

In outward appearance, this expedient view of nature was profoundly pessimistic; but (like Saki's character) many biologists and physicists were sustained by their pessimism. If knowledge set challenges to control, then scientists, not politicians, it was argued, had to be in a position to assert that control. Within the next five years, three arguments were to give this debate special meaning.

First, there were two philosophical responses to Brunetière's accusation that 'we cannot draw from the laws of physics or the results of psychology in any way of knowing anything'. One could, for example, simply agree. For many, the recognition that there was no tenable scientific dogma brought not bankruptcy but liberation, and fed a zeal for a theology of neo-positivism, or neo-empiricism. This neo-empiricism would eventually induce a counterattack, leading (in art) to Futurism and Dadaism and (in biology) to the conceptualism of the 1920s. But for perhaps a decade it sufficed to hold the dyke against critics of scientific rationalism. In fact, it provided an intellectual justification for a broadly unifying band of scholarship, linking the social and the natural sciences - of which the history of science, as a profession, was to be a notable beneficiary.

With neo-empiricism came, moreover, a qualification of mechanism, as a colligative idea underlying the explanation of natural processes. The 'organic' idea, with its traces in Bergsonian vitalism and Weismann's germ plasm, served to give fresh naturalistic justifications for political theories and for the premises of cultural development. In that organic idea, as in social Darwinism,[67] were to be found the rationalisation for a wide variety of state organisations, from socialism to corporate capitalism.[68]

Second, the specialisation and professionalism taking place within science gave special visibility to the philosophical preferences of neo-empiricism. Scientific education acquired a defensive rationale; even 'discovery learning', in H.E. Armstrong's phrase, was to produce respect for the routine processes

---

67 Cf. Bernard Semmel, *Imperialism and Social Reform: English Social-Imperial Thought, 1895-1914* (London: George Allen and Unwin, 1960); Richard Hofstadter, *Social Darwinism in American Thought* (Boston: Beacon Paperback, 1955); G.R. Searle, *Eugenics and Politics in Britain, 1900-1914* (Leyden: Noordhoff International Publishing, 1976).

68 Cf. A.R. Wallace, *The Wonderful Century: Its Successes and Its Failures* (New York: Dodd, Mead, and Co., 1898); J. Marchant, *Alfred Russell Wallace - Letters and Reminiscences* (London: Cassell, 1916), Vol. 2, 181.

of scientific understanding.[69] The inevitable result, in Whitehead's words, was the dullness of the age of the 'professional man',[70] and the routinisation of charisma, which, if anything, deflected science away from the wider Victorian inquiry into man's place in nature.

Third, the bankruptcy debate was given a special symbolic meaning as new discoveries distanced the contestants from the reality of contemporary physics. Becquerel's discovery of radioactivity, and Thomson's measurement of the charge of the electron, became part of a broad transformation in physical theory. Models assuming the continuity of matter and continuous change, fundamental to the tenets of mid-century materialism, were not offset by theories involving particles and discontinuous change. With this new *Gestalt* came a diversity of explanations of physical and biological nature, popularised under such different labels as conventionalism, instrumentalism, and nominalism. With this new diversity came a new modesty, disclaiming the more strident crusades of the Church of Science Militant: 'Men of science, as such, are not to be supposed incapable either of the nobler hopes, nor of the nobler fears to which their fellow men have attained'.[71] With this new modesty came official repudiations of the assertion that science (in the words of the *Grande Dictionnaire Universel* in 1874) sought 'a single law, which will be the true one, accounting for everything'.[72] Henri Poincaré, speaking at the Universal Exposition in Paris in 1900, commended the virtues of humility to his scientific audience. Physicists, he said, were not decipherers of nature's laws, but librarians and cataloguers of experience. Theories and principles are not true or false, but more or less useful.[73] Causal laws, once thought necessary to explain material events, gave way to models based on descriptions of observed or predictable regularities. As *Punch* later quipped, 'All scientists are agreed that there is something wrong with a red-headed man. Of course: it is the colour of his hair.'[74] The search for 'causes' was again left to the theologians.

This recognition of the more modest 'serviceability' of scientific knowledge accorded perfectly well with the mechanical models beloved of the

---

69 H.E. Armstrong, *The Teaching of Scientific Method and Other Papers on Education* (London: Macmillan, 1903), 253-255.

70 A.N. Whitehead, *Science and the Modern World* (Cambridge: Cambridge University Press, 1933), 127.

71 C.G. Knott, *Life and Scientific Work of Peter Guthrie Tait* (Cambridge: Cambridge University Press, 1911), 241.

72 Quoted in John Heilbron, op. cit. note 3.

73 Quoted in ibid.

74 *Punch* (9 October 1918), 225.

Maxwellian generation. Even the mathematic models of French and German physicists were reducible to such teleologically unambitious 'descriptions' of natural events. Both helped to justify the position of physicists against the attacks of 'bankruptcy'. Among biologists and psychologists they forced a wider interest in behaviour and motivation. Even so, most psychologists widely eschewed the claim that 'causes' of behaviour could be determined, or if determined, explained. Freud attempted to make stronger claims, but even his turn to the 'subjective' and the 'irrational' was bounded by an empirical and expedient world view. The contradictions inherent in the Vienna 'secessionists' pointed to precisely this narrow distance between subjectivism and realism.

In retrospect, the bankruptcy of science debate perhaps bore much the same relation to science as the Dreyfus affair bore to anti-Semitism. For the first time in over two generations a sharp light had been thrown on the unstated assumptions of science and its critics. In the closing years of the century, the contest was to prove inconclusive. H.G. Wells refers to the 'ironical silence that follows a great controversy'. There was little mention of it in the English literary press after 1895, but the metaphor of 'bankruptcy' lingered on. As late as 1925, speaking to the British Association, Horace Lamb, the Cambridge mathematician, recalled its current usage, and Beatrice Webb recorded a similar sentiment the following year.[75] C.F.G. Masterman, writing in 1909 on 'Science and Progress', referred to the debate thus:

> The science which was to ally all diseases, the commerce which was to abolish war and weave all nations into one human family, the research which was to establish ethics and religion over science and positive foundation, the invention which was to enable all humanity, with a few hours of not disagreeable work every day to live for the remainder of their time in ease and sunshine - all these have become recognised as remote and fairy visions.[76]

But was this to be all?

The debate demonstrated that the Victorians thought they had solved fundamental problems, when they had merely persuaded themselves that difficult ones were unimportant or insoluble. The problems would not

---

75 Presidential Address, British Association for the Advancement of Science, Southampton, 1925; B. Webb, *My Apprenticeship* (London: Longmans Green, 1950), 90.

76 C.F.G. Masterman, *The Condition of England* (London: Methuen, 1909), 215-215.

disappear; moreover, the issues were both philosophical and highly political. The applications of science-based knowledge were clearer than ever before.

The entire question of public sponsorship of new knowledge was also being debated in many European parliaments. With this new political debate came fears of technology based not on an uninformed Luddism, but on an understanding of the special implications of new knowledge. In 1896, for example, W.E.H. Lecky saw new technology casting fresh shadows on the future of democratic government. The telegraph, he observed, 'had greatly strengthened central government in repressing insurrections, protecting property, and punishing crime', just as the railway had helped centralise politics.[77] He was not alone in these concerns. T.B. Veblen spoke of the 'cross-purposes of science' in a famous essay of 1906.[78] In 1912, E.R. Lankester called for 'mediators' to explain more fully the position of scientific investigators.[79] But these mediators were not necessarily apologists for science. The self-satisfied cult of material progress was not to escape those who, like H.G. Wells, also saw the world from the perspective of the Morlocks.

The bankruptcy of science debate was in many ways a dress rehearsal for the critical examination that would again be forced upon science in the years after World War I. As science abandoned the search for truth, Spenglerian pessimism was perhaps an unavoidable outcome. So, too, was growing public debate about the scientific justification for different economic policies. By 1918 it was clear that professional attitudes, popularised and made public through literary and artistic as well as scientific journals, would no longer permit a single 'scientific orthodoxy'.[80] Idealism and materialism had their political affinities, and such writers as Robert Murray, F.S. Marvin, J.N. Sullivan, and John Strachey frequently crossed pens in debating the appropriate relationship between ideas and action, and the role of 'class consciousness' in science.

On balance, it is wrong to dismiss the 1890s as 'anti-intellectual'. The period witnessed not so much a revolution as a more limited rebellion that struck at certain metaphors and certain mentalities, which it then replaced with

---

77 W.E.H. Lecky, *Democracy and Liberty* (London: Longmans, 1896), 209.

78 T.B. Veblen, *The Place of Science in Modern Civilisation* (New York: B.W. Huebsch, 1919), Chapter 1; first published in *American Journal of Sociology*, 11, (1906), 585-609.

79 E.R. Lankester, *Science from an Easy Chair* (London: Methuen, 1914), Chapter 31. Lankester is one of the few English writers to use the phrase 'bankruptcy of science' in recalling the debates of his time.

80 Roy MacLeod and Kay MacLeod, 'The Social Relations of Science and Technology, 1914-1939', in C. Cipolla, (ed.), *Fontana Economic History of Europe* (London: Collins, 1976), Vol. 5, part 1, 301-335.

others, that were often not greatly superior. Whitehead once complimented the nineteenth century on discovering the method of invention. The late-Victorian period also invented a method of qualitative assessment, of refusal to accept unchallenged even the most pervasive of intellectual assumptions. As Oliver Lodge commented in 1913, the new century had already proved to be a period of rapid progress, combined with fundamental scepticism.[81] That may be the principal message of the bankruptcy debate today.

Positivism and the 'creed of science' survived this rush of 'creditors'. Indeed, the claims of 'bankruptcy' were much exaggerated, as the merchants of mechanistic, reductionist science continued to transact business throughout the academic marketplace.[82] But never again could orthodoxy rest quite secure. Time and time again, the twentieth century would vindicate the right to make rational inquiry into the business of inquiry itself.

From the maelstrom of the 1890s came two major developments. First, some scientists began to pay closer attention to metaphor and message, as science's public image of practical power and iconoclasm was marred by the suggestion of ethical dissembling and addiction to the opiates of unaccountable inquiry.[83] Second, there came fresh models of the natural and human universe, in particular, ones which sought new syntheses and afforded new sentiment to science.[84]

Beyond this, the bankruptcy of science debate also showed the inability of scientists and laymen to ignore the political questions implicit in competing ideas about man, nature, and society.[85] For the future, any reconciliation of

---

81 Oliver Lodge, Presidential Address to the British Association, Birmingham, 1913.

82 Hughes, op. cit. note 7, 39.

83 For the range of responses, see G. Berdoe, 'Humane Methods in Medicine', *The Humane Review*, II, (1901-1902), 97-111; 'Ethics and Science', in Julie Wedgewood, *Nineteenth-Century Teachers and Other Essays* (London: Hodder and Stoughton, 1909), 304-323; and Sir Richard Gregory, *Gods and Men* (London: Stuart and Richardson, 1949), 191 and 204-209; Cf. C.H. Waddington, *The Scientific Attitude* (Hammondsworth, England: Penguin, 1941).

84 Cf. H.G. Wells, *The Discovery of the Future* (London: T.F. Unwin, 1902); Gerald Holton, 'Modern Science and the Intellectual Tradition', in *Thematic Origins of Scientific Thought: Kepler to Einstein* (Cambridge, MA: Harvard University Press, 1973), chapter 14; and David Bohm's brilliant *Wholeness and the Implicate Order* (London: Routledge and Kegan Paul, 1981).

85 Within the framework of recent philosophy of *social* science, see esp. Alasdair MacIntyre, *After Virtue* (London: Duckworth 1981); Russell Keat, *The Politics of Social Theory: Habermas, Freud, and the Critique of Positivism* (Oxford: Blackwell, 1981); and David Thomas, *Naturalism and Social Science: A Post-Empiricist Philosophy of Social Science* (Cambridge: Cambridge University Press, 1980).

science with humanism required a faith in the triumph of hope over expectation. Today, we still mourn the difficulty of joining, through our educational system, the instrumental creed of technical competence with a passion for humane understanding. Looking to the 1890s from the 1980s, freedom from a belief in the unlimited power of science compels us to question authority allegedly based on science and chastens us to work with, rather than apart from, our fellow men and women. In this spirit, we may yet avoid fashioning merely improved means to unimproved ends.

IV

Evolutionism, Internationalism and Commercial Enterprise in Science: The International Scientific Series 1871–1910

Introduction

Despite the considerable role of publishing in the history of scientific ideas, historians of science have paid the subject remarkably little attention. Most would agree that scientific books and journals are more than merely cultural artifacts. Yet few have studied the history of publishing *per se* as an essential element in the development of the modern scientific enterprise.

When one turns to the history of publishing, however, difficulties immediately arise. For many reasons, the history of scientific publishing in general, and of international scientific publishing in particular, has been neither popular nor easy. For here, one must deal with more than the antiquarian's history of individual books, or authors, or ideas. Rather, one must examine a complicated pattern of economic and institutional, as well as intellectual factors; the promotion of cultural values, as well as the pursuit of ideas. The historian of scientific publishing must concern himself with the image, as well as with the content, with commercial prospects, as well as philosophical positions, with the vagaries of fashion, as well as with the codification of new knowledge.

These difficulties, abundantly clear today, first became apparent in the mid-nineteenth century. Science has traditionally enjoyed an ideology of 'internationalism', and scientific publishing, since the sixteenth century, an international market place. Men of science, editors and men of letters have together contributed towards its cultural ascendency. But around the middle of the last century, several 'vested interests' came into play. First, there emerged, throughout Europe, a new and growing class of scientists for whom publication was commercially and institutionally, as well as intellectually, important. Academic hegemonies, no longer simply matters of patronage, could be established (or undone) by the control of individual journals and key texts. And publishers, responding to new institutional opportunities, recognised the new readership this implied. Moreover, throughout the Western world, governments began to encourage expansion in education. Wherever this occurred, especially in secondary education, there arose a demand for books. The natural sciences took their place in this expansion, especially where, as in Germany and France, scientific education became a token of national economic and cultural prestige.

Within Europe, but more especially within Britain and America, there were other impulses towards expansion in scientific publishing. The 'fashionable' interest in science, so prominent a feature before 1830, now gave way to a 'popular' interest among a larger, literate audience, seeking in science both rational entertainment and useful knowledge. Schools, mechanics' institutes, public libraries and field clubs replaced the Literary and Philosophical Societies, as the market place for science books. Increasingly, publishers and journalists, and popularizers of many descriptions, came to recognise the opportunities presented. The message of self-help through self-knowledge, popularised by Samuel Smiles and others, contributed to an atmosphere receptive to the discussion of new ideas. And by the 1860s, in the midst of the Darwinian debate, publishers could build upon the debates surrounding the concept of scientific naturalism. By the 1870s, a growing belief in the existence of underlying principles of uniformity in nature had begun to nourish a new hegemony within European science. Over the next two decades, that hegemony was increasingly mediated and represented by a generalised commitment to the principles of 'evolution' – principles understood to operate at all levels, through the relations of man, nature and society.

In retrospect, scientific publishing – particularly international publishing – was decisive in establishing this hegemony, and in supplying its mediations with the reading public. In more practical, if more prosaic, terms, publishers found in science a respectably marketable commodity which had the added appeal of a fashionable 'modernity'. With the growth of new knowledge, 'specialisation' seemed unavoidable, and infinitely sustainable. At the same time, the opposing force, towards 'general' comprehensive textbooks, was equally strong. Neither market need be neglected. For many publishers, the ideal way, commercially speaking, to recapture the 'unity of science', to restore the 'circle of the sciences' and yet to recapture the 'specialist' reader, lay through the publication of books in 'series'. Indeed, formal series of books acquired growing popularity in both Europe and America. Whether 'classbooks', primers or standard texts, they closely embraced the entrepreneurial interests of publishers, scientists and educators alike. As a commercial activity, they escaped risk. Although not intended to eclipse single volumes in Britain (which represented 90% of titles produced by leading houses), they could expect a regular library institutional sale. For men of science, they conveyed both a broad editorial vision, and a representation of science as culturally more than the sum of its increasingly fragmented parts. For educators and teachers, they conveyed an image of order, and a sense of stability at a time when the restless advance of scientific thinking left few beliefs undisturbed.

The economic and intellectual importance of the scientific series, and the relevance of this *genre* to the history of scientific publishing, is perhaps best illustrated by an extended example, and by a close study of the factors involved in its particular evolution. The case is that of the *International Scientific Series* – of all Victorian attempts at codifying and popularising scientific knowledge in a systematic fashion to a wide reading public, perhaps the most famous. The ISS, whose 'familiar red covers were a guarantee of sound material within' (Eve and Creasey, 1945, p. 154), appeared in over 120 titles between 1871 and 1910, in four languages, in the U.S.A. and five countries of Europe. Conceived by Americans, nourished by Englishmen, the ISS was sustained by the new 'scientific movement' sweeping Europe and America in the late 1860s and early 1870s. It was blessed by the 'Young Turks' of science (cf. Huxley, 1918, p. 541), and was dedicated to the proposition that science as the 'new learning' had a central role to play in social progress and international prosperity. For a generation, the series endured; ultimately, it died – and in the trajectory between its birth and death one finds revealed the enormous hopes, and equally enormous disappointments involved in the search for order unity within the natural and social sciences.

This history begins with the entrepreneurial vision of Edward Livingstone Youmans (1821–1887).

EDWARD YOUMANS COMES TO EUROPE

Youmans was born in Albany, New York, in 1821 and conquered childhood blindness to become one of the country's most prolific scientific popularizers. Reading Chambers and Lyell in the 1840s, he became a confirmed evolutionist, and, joining the firm of Appleton in New York, he pressed for serial publication of new work in palaeontology and geology. In the 1860s, when Darwinian theory arrived in America, Youmans cast himself in the role of herald archangel – pressing Appleton's to take up the Darwinian cause. Previously, Charles Appleton – amid intense rivalry with Harpers, Rand McNally and the Boston houses – had spared no expense to capture the 'world's outstanding works in science' for quite candid commercial reasons:

> When the growing interest in educational matters called for new textbooks, D. Appleton and Co. provided them. When a new cyclopedia was urgently needed, then Appleton's were ready with the best one ever prepared; and when interest in the development of scientific thought created a demand for a new literature, the Appleton house provided it. (Madison, 1966, pp. 72–73)

When it came to capturing scientific authors, Appleton had at his right hand Youmans, who became known among publishers as the most dynamic force in the advancement of scientific thought in America (Overton, 1925, p. 35; cf. Fiske, 1894).[1]

Long before the publication of the *Origin of Species,* Youmans found a working form of the evolutionary metaphor in the writings of Herbert Spencer. In 1856, Youmans read Spencer's *Principles of Psychology,* and, in 1859, his *Social Statics.* In

1860, soon after seeing the prospectus of the philosophical series Spencer proposed to issue by subscription, Youmans offered to become Spencer's American agent. In 1861, Youmans met Spencer in London, returned to New York, and began proselytising in America for the grand Spencerian concept of evolutionary development. 'In no part of the English world has Herbert Spencer's philosophy met with such a general and cordial reception as in the United States', John Fiske later recalled. It was to Youmans that this reception was largely due (Fiske, 1894).

Youmans' imagination was particularly fired by the prospect of diffusing Spencerian evolutionism across international frontiers. As literary acolyte to Spencer, he equated his task with that of international and progressive social development (Duncan, 1908, p. 178). Youmans himself was a committed Anglophile. During the mid-1860s, he made repeated trips to the scientific circles of London. He became a frequent guest of the famous 'X-Club' (cf. MacLeod, 1970), where he met Tyndall, Huxley, Lubbock and several of his future authors (Huxley, 1900, p. 258). In 1867, Youmans arranged for Appleton's to publish Huxley's *Lessons on Elementary Physiology* and, while in England, he edited reprints of a famous series of addresses *(Of Modern Culture: Its True Aims and Requirements)*, which included addresses on the claims of scientific education.[2] Youmans believed he could tap in America a ready commercial market for English intellectual goods. Despite the growing and pervasive German influence on American university life, C.S. Pierce recalled that 'the type of our thought was decidedly British'.[3] Whitelaw Reid, in an address at Dartmouth in 1873, summarised the position thus:

> Ten or fifteen years ago, the staple subject here for reading and talk, outside study hours, was English poetry and fiction. Now it is English science. Herbert Spencer, John Stuart Mill, Huxley, Darwin, Tyndall have usurped the places of Tennyson and Browning, and Matthew Arnold and Dickens. (*Scribner's Monthly*, 1873)

Happily for Youmans, Appleton, like his English contemporaries, particularly John Murray and Alexander Macmillan, was anxious to dominate this market. But there were risks. Economic difficulties had undermined the short-lived *Appleton's Journal* (edited by Youmans) in 1869. Margins had to be calculated carefully; investment sustained over initial periods of loss. Moreover, there was the fear of competitive pirating, which made profit margins unpredictable. For authors and publishers alike, lack of copyright protection was one of the most serious omissions in prevailing arrangements. Until the 1880s, the international economic rights of English and other European authors and publishers were safeguarded by only a small body of copyright laws and by a network of informal agreements governing reprinting and translation. By the late nineteenth century, however, the growth of a substantial demand for scientific books in Europe and America placed authors in an unfavourable position. Even when European copyright arrangements were codified by the Berne convention in 1886, publishers in the United States, which was not a signatory, remained a threat.[4] It was well known during the 1830s and 1840s for English works of literature and fiction to be regularly 'pirated' by American houses. In the 1870s and 1880s, there was every reason to expect that scientific works would share the same fate.[5] In capturing the market for British and European scientific books in America, a new policy was necessary. With sybilline acumen, Appleton became generally 'much more inclined to friendly cooperation and mutual concession than to barbarous competition' (cf. Unwin, 1926). In particular, and to secure exclusive permission to publish their works, Appleton made it 'a point of honour to treat foreign authors as though they were legally entitled to some equitable rights' (Huxley, 1920, p. 283).

Youman's struggle on behalf of Spencer and Huxley, and his own experience with *Appleton's Journal*, had taught him the financial importance of attracting consistently the best writers. In 1870, with the cooperation of the X-Club and its leading English men of science, Youmans therefore proposed to Appleton the idea of launching a new 'International Scientific Library' designed to contain 'the best work of every important scientific thinker of the day in all countries'. In an age of 'specialists', such a series could

> bring so much of each separate science as might be thought necessary for the ordinary layman within the compass of a single volume, which, while being accurate, should also be in a 'tongue understanded of the people'. (Kegan Paul, 1899, p. 279)

The plan to secure proper royalties and recognition for his authors was one of eloquent simplicity. By prior arrangement with participating publishers in several different countries, all authors would receive a fixed percentage royalty on their books, and would receive that royalty in every country in which their books were sold. In the absence of legislated international copyright agreements, this would virtually guarantee individual rights on a national basis.

Appleton approved, and Youmans sailed to England in June 1871 for a decisive visit of six months. Furnished with introductions through the X-Club and a letter from Spencer (which stressed Appleton's success in securing fair contracts in America for English men of science)[6] Youmans began his round. His task was not straightforward. Some potential authors were suspicious of 'popular' series, while others 'so eminent as specialists as to be well-nigh indispensable', were too busy to contribute. And 'not a few bluntly stated their opinion that the series would not "go"' (Fiske, 1894, p. 275).[7] Despite the 'rapt incredulity' of his listeners, Youmans persevered. 'Its importance grows upon one' he wrote his sister in July, 1871, 'and when I think of the quality and delicacy of the enterprise, and of the important consequences that may flow from it ... I am anything but dissatisfied with the aspect of things' (Fiske, 1894, p. 276).

Support steadily materialised. '[Alexander] Bain is all right and will be helpful', Youmans soon wrote to his sister from London, and 'I have this morning had a very interesting interview with [Francis] Galton. He thinks well of the scheme, and, strange to say, is now writing upon the very subject that I had assigned to him' (Fiske, 1894, p. 276). By July, Charles Darwin even insisted that the subject be brought up at the forthcoming British Association meeting in Edinburgh, for endorsement by all British men of science there assembled.

As it happened, the Edinburgh meeting of August 1871, was of critical importance to the scheme. Moving with Spencer's help through the 'bewildering crush', Youmans held a special session on the 'fringe' of the Edinburgh meeting, which included J.D. Hooker, T.H. Huxley, W.B. Carpenter, W.K. Clifford, Balfour Stewart and Alexander Bain, together with Dr. William Lindsay, the botanist of Perth. The session decided that Huxley, Tyndall and Spencer should form an advisory committee, representing the biological, physical and 'human' sciences. The meeting then turned to enlisting prospective authors. Edward Smith 'is the authority on diet', and Smith was duly registered. 'Tylor is great on primitive man – greater than Lubbock' although 'he may not be able to leave his present publisher'. His name, too, was inscribed (Fiske, 1894, p. 279).

After the Edinburgh meeting, Youmans returned to London with Spencer to meet Appleton who had just arrived from New York, and to conclude arrangements for an English edition with Henry King – a 'wide-awake wholehearted fellow' of 65 Cornhill Street. King, formerly a well-known bookseller of Brighton, and a partner of George Smith (of Smith, Elder and Co.) between 1853 and 1868, had recently begun his own publishing house, and was looking for a series which would help 'make his name'. King's optimism gave Youmans great hope. 'They will drive it with energy', Youmans declared, 'and are bound to make a success ... The only apparent remaining difficulty is to transfer authors to the new house. Several are sure, some doubtful, and there are four hopeless, but their places can be supplied' (Fiske, 1894, p. 284).[8]

In a 'Private and Confidential' letter circulated in August 1871, Youmans sought the attention of 'English Scientific Writers' for his project of 'monographs or elaborate essays' in a 'form suited for wide circulation'. Youmans drew particularly on his American market, which, he claimed, 'would give efficient support to such an enterprise'. 'That country', he said, 'now contains nearly 40 millions of inhabitants and they are eminently a reading and book-buying people; education is extending and improving, and the interest in science, especially its progressive aspects, is also becoming more earnest and more general' (Huxley Papers, 1871).

By mid-October 1871, Youmans had great expectations. In five years, he predicted, the series would run to seventy-five or one hundred volumes. 'It will be the world's popular cyclopedia of reading science. Spencer is delighted, but half-bewildered; every once in a while he breaks out, "who would have thought such a result would arrive from your first beginnings with me?" ... He says this whole movement is going to revolutionise the position of English authors' (Fiske, 1894, p. 285).

In November, Spencer accompanied Youmans to Paris, and helped him persuade Henri Paul Bailliere to accept the scheme.[9] French scientific publishing presented special problems. It was widely believed that scientific books could sell, but it was known that circulation was slow and distribution was limited, so new ideas from abroad could only by degrees, and with difficulty, make their way.[10] Still there were important questions of prestige at stake. Youmans' proposal reached Paris in the wake of Germany's victory – a victory seen in no small measure as a triumph for German science. The *Revue Scientifique*, founded in 1871, was convinced that 'c'est la science et l'esprit scientifique qui ont vainçus'. Bailliere saw the new 'Bibliothèque Scientifique Internationale' as a means of stimulating the resurgence in science which Pasteur and the Third Republic righteously sought. In November 1871, 'France is committed', Youmans rejoiced, 'and I have just received a French note from the publisher pledging ten books from the ablest men in France' (Fiske, 1894, p. 286). Again, with Spencer's prompting, Youmans went on to Berlin, but this time alone. Germany was not quite so easy to win. Youmans – alas – spoke no German, and Spencer had few German friends. Moreover, as Youmans explained 'France was dispatched quickly because Paris *is* France, but it is otherwise here. The men are scattered ... and [while] there is great boasting of "universal education", yet there is no place in the world where there is on the part of the learned such contempt for popular education, or such shallow and worthless trash as they write when they attempt to write for "the people".' Youmans continued sadly:

> Germany is not going to suit me. They are too cocky – 'We don't want your translations; we can make our own books; the talent is here', is the way publishers talk. On the other hand I find that scientific men here have a contempt for 'popularisation' more intense than anywhere else. They ostentatiously despise it and the countries that tolerate it. It looks like a very bad market for my pigs, but I am going to know more about it before I get through. (Fiske, 1894, pp. 286–287)

Above all, there was evidently no great love for what was called 'English philosophy', and astonishingly, 'no one had heard of Spencer'.[11]

Youmans moved from Berlin to Leipzig in search of a publisher, with growing impatience. 'I feel I am at the extremist point of my Quixotic career in Leipzig', he wrote:

> The Dutch [i.e. the Germans] are prodigiously slow ... They continue as ever to have their doubts, and I feel it awful hard work to wait. Still, it will have to be done but once, like gravitation and evolution, and so I continue to be patient. (Fiske, 1894, p. 289)

By December, however, the Germans had succumbed, and Brockhaus of Leipzig agreed to take the series. Returning to England, Youmans' views of Germany underwent a sea change, not least because he felt 'Germany is more ripe for the movement than even England; its best men can be procured' (Fiske, 1894, p. 290).[12] In fact, in Germany, as in France, an 'inner circle' of English scientific men had done the trick. In the end, what Spencer's reputation had wrought in England and France, Darwin's magic had worked in Germany. 'The whole nation is pervaded with religious skepticism, and they are without philosophic guidance', Youmans wrote. That guidance Darwinism had begun to supply.

By the winter of 1871, the series had 'caught on'. In mid-December, Youmans dined with Spencer, Tyndall and Huxley in London, and found the atmosphere

> precious different from my first dinner there six months ago. They are fairly in harness and trot quite smoothly. They took hold of work as a matter of course, and are going to make a pretty good committee. It seems to be universally agreed that we have a great thing in prospect and well under way, and which can hardly fail to result in large advantages to many authors and to the public also. (Fiske, 1894, p. 293)[13]

Huxley and Tyndall agreed to advise Youmans on condition that it would not involve 'the slightest correspondence or business, or anything more than the giving of an opinion now and then in regard to the competency of writers' (Duncan, 1908, p. 187). In editorial matters Youmans looked to Spencer alone for guidance.

In February 1872, arrangements with Messrs. Bailliere and Brockhaus were concluded. By the terms of agreement, a committee of 'not less than three' men of science in each country would decide what works should be admitted into the Series, and the publishers of each country would then publish 'according to the custom of their respective houses'.[14]

In France, most work fell upon Emil Alglave[15], a young lawyer and economist, who had begun the *Revue des Cours Scientifique* (later the *Revue Scientifique*) in 1864. In Germany, the committee of three – all university professors – was, at first, to include Rudolf Virchow of Berlin and Jacob Rosenthal (then of Berlin, but later of Erlangen). Eventually, however, the German committee included Rosenthal, I.R. Tzermat ('Chermac' in Youmans' letters) of Leipzig[16], and Oscar Schmidt of Strasbourg. All translations were to be issued in crown octavo. A first edition of 1250 copies, and all subsequent editions, would be issued at the discretion of each respective house. Each house would keep independent accounts; works would be exchanged in stereotype, and all illustrations in electro-type clichés. Books not accepted for the series in a given country could be exchanged with another participating publisher on the same financial terms. Volumes in the series would be sold for five shillings in Britain, \$1.50 to \$1.74 in America, 6RM in Germany, and six francs in France. Author's contracts were to be negotiated individually, but were usually based on a fee of £50 on publication of the first 1250 copies.[17] Royalties for 'foreign authors' were fixed at 7½% of the published price of works translated. The costs of translation were to be defrayed by each publisher. By far the greatest blessing, however, was the implicit provision for international distribution, which, according to Spencer, 'practically amounted to international copyright' (Duncan, 1908, pp. 158–159). Similar arrangements were made with Dumolard Fratelli in Milan and Znanie in St. Petersburg, for reproduction in Italian and Russian.

The series was advertised in England by Henry King in June, 1872. Spencer modestly proposed that 'each of the volumes should ... [deal] with some part of science capable of being cut out from the rest and within the limits of which there had been recent developments of importance' (Spencer, 1904, p. 230). More grandly, King claimed these 'Popular Treatises' would embody the 'latest investigations in the various departments of Science at present most prominently before the world' written by 'many of the most distinguished professors in England, America, Germany and France'.

In fact, King's enthusiasm defied the sceptics. 'It is by no means easy to say how far a scientific subject may be fairly popularised', warned the *Athenaeum*. 'To entice the public to swallow a dose of science a good deal of sugaring is no doubt necessary ...' But Youmans disagreed. While not specially intended for the 'instruction of beginners', the ISS titles were intended to 'address the *non-scientific* public; while they would be as far as possible explanatory in character and free from technicalities'[18], the object of each was to 'bring his subject as near as he can to the general reader'. In New York, Appleton's advertisement proclaimed that the series would

> give authentic yet popular expression to the latest and growing thought on the leading subjects of progressive inquiry. The various branches of physical, biological, psychological and social science, which have undergone marked revision within a record period, will be so presented as to help a better understanding of the phenomena of human nature and the economy of life. (Madison, 1966, p. 73)[19]

In France, it was determined from the outset that the series would not be limited to the physical and natural sciences; all titles 'empruntent les methodes d'observation et d'expérience qui ont rendues si fécondes depuis deux siècles'. The *Bibliothèque Scientifique Internationale* asserted it was not 'une entreprise de librairie ordinaire'. Its object, to popularise science in all its aspects, was no less important than the reflection that

> On peut ainsi, sans quitter la France, assister et participer au mouvement des esprits en Angleterre, en Allemagne, en Amerique, en Italie, etc., tout aussi bien que les savants memes de chacun de ces pays.[20]

In Germany, Brockhaus began the series in 1873, and made the 'Internationale Wissenschaftliche Bibliothek' the most important of its encyclopedic endeavours.

In America, then witnessing 'large increases' in popular scientific publication (Newcomb, 1874), and helped by Tyndall's lectures in 1872–1873, the series took hold.

THE SERIES IN THE SEVENTIES

The 1870's were perhaps the most exciting years of the ISS. In America, Appleton's titles probably encountered stiff opposition from 'Van Nostrand's Science Series', edited by Prof. George Plympton, 127 titles of which appeared (in green cloth), price 50 and 75 cents, between 1870 and 1902. The titles in Van Nostrand's list, chiefly concerned with industry and engineering, sold steadily as a series until the 1920s. Other American rivals were Henry Holt's 'American Science Series' which successfully published American authors, including William James' *Principles of Psychology* (1890), and Holt's 'Nature' series which began in 1906, and which promised 18 volumes.

In England, the ISS had the advantage of originality and surprise, coupled with the virtue of offering world-famous authors to a middle-brow audience at moderate prices. Given this audience, the series had the force of a new testament to the age, codifying the 'Religion of Science' and belief in scientific method which Beatrice Webb described as 'the most salient element in the mid-Victorian Time Spirit' (Webb, 1926, p. 123). It was Spencerian in range, as much as an inspiration – as the evolutionary metaphor was amplified from botany to speech, from music to political organisation. Quoting G.H. Lewes, Beatrice Webb (1926, p. 115) later recalled the evolutionary optimism of the 1870s: 'Onward and ever onward, mightier and forever mightier, rolls this wondrous tide of discovery'. And onward rolled the ISS.

In June 1872, King issued a list of 34 prospective titles, and between 1872 and 1879 produced 27 books, or an average of four per year. The first title was John Tyndall's *Forms of Water in Rain and Rivers, Ice and Glaciers* [1][21], a collection of essays begun as six lectures for children at the Royal Institution in 1871. The second was Bagehot's *Physics and Politics* [2], which had already appeared as a series in the *Fortnightly*. It went into seven languages by 1888 (Barrington, 1914). Dr. Edward Smith's classic guide to nutrition, *Foods* [3], and Alexander Bain's immensely influential *Mind and Body* [4], both promised at Edinburgh in 1871, appeared 'without much delay' in 1873 (Bain, 1904, p. 321)[22]. By 1910, *Mind and Body* had gone through eleven editions.

In 1871, Spencer, who had not at first intended to contribute to the series himself, was persuaded by Youmans to write a volume on the study of sociology. Successive chapters were first to appear both in the *Contemporary Review* of London, and in an American monthly. Meanwhile, with his usual gift for seizing opportunities, Youmans, in 1872, took the step of creating in New York the *Popular Science Monthly*[23], an important 'class periodical' devoted to 'broadening conceptions of science as analysis of mind'; the *PSM* was to be in the periodical press what the ISS would be in the world of books. As arranged, Spencer's first chapter appeared in the opening issue (Duncan, 1908, p. 160).[24] His *Study of Sociology* [5] was among the most successful volumes in the ISS, going through 22 English editions by 1914. In French translation (as *La Science Sociale*), it went through 14 editions by 1914, making it the most popular foreign work in the French series. Spencer's book brought the series international fame. In America, William Graham Sumner used the *Study of Sociology* as a text at Yale despite the fact that the President, Noah Porter, forbade it as 'antitheistic'. Two new titles – Balfour Stewart, *The Conservation of Energy* [6], and J. Bell Pettigrew, *Animal Locomotion* [7] – brought the series in 1873 to a total of seven; by 1874, when Charles Kegan Paul, formerly a master at Eton, joined Henry King as literary adviser, the series was well under way. It had, in fact, become the largest undertaking in the firm (Mumby, 1934, pp. 183–184), and gave Kegan Paul, Trench and Trubner its first major push to prominence.

Despite these early successes, the early 1870s were not untroubled. A few authors never produced, and Huxley first postponed his contribution, then changed his subject completely.[25] In America, where there was a large increase in scientific publishing (Newcomb, 1874), the series demanded constant care.[26] In 1873, amid murmurings about the alleged 'exclusion' of American authors (Duncan, 1908, p. 159)[27] Youmans fell seriously ill, badly strained from overwork (Fiske, 1894, p. 321), and business arrangements in England and America were temporarily left to drift. Moreover, in March 1874, what Youmans delicately described as a 'difference of opinion' arose between the English editorial committee, Henry King and Appleton (Spencer, 1904, pp. 258–278). The exact nature of these differences is not clear, but by the summer of 1873 we find Spencer asking Youmans to visit London to sharpen the edge of the Englishmen's amiable, but unbusiness-like, attitude. The series, wrote Spencer 'evidently wants a spur – some of the authors are lagging, and it is quite time that measures were taken for finding successors to them ...' (Duncan, 1908, p. 187). But Spencer's 'hard line' was contrary to the attitude of Huxley and Tyndall, who had deliberately refused to press authors unduly. Youmans felt bound to remind Spencer of their gentlemanly terms of agreement; he did, however, share Spencer's concern. 'The "Series" seems to be in a very bad way (internationally) and I don't know but we shall have to let it go; it was a quixotic project, and I doubt if it is worth much further attention' (Duncan, 1908, p. 187). The French series in particular needed a push; of the first ten volumes to appear in France, seven were by British authors. Emile Alglave, teaching first in Douai and then in Lille, lived too far from Paris (psychologically and geographically) to command intellectual assent.

Despite all these hazards and shortcomings, the ISS prospered. Several learned weeklies, notably the *Saturday Review* and the *Athenaeum,* with middlebrow, educated readerships, regularly reviewed its volumes.[28] In time-honoured fashion, sharp differences of opinion among reviewers did not hurt sales, nor authors' reputations. Spencer's *Study of Sociology*, which caused the *Saturday Review* to wonder whether it was 'worthwhile for a philosopher to spend time in popularising his own ideas' (*Saturday Review*, 1873), was warmly received in *Nature*:

> Nowhere, perhaps, than in sociological phenomena is the truth of the doctrine of evolution, the central doctrine of Mr. Spencer's philosophy, more strikingly displayed, and nowhere subject to proper limitations is it likely to prove more serviceable. (*Nature*, 1874)

In May 1874, the *Spectator* welcomed Henry Maudsley's *Responsibility in Mental Disease* [8] which argued that, while heredity disposed certain individuals towards both crime and insanity, the social influences of bad housing and poor education played an important part. The *Spectator* approved of its bearing on the 'welfare of the community and the instruction and improvement of the rising generation' (*Spectator*, 1874).

Perhaps the greatest boost to sales and notoriety, however, came the following year, when there appeared the second most successful title of the entire ISS (and the most popular American volume) – John William Draper's famous *History of the Conflict between Religion and Science* [13]. Following the 'victorious certainty' of Tyndall's Belfast address, Draper's book led to a terrific fanfare against materialism, agnosticism and iconoclasm in general. Draper went through twelve editions in English and was translated into French and German.[29]

By the mid-1870s, reforms in national scientific education were attracting great attention (MacLeod, 1969). The 'endowment of research' and the encouragement of scientific education involved demands for reform. Several publishers were alive to the implications of these new developments for the educational book market.[30] Macmillan began the 'Science Primer' series, with texts written by T.H. Huxley, Henry Roscoe, Balfour Stewart and Archibald Geikie – whom the *Publishers Weekly* called 'the very best men for the best work' (*Publishers Weekly*, 1873). By 1876, the *Academy* could remark that 'The cause of science is not likely to suffer in our schools for want of suitable textbooks'. Between five and ten series of textbooks were available, each attempting to set out 'the orderly progression of ideas from general to particular, which seems the watchword of the age'.[31] The publication of these textbooks could be very lucrative. Stanley Jevons, who wrote a *Primer of Logic* for Macmillan's *Science Primer* series in 1876, said 8000 copies of his volume were sold in the first six months of publication. But as A.J. Mundella observed to John Lubbock, mere textbooks were not enough: 'They are merely *crams* and burden the memories of children with a jargon of technical phraseology to which they attach no meaning' (Lubbock Papers, 1880). What was needed were more sophisticated general books for the enlightened, interested student. In this development the ISS, in many respects, led the way. Although not specifically intended as textbooks, some of them (notably Huxley on the *Crayfish* [28], published in 1880) were so used, and were soon to be found in the libraries of the Headmasters Conference schools.

On the lives of its authors, the series, and others like it, had a noticeable effect. W.S. Jevons' *Money and the Mechanism of Exchange* was published in 1875 [17], and galloped through 22 profitable editions. Meanwhile, his *Elementary Lessons in Logic*, for Macmillan's Science Class Books, sold 3350 copies. In November 1874, he wrote to his sister that the *Lessons* were earning him £70 per year; his conclusion: 'I think I am going to write more school or college books' (Jevons, 1886).

After Henry King's dealth in 1876, the ISS (which had passed to Charles Kegan Paul) was taken into the new firm of Kegan Paul, Trench and Co. Shortly afterwards, failing health forced Youmans to release more and more of the vital work of organisation to others. Although he lived until 1887, by 1880 he had ceased to be the animating, coordinating influence. In England, Alfred Trench, a younger son of the Archbishop of Dublin, attempted to develop the series in an enlightened fashion. As Kegan Paul later recalled, Trench happily took advice both from his father on 'agnostic books' and contributions from men of science on 'advanced' subjects:

> He thought the day was long past in which questions discussed in those books could be shelved; that our attitude should be that when they were treated in a reverent and serious spirit, we should by no means refuse to publish the works of a freethinking or agnostic type. (Mumby, 1934, p. 189)

Kegan Paul's reputation for liberal eclecticism won him the patronage of 'some of the leaders of the advanced liberal movement in England' (*Critic,* 1883), who often wrote for the *Nineteenth Century* – a review which performed the same function for Kegan Paul that *Macmillan's Magazine* (from 1880) performed for Macmillan. Profits apart, the ISS helped to consolidate this liberal, moral and internationalist image of science increasingly preferred by the leading secular middle-brow publishers of London.

THE SERIES IN THE 1880's: CONSOLIDATION AND EXPANSION

In 1882, G.J. Romanes observed in *Nature* that the ISS 'both on account of its popularity at home, and of its well organised machinery for securing rapid translations abroad, is the most suitable place for publishing results which are in so eminent a degree of interest to general readers (*Nature*, 1882). During the 1880s, the series issued 43 new titles, twice as many as in the period 1872–1879 (see Table 1), together with continuing editions of older ones – Adolph Wurtz, Charles Young, Jean-Louis de Quatrefages, Nicolas Joly, Robert Scott – works often evolutionary in temperament, progressive in tone, comparative in style.

Table 1
Distribution of ISS titles (English edition) by country of authorship

	1872–79	1880–89	1890–99	1900–09	1910+	Total titles	Total authors	Authors of different nationalities as percentage of total
U.K.	14	19	15	4	2	54	45	52.9
U.S.A.	4	4	2	–	–	10	18	21.2
France	4	7	2	3	1	17	9	10.6
Germany	4	6	–	–	–	10	5	5.9
Italy	1	2	–	–	–	3	4	4.7
Others	–	2	1	1	–	4	4	4.7
	27	40	20	8	3	98	85	100.0

The series perhaps reached its apogee with the appearance of Sir John Lubbock's immensely popular *Ants, Bees and Wasps* [40] in 1882. The book – consisting of papers read to the Linnaean Society and the Royal Institution on insect psychology, labour and community life – enjoyed the added advertisement of Lubbock's Presidency of the British Association in 1881.

During the 1880s, the series increasingly devoted itself to exploring the implications to be drawn from Darwinian biological theory. Not surprisingly, the series gained added visibility whenever noted Darwinians, including Ray Lankester and George Romanes, reviewed its titles in *Nature*.[32] The effect was, for a dozen years at least, highly integrative. In 1887, at the Queen's Jubilee, Grant Allen contrasted the 'fragmentary and disjunctive' science of 50 years previously with the spectacle of modern science as a

> simple, unified and comprehensive cosmos, consisting everywhere of the same prime elements, drawn together everywhere by the same great forces, animated everywhere by the same constant and indestructible energies, evolving everywhere along the same lines in accordance with the self-same underlying principles. (Allen, 1887)

The series illuminated this sense of a unified, comprehensive evolutionary dynamic. In the purest form of flattery, the idea of the ISS found one important imitator in Havelock Ellis, who in 1888 felt that, despite its success in leading 'advanced thought' for over seventeen years, the 'old (sic) International Scientific Series'

> in which such memorable books appeared, had now reached the state of senility. A new series on more modern lines was needed, with more attention to those sciences and those aspects of science which were now coming to the front. (Ellis, 1967, pp. 170–171)

One could not keep a good idea down.

In 1889, Ellis began the 'Contemporary Science Series' (opening with Patrick Geddes and J. Arthur Thomson's *The Evolution of Sex*) published by Walter Scott, which carried on for 25 years and 48 distinguished volumes (Ellis, 1967, pp. 170–171). Meanwhile, Appleton's themselves began a new 'International Education Series' (edited by William T. Harries) which produced 64 volumes and was sold in 1890 to the American Book Company (Madison, 1966, p. 76). In America, Appleton's leadership was contested by Putnam's *New Progressive Science Series*, edited by J. McKeen Cattell from 1898. Putman's series (which included Simon Newcomb's *The Stars*, C.A. Young's *The Earth as a Planet* and T.C. Mendenhall's *The Measurement of the Earth*)[33] even received a warm reception in England. Indeed, so successful was the idea that, by 1890, some thought that there was a surfeit of scientific series. As Tyndall wrote despairingly to Lubbock:

> In the midst of Napoleon's victories, the French were known to cry out 'Assez Buonaparte', and in the midst of the triumphs of science one is also tempted to cry 'Assez'. Is there to be no end to scientific books and scientific research? (Lubbock Papers, 1889)

THE SERIES IN THE NINETIES – DECLINE AND FALL

In fact, as Tyndall surmised, there were factors at work which would hamper its progress. Beginning in the mid-1880s, the series moved into hard times. After its rapid growth in the 1870s, it had declined sharply by the 1890s, and scarcely managed to survive into the twentieth century. Evidence suggests that contemporary series generally did not follow this pattern, but the ISS was vulnerable on several points. In retrospect, its editors and publishers failed to anticipate these difficulties, and to respond to new market opportunities. But this is an easy judgement. To contemporaries, the position was much more complicated.

If the late 1880s were the watershed of the series they also signalled the beginning of what Jerome Buckley has called 'a revolt from reason'.[34] Evolutionism had not won man glories of the spirit or the satisfaction of human kindness, and materialism and the 'tyranny of science' had made profound enemies.[35] Tennyson summarised his own sensation of disillusionment and despair with science and its mystique in *Lockesley Hall, Sixty Years After*. Beatrice Webb described her feelings in 1882 when she explained that she had tried the 'religion of science and [had] found it wanting'. Science could never be, in her terms, the religion of a 'suffering humanity' (Webb, 1926, p. 80). Even William Graham whose volume on *Socialism, New and Old* [70] was published in the ISS in 1890, delivered a heavy blow against the aggressions of 'modern science' in the realm of politics and religion.

Moreover, by the turn of the century, the concept of unified positive knowledge was under attack (cf. Strong, 1906), as A.J. Balfour demonstrated in his Presidential Address to the British Association in 1904. The 'revolt against positivism' incorporated a revolt against the evolutionary naturalism that had served as the conceptual 'character' of the ISS. When Grant Allen died in 1899, his biographer admitted that he 'perhaps ... too confidently set up Darwin and Mr. Herbert Spencer in the place of his lost Hebrew prophets (Le Gallienne, 1899).

And both in France and England, Brunetière and Ray Lankester declaimed against the 'bankruptcy of science'.

It is difficult to assess the effect of these changing beliefs on reading fashions; any effect upon particular series was, to say the least, indirect. Nevertheless, the climate had become distinctly unfavourable. In England, the widening separation between science and religion, literature and philosophy, now mocked the 'unified spirit' of *la science totale* and *Wissenschaftlichkeit* of the 1870s. The British reading public seems to have moved away from the mid-Victorian fascination with natural history and improving texts. Science became romanticised, as in the novels of W.D. Howells and H.G. Wells; or made the agent of detection or crime, as in R.L. Stevenson's *Jekyll and Hyde* (1898). 'Popular science' became operational knowledge, and those several series praising the 'heroes of science'[36] which were successful, looked back to an arcadian vision of Victorian achievement. C.P. Putnam and Sons began a 'Leaders in Science' series[37] (edited by S. Chapman) in 1919. But these new publishing ventures thrived on history; glorifying the past, rather than cultivating a future. The cumulative effect was to produce a climate which, critical of many former certainties, did not augur well for the Series, or its contemporaries.

By the 1890s, the ISS met a much less favourable reception in the 'quality' weeklies of London. The ambitious scope of the series did not agree with everyone. Indeed, the influence and authority of the 'familiar black and red covers' as one put it, invested each book with a 'considerable capacity for harm' (*Athenaeum,* 1890). When a book was simply 'bad' (as, in one case, the *Athenaeum* described Poulton's *Colours of Animals* (1890, [68]), it could make a very much worse impression than an ordinary bad book. Moreover, as G. Wright's *Man and the Glacial Period* (1893, [72]) demonstrated, a book's mere presence in the series could lend a misleading air of authority to an otherwise controversial interpretation.[38]

Sometimes, this impression of authority was belied by the facts. As Kegan Paul, with a trace of bitterness, recalled, 'some of the most distinguished contributors gave merely old magazine articles, while others were not so well known to the world in general as men of scientific importance' (Kegan Paul, 1899, p. 280). True, T.G. Bonney's *Ice Work, Present and Past* (1896, [78]) was welcomed as one of the best books on glacial geology (*Athenaeum,* 1896), and as a model of what the series should be. 'An exposition of the results of modern research and modern thought which shall be an unrestricted and international character, written without offence to prejudices or persons...' (cf. *Nature,* 1896). But while *What is Electricity?* (1897, [82]) was considered 'pleasant reading', F.W. Eldridge-Green's *Memory and its Cultivation* (1897, [83]) was considered out-of-date, allegorical and polemical in the worst sense (*Nature,* 1891). M.G. Cooke's *Introduction to Freshwater Algae* (1890, [69]) took its place as a work of reference next to William Phillip's *Manual of the British Discomycetes* (1887, [61]); but Thomas Stebbing's *History of the Crustacea* (1893, [74]) was rejected as a disappointing failure in scientific argument (*Nature*, 1893)[39]. Harry Wallis Kew's *Dispersal of Shells* (1893, [75]) was taken as a most useful guide to the transmutation of species, but George Henslow's *Origin of Plant Structures* (1895, [77]) was criticised by C.A. Barber in *Nature* for its neo-Lamarckian bias (*Nature*, 1895).

There were also grounds for financial concern. The new titles were not selling as well as the older ones. In his eagerness to accept the series in 1871, Henry King had offered what struck Kegan Paul as liberal terms, possibly 'too liberal indeed to make the series a great financial success' (Kegan Paul, 1899, p. 280) – unless there were several editions. Many of the works, moreover, including Huxley's excellent study of *The Crayfish* [28], had expensive illustrations. Finally, there were the usual difficulties of translation. Mrs. Kegan Paul did many of the translations from French (for which she was paid £25 a time), but the office could not easily cope with other languages. As Kegan Paul woefully recounted, H. Vogel's *Light and Photography* (1875, [15]) had to be translated from German and only reached him in proof. But the translation was misplaced, and, since the whole edition of 1250 had to be condensed before publication, a new translation had to be made. This translation was criticised in *Nature* until a second, and 'thoroughly revised' edition could be prepared (*Nature*, 1875; 1876). Even then, the *Academy* slammed it as a 'flagrant case of book-making without sufficient capital – absolutely useless as a scientific or practical textbook or manual' (*Academy*, 1876).

Amidst these difficulties, the publishers in London grew increasingly unable to deal with scientific content. Kegan Paul's reaction to a loss on Lubbock's *Seedlings* (1896, [79]) was to argue for issuing a 'readable edition of the book from which the purely scientific matter is omitted' (Lubbock Papers, 1893). After Youmans' death in 1887, the series lacked a strong 'scientific editor'. The ISS had been weighted strongly in favour of the biological and social sciences (reflecting, probably, Spencer and Huxley). In omitting volumes on mathematics or astronomy, in giving slight treatment to physics, or the 'new chemistry' (only 10 of the 98 titles are in these fields), the series by the 1890s gave a misleading impression of the breadth of modern science.

To compound matters, Kegan Paul misjudged his market. Writing to Lubbock in 1891, he confessed his fears as to the ability of a new series, 'Modern Science', to pay its way:

> The books, one and all, as far as we have seen them, seem admirable, but not such as the public are likely to buy in large numbers, and we doubt whether there is a sufficient number of real students and scholars to purchase the very large quantities necessary to make 'Modern Science' a paying speculation. (Lubbock Papers, 1891)

In fact, as the experience of Macmillan had shown, science was a gilt-edge investment. There is no evidence that reducing the price of books in science helped sell them – a factor American publishers put to their advantage. Kegan Paul had difficulty with content, in large part because he had difficulties with his authors. If good science books were becoming difficult to sell, good authors were also becoming more difficult to find. The proportion of F.R.S.'s (including foreign members) among authors in the English series had hovered around 30% in each decade between 1872 and 1899, but after 1900 this proportion fell to eight per cent. The proportion of British authors sufficiently distinguished to appear eventually in the *DNB* also fell sharply (see Table 2).

Within the ranks of scientific writers, the age of the 'great generalists' – Huxley, Tyndall and Spencer – had passed. Their successors were far less distinguished, and sold less well. Increasingly, publishers turned to specialised texts for professional use, and for the teaching of specific disciplines.[40] This movement towards specialisation, towards institutional development, and towards 'national efficiency' tended, if anything, to distract readers (and publishers) away from general series which emphasised popular scientific culture, and which were directed to an international audience.

This situation worried W.C.D. Whetham, whose *Recent Development of Physical Science*, which first appeared in August 1904, actively looked to the

> possibility of treating the wider and deeper generalisations of natural science as fit subject matter for current thought and literature ...

Yet, Whetham believed, there was not a sufficient and 'proper understanding of the scientific spirit and scientific method even among the more educated portion of the community'. The problem was acute for scientists, too. As a contemporary zoological reviewer for *Nature* bewailed:

> The accumulation of facts in zoology during the last quarter of a century has been so overwhelming that it becomes most difficult even to gauge the capacity of a book ... which deals broadly with the whole field. (*Nature*, 1903)

Welded to the examination system, science books became increasingly textbooks – in the words of a late-Victorian reformer – mere 'collections of test-cards framed

Table 2
Distinctions among authors of the ISS (English edition)

	1872–79	1880–89	1890–99	1900–09	1910+	Totals
Number of titles	26	43	17	10	2	98
Number written by authors mentioned in DNB	12	10	4	–	1	27
Number written by FRS's (UK and Canada)	6	11	5	–	1	23
Number written by FRS's (Foreign members)	2	2	–	–	–	4

on models of those set by inspectors of schools' (Dyer, 1893, p. 17). The predictable result was evident as early as the late 1880s. A reviewer in *Nature* observed:

> It is by no means an uncommon misfortune to find that textbooks are not written by persons the most competent or with the widest experience; hence the results are very far from satisfactory, and no one expresses much gratification. (*Nature*, 1888)

The end of the ISS came gradually, but ineluctably. In Germany, the series stopped in 1889. In America, where records have vanished, the series seems to have ended by 1910, and in England, the last new title appeared in 1911. By that time, clearly, the era of international peace which had given the cooperative aim of the series its legitimacy, had collapsed – and with it passed, for a time, heroic appeals to the scientific book as a symbol of 'free trade in ideas' (see Table 3).

Faltering, the English edition was suspended between 1899 and 1902, whilst advertisements in *Nature* offered to re-issue the series 'at a greatly reduced price'.[41] Only one volume, H.M. Vernon, *Variations in Animals and Plants* (88) appeared in 1903, and the series lapsed again between 1904 and 1907. The future prospects of the ISS were unpromising; yet, on the whole, it had proved profitable.[42] Of the ninety-eight titles issued by 1911, no less than seventy-four had gone through more than one edition. Although superseded by Andrew White's *History of the Warfare of Science with Theology in Christendom,* Draper's *History of the Conflict between Science and Religion* alone sold over 25000 copies; it enjoyed fifty printings by the early 1930s, and, in 1950, it was still on Appleton's list.[43] Draper's *Conflict between Religion and Science* was also reprinted by the Vanguard Press in New York, in 1926; Robert Thurston's *History of the Growth of the Steam Engine* was last reprinted by Cornell University Press in 1939; Abercombie's *Weather* was last reprinted in 1946, and Spencer's *Study of Sociology* was last reprinted by Dover Paperback in 1961.

In France, the collective ideal survived the century. This was probably due to a combination of factors, about which one has only the slimmest evidence. On the one hand, the number of French series – scientific, literary and philosophical – in which scientific titles could appear, continued to grow. Declaring the quarter-century ending in 1910, 'la grande periode de specialisation de la medicine et des sciences', Masson et Cie alone published 1370 scientific and medical titles between 1889 and 1900, and began several new series, including a 'Bibliothèque de la Nature' (Masson et Cie, 1954). Even Tyndall's French publishers, Gauthier-Villars, began their own series of 'Actualités Scientifique'. In France, the ISS also lived on, strengthened by traditional symmetries between the natural and social sciences, by an evident accommodation between intellectual demand and commercial interests. One may speculate that the ghost of Spencer, mediated by Durkheim, retained some hold on French intellectual life. The new Presses Universitaire de France, succeeding Felix Alcan, began in 1920 *La Nouvelle Biblio-*

Table 3

Distribution of ISS titles (English, French and German editions by country of authorship, at date of authorship)

Authors	1872/79 Edition			1880/89 Edition			1890/99 Edition			1900/09 Edition			1910/14 Edition			Total		
	UK	Fr.	Ger.	UK	Fr.	Ger.	UK	Fr.	Ger.	UK	Fr.	Ger.	UK	Fr.	Ger.	UK	Fr.	Ger.
UK	14	11	11	19	7	9	15	1	*	4	2	*	2	1	*	54	22	20
USA	4	2	3	4	3	4	2	–	*	–	–	*	–	–	*	10	5	7
Fr.	4	8	5	7	13	3	2	20	*	3	10	*	1	9	*	17	60	8
Ger.	4	6	10	6	4	6	–	–	*	–	3	*	–	–	*	10	13	16[a]
Italy	1	2	3	2	3	2	–	–	*	–	2	*	–	–	*	3	7	5
Other	–	1	2	2	1	4	1	3	*	1	–	*	–	–	*	4	4	6
	27	30	34	40	31	28	20	24	–	8	17	–	3	10	–	98	112	62[b]

[a] Contributors from Strasbourg are counted as German in the period 1871–1914.

[b] Brockhaus announced the publication of 68 titles in all, but only 62 can be traced.

thèque Scientifique Internationale. A half century later, this PUF series continues. In content, there is little to suggests its origins, but in title at least, it keeps alive a semblance of its original ideology.

If the survival of the series in France owes much to the particular supportive characteristics of the French publishing industry, its subsidies, and its selling strategies, in England and America the series was arguably much more sensitive to the changing pressures of the market. As such, the end of the ISS signalled the end of an epoch in English and American scientific publishing, and possibly a turning point in the organisation of science. With the end of the Great War, the concept of the international series was occasionally taken up again, but with very little of the same unifying literary and humanitarian spirit. In 1920, Kegan Paul (now Routledge and Kegan Paul) launched a new series – the 'International Library of Psychology, Philosophy and Scientific Method' (edited by C.K. Ogden) – the first volume of which was G.E. Moore's *Philosophical Studies.* By 1934, there were nearly 120 books in the series, which was described as an 'heroic attempt to incorporate all current tendencies of international thought' in the field.[44] Between the wars, Routledge also published the famous *Future of Science* series. But these ventures now spoke of a vastly altered conception of science, of specialised knowledge and of science and knowledge no longer 'unified' by a Spencerian evolutionary, perspective.

The ISS had fostered, commercially and intellectually, a scientific consciousness. With poetic justice, having failed to adapt itself, or to find an ecological niche, it lost its selective advantage in the struggle for survival. Its history reflects the tensions within the mid-Victorian synthesis, and the pressures of commercial competition, intellectual fragmentation and specialisation. Without doubt, by the turn of the century, the vested interests, public demands, and intellectual justifications for scientific publishing were no less pressing – but their force and direction had changed. This, above all, the ISS, and other series to which we can generalise, amply illustrate. An attempt to offer a new 'common context' had ultimately failed. In England and America, at least, the circle of the sciences had been broken – and publishers, professors, and researchers, moved in to collect the pieces.

Acknowledgements

I am grateful for the kind cooperation of Mrs. Margaret Selby, of Routledge and Kegan Paul Ltd., in making available to me records and ledgers of the English edition; and for the assistance of the Lilly Library of Indiana University, in allowing me to consult the Appleton papers in their possession. I am also indebted to Fraulein Schlecht of F.A. Brockhaus, formerly of Leipzig, and now of Wiesbaden; to M. Hervé Lauriot Prevost of Presses Universitaires de France, Paris; and to Mr. Burton Frye of Appleton-Century-Crofts, New York, for helping in my enquiries. I am also grateful for the assistance of M. Clemens Heller and his staff at the Maison des Sciences de l'Hommes, Paris. This research was supported in part by a grant-in-aid from the American Council of Learned Societies, and by the SSRC/CNRS Exchange programme.

Appendix I. The International Scientific Series

English Editions: Kegan Paul, Trench, Trubner & Co.

(Detailed information on illustrations is not available for all titles.)

[1] TYNDALL, John	*Forms of Water: in Clouds and Rivers, Ice and Glaciers* (1871) 192 pp.; 25 illus.; (13 editions)
[2] BAGEHOT, Walter	*Physics and Politics, or Thoughts on the Application of the Principles of Natural Selection and Inheritance to Political Society* (1872) 228 pp.; (13 editions)
[3] SMITH, Edward	*Foods* (1873) 485 pp.; 156 illus.; (10 editions)
[4] BAIN, Alexander	*Mind and Body: The Theories of their Relation* (1873) 196 pp.; 4 illus.; (10 editions)
[5] SPENCER, Herbert	*The Study of Sociology* (1873) 446 pp. (22 editions)
[6] STEWART, Balfour	*The Conservation of Energy, being an Elementary Treatise on Energy and its Laws* (1873) 180 pp.; 14 illus.; (9 editions)
[7] PETTIGREW, J. Bell	*Animal Locomotion, or Walking, Swimming and Flying, with a Dissertation on Aeronautics* (1873) 264 pp.; 130 illus.; (4 editions)
[8] MAUDSLEY, Henry	*Responsibility in Mental Disease* (1874) 313 pp.; (5 editions)
[9] COOKE, Josiah P.	*The New Chemistry* (1874) 333 pp.; 31 illus.; (11 editions)
[10] AMOS, Sheldon	*The Science of Law* (1874) 417 pp.; (9 editions)
[11] MAREY, Etienne Jules	*Animal Mechanism* (1874) 283 pp.; 117 illus.; (4 editions)
[12] SCHMIDT, Eduard Oscar	*The Doctrine of Descent and Darwinism* (1875) 334 pp.; 26 illus.; (8 editions)
[13] DRAPER, John William	*The History of the Conflict between Religion and Science* (1875) 373 pp.; (24 editions)
[14] COOKE, Mordecai Cubitt (Ed. M.J. Berkeley)	*Fungi: Their Nature. Influence and Uses in their application to Art, Science and Industry* (1875) 299 pp.; illus.; (6 editions)
[15] VOGEL, Hermann	*The Chemistry of Light and Photography* (1875) 302 pp.; 100 illus.; (1 edition)
[16] WHITNEY, William Dwight	*The Life and Growth of Language* (1875) 526 pp.; (7 editions)

[17] JEVONS, Stanley	*Money and the Mechanism of Exchange* (1875) 349 pp.; (22 editions)
[18] LOMMEL, Eugene	*The Nature of Light, with a General Account of Physical Optics* (1875) 356 pp.; 188 illus.; (7 editions)
[19] VAN BENEDEN, Pierre J.	*Animal Parasites and Messmates* (1876) 274 pp.; 83 illus.; (4 editions)
[20] SCHÜTZENBERGER, Paul	*Fermentation* (1876) 331 pp.; 28 illus.; (5 editions)
[21] BERNSTEIN, Julius	*The Five Senses of Man* (1876) 304 pp.; 91 illus.; (7 editions)
[22] BLASERNA, Pietro	*The Theory of Sound in its Relation to Music* (1876) 187 pp.; numerous illus.; (7 editions)
[23] LOCKYER, J. Norman	*Studies in Spectrum Analysis* (1875) 258 pp.; numerous illus.; (6 editions)
[24] THURSTON, Robert Henry	*A History of the Growth of the Steam Engine* (1878) 490 pp.; numerous illus.; (5 editions)
[25] BAIN, Alexander	*Education as a Science* (1879) 453 pp.; (10 editions)
[26] DE QUATREFAGES, J.L. Armand	*The Human Species* (1879) 498 pp.; (6 editions)
[27] ROOD, Ogden, N.	*Modern Chromatics* (1879) 330 pp.; 130 illus.; (4 editions)
[28] HUXLEY, Thomas Henry	*The Crayfish: An Introduction to the Study of Zoology* (1880) 371 pp.; 82 illus.; (7 editions)
[29] BASTIAN, H. Charlton	*The Brain as an Organ of Mind* (1880) 708 pp.; 184 illus.; (5 editions)
[30] WURTZ, Charles Adolphe (transl. E. Cleminshaw)	*The Atomic Theory* (1880) 344 pp.; (7 editions)
[31] SEMPER, Karl	*The Natural Conditions of Existence, as they affect Animal Life* (1880) 472 pp.; 108 illus.; (5 editions)
[32] ROSENTHAL, Isidor	*General Physiology of Muscles and Nerves* (1880) 324 pp.; 75 illus.; (4 editions)
[33] LE CONTE, Joseph	*Sight: An Exposition of the Principles of Monocular and Binocular Vision* (1881) 275 pp.; numerous illus.; (3 editions)
[34] SULLY, James	*Illusions: A Psychological Study* (1881) 372 pp.; (4 editions)
[35] JUDD, John W.	*Volcanoes* (1881) 383 pp.; 96 illus.; (6 editions)
[36] MORSELLI, Henry	*Suicide: An Essay on Comparative Moral Statistics* (1881) 388 pp.; illus.; (3 editions)
[37] LUYS, Jules Bernard	*The Brain and its Functions* (1881) 327 pp.; illus.; (4 editions)
[38] VIGNOLI, Tito	*Myth and Science* (1882) 370 pp.; (4 editions)
[39] YOUNG, Charles Augustus	*The Sun* (1882) 321 pp.; numerous illus.; (5 editions)

[40] LUBBOCK, Sir John — *Ants, Bees and Wasps: A Record of Observations on the Habits of the Social Hymenoptera*
(1882) 436 pp.; 5 illus.; (17 editions)

[41] ROMANES, Georges J. — *Animal Intelligence*
(1882) 520 pp.; (8 editions)

[42] STALLO, Johann Bernard — *The Concepts and Theories of Modern Physics*
(1882) 313 pp.; (4 editions)

[43] RIBOT, Théodule Armand — *Diseases of Memory*
(1882) 209 pp.; (4 editions)

[44] JOLY, Nicolas — *Man before Metals*
(1883) 365 pp.; 148 illus.; (6 editions)

[45] AMOS, Sheldon — *The Science of Politics*
(1883) 490 pp.; (9 editions)

[46] SCOTT, Robert H. — *Elementary Meteorology*
(1883) 408 pp.; numerous illus.; (8 editions)

[47] VON MEYER, Georg Hermann — *The Organs of Speech, and their Application in the Formation of Articulate Sounds*
(1883) 349 pp.; 47 illus.; (2 editions)

[48] SIDGWICK, Alfred — *Fallacies: A View of Logic from the Practical Side*
(1883) 375 pp.; (3 editions)

[49] DE CANDOLLE, Alphonse — *The Origin of Cultivated Plants*
(1884) 468 pp.; (2 editions)

[50] ROMANES, Georges J. — *Jelly Fish, Star Fish and Sea Urchins, being a Research on Primitive Nervous Systems*
(1885) 323 pp.; illus.; (2 editions)

[51] CLIFFORD, William K. — *The Common Sense of the Exact Sciences*
(1885) 271 pp.; 100 figs.; (5 editions)

[52] WARNER, Francis — *Physical Expression: its Modes and Principles*
(1885) 372 pp.; 51 illus.; (2 editions)

[53] HARTMANN, Robert — *Anthropoid Apes in their Relation to Primeval Times*
(1885) 308 pp.; 51 illus.; (2 editions)

[54] SCHMIDT, Eduard Oscar — *The Mammalia in their Relation to Primeval Times*
(1885) 308 pp.; (2 editions)

[55] POSNETT, H. Macaulay — *Comparative Literature*
(1886) 402 pp.; (1 edition)

[56] MILNE, John — *Earthquakes and other Earth Movements*
(1886) 376 pp.; 38 figs.; (5 editions)

[57] TROUESSART, Edouard L. — *Microbes, Ferments and Moulds*
(1886) 314 pp.; 107 illus.; (3 editions)

[58] HEILPRIN, Angelo — *Geographical and Geological Distribution of Animals*
(1887) 435 pp.; (2 editions)

[59] ABERCROMBY, Ralph — *Weather: A Popular Exposition of the Nature of Weather Changes from Day to Day*
(1887) 472 pp.; 96 figs.; (6 editions)

[60] BINET, Alfred and FERE, Charles — *Animal Magnetism*
(1887) 378 pp.; (5 editions)

[61] PHILLIPS, William — *Manual of British Discomytes with Descriptions of all the Species of Fungi hitherto found in Britain included in the Family, and illustrations of the Genera*
(1887) 446 pp.; (2 editions)

[62] Levi, Leone	*International Law: With Materials for a Code of International Law* (1887) 346 pp.; (2 editions)
[63] Dawson, Sir J. William	*The Geological History of Plants* (1888) 290 pp.; illus.; (1 edition)
[64] Henslow, George	*The Origin of Floral Structures through Insect and other Agencies* (1888) 349 pp.; 88 illus.; (2 editions)
[65] Lubbock, Sir John	*On the Senses, Instincts, and Intelligence of Animals with Special Reference to Insects* (1888) 292 pp.; over 100 illus.; (6 editions)
[66] Starcke, Carl Nicolaj	*The Primitive Family in its Origin and Development* (1889) 315 pp.; (2 editions)
[67] Lagrange, Ferdinand	*Physiology of Bodily Exercise* (1889) 395 pp.; (3 editions)
[68] Poulton, Edward Bagnall	*The Colours of Animals, their Meaning and Use, Especially considered in the Case of Insects* (1890) 360 pp.; 67 illus.; (2 editions)
[69] Cooke, Mordecai Cubitt	*Introduction to Fresh-Water Algae* (1890)
[70] Graham, William	*Socialism: New and Old* (1890) 416 pp.; (5 editions)
[71] Edridge-Green, Frederick William	*Colour-Blindness and Colour-Perception* (1891) 311 pp.; 3 illus.; (1 edition)
[72] Wright, G. Frederick	*Man and the Glacial Period* (1892) 385 pp.; illus.; (2 editions)
[73] Maunde-Thompson, Edward	*Handbook of Greek and Latin Palaeography* (1892) 343 pp.; (3 editions)
[74] Stebbing, Thomas R.R.	*A History of Crustacea: Recent Malacostraca* (1893) 466 pp.; illus.; (1 edition)
[75] Kew, Henry Wallis	*The Dispersal of Shells: An Inquiry into the Means of Dispersal Possessed by Fresh Water and Land Mollusca* (1893) 291 pp.; illus.; (1 edition)
[76] Lefevre, André	*Race and Language* (1894) 424 pp.; (1 edition)
[77] Henslow, George	*The Origin of Plant Structures by Self-Adaptation to the Environment* (1895) 256 pp.; (1 edition)
[78] Bonney, Thomas G.	*Ice Work, Present and Past* (1896) 295 pp.; illus.; (2 edition)
[79] Lubbock, Sir John	*A Contribution to our Knowledge of Seedlings* (1896) 288 pp.; 282 figs. (1 edition)
[80] Parry, Sir Charles Hubert E.	*The Evolution of the Art of Music* (1896) 342 pp.; (1 edition)
[81] Angot, Alfred	*The Polar Aurora* (1896) 264 pp.; illus.; (1 edition)
[82] Trowbridge, John	*What is Electricity?* (1897) 315 pp.; illus.; (1 edition)
[83] Eldridge-Green, Frederick William	*Memory and its Cultivation* (1897) 311 pp.; (1 edition)

[84]	VINCENT, Ralph Harry	*Elements of Hypnotism* (1897) 271 pp.; 17 illus.; (2 editions)
[85]	MILNE, John	*Seismology* (1898) 338 pp.; 52 figs.; (1 edition)
[86]	LUBBOCK, Sir John	*Buds and Stipules* (1899) 239 pp.; 340 figs.; (1 edition)
[87]	DEMOOR, Jean; MASSART, Jean; VAN DER VELDE, Emile (Transl. Mrs. Chalmers Mitchell)	*Evolution by Atrophy in Biology and Sociology* (1899) 322 pp.; (1 edition)
[88]	VERNON, Horace Middleton	*Variation in Animals and Plants* (1903) 415 pp.; (1 edition)
[89]	BINET, Alfred	*The Mind and the Brain* (1907) 280 pp.; (1 edition)
[90]	POINCARÉ, Lucien	*Evolution of Modern Physics* (1907) 344 pp.; (1 edition)
[91]	LE BON, Gustave	*The Evolution of Forces* (1908) 388 pp.; 42 figs.; (1 edition)
[92]	MAKOWER, Walter	*The Radio-Active Substances; their Properties and Behaviour* (1908) 301 pp.; (1 edition)
[93]	COMBARIEU, Jules	*Music: Its Laws and Evolution* (1910) 334 pp.
[94]	DEPERET, Charles	*The Transformations of the Animal World* (1909) 360 pp.; (1 edition)
[95]	MACNAMARA, N. Charles	*Human Speech: A Study in the Purposive Action of Living Matter* (1908) 284 pp.; 44 illus.; (1 edition)
[96]	GARRETT, Albert E.	*The Periodic Law* (1909) 294 pp.; (1 edition)
[97]	MACNAMARA, N. Charles	*The Evolution and Function of Living Purposive Matter* (1910) 298 pp.; illus.
[98]	TUTTON, Alfred E.H.	*Crystals* (1911) 301 pp.; 120 illus.

Note: The edition number in parenthesis refers throughout to the number of editions reached by 1909. This, in most instances, was the final edition.

Appendix II. Dates of First English Editions

1872	Tyndall [1], Bagehot [2]
1873	Smith [3], Bain [4], Spencer [5], Stewart [6], Pettigrew [7]
1874	Maudsley [8], Cooke [9], Amos [10], Marey [11]
1875	Schmidt [12], Draper [13], Cooke and Berkeley [14], Vogel [15], Whitney [16], Jevons [17], Lommel [18], Lockyer [23]
1876	van Beneden [19], Schützenberger [20], Bernstein [21], Blaserna [22]
1877	

1878	Thurston [24]
1879	Bain [25], Quatrefage [26], Rood [27]
1880	Huxley [28], Bastian [29], Wurtz [30]
1881	Semper [31], Rosenthal [32], Le Conte [33], Sully [34], Judd [35], Morselli [36], Luys [37]
1882	Vignoli [38], Young [39], Lubbock [40], Romanes [41], Stallo [42], Ribot [43]
1883	Joly [44], Amos [45], Scott [46], Meyer [47], Sidgwick [48]
1884	De Candolle [49]
1885	Romanes [50], Clifford [51], Warner [52], Hartmann [53], Schmidt [54]
1886	Posnett [55], Milne [56], Trouessart [57]
1887	Heilprin [58], Abercrombie [59], Binet [60], Phillips [61], Levi [62]
1888	Dawson [63], Henslow [64], Lubbock [65]
1889	Starcke [66], Lagrange [67]
1890	Poulton [68], Cooke and Berkeley [69], Graham [70]
1891	Eldridge-Green [71]
1892	Wright [72], Thompson [73]
1893	Stebbing [74], Kew [75]
1894	Lefevre [76]
1895	Henslow [77]
1896	Bonney [78], Lubbock [79], Parry [80], Angot [81]
1897	Trowbridge [82], Eldridge-Green [83], Vincent [84]
1898	Milne [85]
1899	Lubbock [86], Demoor et al. [87]
1900	–
1901	–
1902	–
1903	Vernon [88]
1904	–
1905	–
1906	–
1907	Binet [89], Poincaré [90]
1908	Le Bon [91], Mackower [92], Macnamara [95]
1909	Deperet [94], Garrett [96]
1910	Combarieu [93], Macnamara [97]
1911	Tutton [98]

Notes

[1] Youmans was described as one of Appleton's 'greatest contributions to publishing and American culture' (Madison, 1966, p. 31).

[2] The book, published by Macmillan, contained one lecture ('Observations on the Scientific Study of Human Nature') which Youmans had himself given to the London College of Preceptors.

[3] Quoted in C. Wright Mills, *Sociology and Pragmatism, the Higher Learning in America*, New York, 1964, p. 93.

[4] Provision for international copyright for British works was made by the International Copyright Acts (1837 and 1881), as elaborated by the Copyright Act, 1911. Before the Chase Act of 1891, which took 53 years and 12 Bills through Congress, copyright in the United States could only be obtained by American citizens; cf. E.J. MacGillvray, *Treatise upon the Law of Copyright in the United Kingdom*, London, 1902. George Putnam appears to have been the first American publisher to pay royalties to British authors, beginning in 1848, but many of the better houses formed a kind of 'unofficial copy-

right'. American publishers often argued that American readers were poor and that English authors could live on their English sales. There was also some fear that, if copyright were granted to English authors, pirating might pass to English control; cf. George S. Gordon, *Anglo-American Literary Relations*, Q.U.P., 1924, pp. 83–86 (cf. also Madison, 1966, pp. 25–29; 53–57).

[5] Youmans, himself, was not entirely innocent of this – as his record in publishing material previously published in *Nature* in the *Popular Science Monthly* (begun in 1872), proves.

[6] Spencer, *Autobiography*, ii, p. 227, Spencer to Youmans, 4 July 1871, also endorsed by Tyndall and Lubbock.

[7] These pages are indebted to Fiske (1894, Ch. XIII, pp. 266–294) which offers the best account of the origins of the series through the letters of Youmans to his sister.

[8] Leonard Huxley intriguingly describes H.S. King as politically a 'stout conservative'; what bearing, if any, this may have had on the future of the ISS is not clear: L. Huxley, *The House of Smith, Elder*, London, 1923.

[9] H.P. Bailliere (1840–1905), son of Jean-Baptiste Bailliere (1797–1885), also created the famous publishing house of Bailliere et Cie. By 1871 Maison Bailliere had built a wide reputation in scientific and medical texts and bibliography. Through Gustave-Germer Bailliere (a cousin of Henri-Paul), the Maison had already begun the 'Bibliothèque de Philosophie Contemporaine' (1868), and had acquired (with the editorship of Emile Alglave) *La Revue des Cours Scientifique*. This was reissued in 1872 as *La Revue Scientifique*.

[10] Cf. Evidence to the Royal Commission on Copyright, 1878, [c. 2036–1], Q 2526.

[11] It happened, however, that the Germans became more interested in Spencer's title than in any other, and Youmans thought this must lead to a prompt demand for the *Synthetic Philosophy*.

[12] In fact, scientific publishing was flourishing in Germany. W. Crookes found that from a sample of 31 scientific books published in the first fortnight of September 1873, 19 were German, 8 were French and only two were British (W. Crookes, 'Endowment of Scientific Research', *Quarterly Journal of Science*, **VI** (N.S.), (1876), 469). An excellent example of the success of German scientific publishing at this general level can be found in the *Geschichte der Wissenschaften in Deutschland*, Königliche Academie der Wissenschaften, München, which published 24 volumes between 1864 and 1913; cf. G. Sarton, *Horus: A Guide to the History of Science*, New York, 1952, p. 124.

[13] Despite comment to the contrary, the vision of European cooperation seems as much attributable to Spencer as to Youmans himself: A. Appleton, *The House of Appleton-Century*, New York, 1936, pp. 13–15.

[14] Memorandum of Agreement between H. King and G. Bailliere, 5 February 1872, in possession of Routledge and Kegan Paul.

[15] Emile Alglave (1842–1928) was, in 1870, *professeur agrégé* in the Faculty of Law in the University of Paris, and later an influential writer on the reform of import duties and alcohol regulations; cf. *Dictionnaire de Biographie Française*, Paris, 1933.

[16] Tzermat died in September 1873, and his place was taken by Professor H. Marquardsen of Erlangen; cf. Heinrich Brockhaus, *Die Firma F.A. Brockhaus von der Begründung bis zum 100 jährigen Jubiläum, 1805–1905*, Leipzig, 1905, pp. 306–307.

[17] Memorandum of Agreement cited in Note 14. It was also specified that no national committee could admit to its series a work of another country which did not come into the list sanctioned by the committee of that country.

[18] H.S. King and Co., First List Advertisement, June 1871.

[19] See: *The Publishers and Stationers Weekly Trade Circular* (afterwards *Publishers Weekly*), *I* (25 January 1872), 47; a further advertisement, listing the titles to be published, featured in the same issue (p. 72), with the description: '(These Books) will be not so much for beginners in science as to acquaint those who have had at least a fair general cultivation in the latest developments of the subjects treated'.

[20] Part of the advertisement for the series which appeared, for example, in the French edition of Balfour Stewart, *The Conservation of Energy* (1899).

[21] For a chronological list of authors in the English edition, see Appendix I; numbers in brackets refer to the order of publication (see Appendix II).

[22] For Bain's contributions to associationism and sensory physiology, see R. Young, *Mind, Brain and Adaptation in the Nineteenth Century*, Oxford, 1970, Ch. 3. Most of Bain's works were published not by Kegan Paul, but by John Parker, later Longmans.

[23] The *PSM*, in periodical form, reflected the same concept; cf. *PSM*, **I** (May, 1872), 113. The *Popular Science Monthly*, with which Appleton intended to bring to America the scientific thought of Europe, became 'the most widely read and influential periodical of its kind' (Madison, 1966, p. 74). It was sold in 1909, when Appleton's suffered a severe financial crisis, and was edited by J. McKeen Cattell (then editor of *Science*), until 1915, when it was divided into the *PSM*, and the *Scientific Monthly;* the former thereafter became a 'popular' journal in a different sense, while the latter survived in its original form until 1950, when it was incorporated into *Science.*

[24] Youmans' impatience to secure Spencer as an author was a precipitating factor in the genesis of the *PSM;* cf. A.J. Meadows, *Science and Controversy: A Biography of Sir Norman Lockyer*, London, 1972, pp. 30–31.

[25] Huxley's famous study of *The Crayfish* was eventually published in 1880 [28], and went through seven editions.

[26] This was the common plight of all series. Henry Holt's attempt to develop the rival 'American Science' Series required 'great patience, constant concern and adroit diplomacy. Authors failed to meet deadlines, or backed out suddenly, or wrote on a level unsuitable for the intended reader' (Madison, 1966, p. 98). Recent enquiries of Holt have failed to elicit details of this series, which would undoubtedly provide interesting parallels.

[27] Joseph Cooke's *The New Chemistry* (1874) was the first American volume in the series.

[28] A search has been made for each of the 98 English volumes in the following journals: *Academy, Spectator, Nature, The Saturday Review* and the *Athenaeum.* The *Academy* rarely reviewed the ISS, the *Spectator* reviewed only four and *The Saturday Review*, only seven; *Nature*, with 35 was better, but this analysis gives only a partial impression of the series' reception as a whole. Searches in more specialist journals would undoubtedly reveal more detailed reactions, and although the general impact would probably remain difficult to assess, the methodology of Alvar Ellegård would almost certainly yield interesting conclusions.

[29] Part of Tyndall's Belfast Address had been based on Draper's earlier book on European culture. In France, Draper was denounced in a series devoted to Actualités Scientifiques as 'une expression passionée de haine, une declaration violente de guerre, une declaration pleine de colère, basées sur une interpretation partiale, arbitraire et fausse des faits de figures de l'histoire': L'Abbé Moigno, *La Foi et la Science: Explosion de la Libre Pensée.* Youmans' negotiations with Draper are described in Donald Fleming, *John William Draper and the Religion of Science.* Philadelphia, 1950, pp. 123–126.

[30] As early as 1854, the Royal Society of Arts held an educational exhibition with a three-volume Prospectus which advertised science texts likely to be accepted by the Science and Art Department. Among the contemporary rivals were Longmans' 'Textbooks of Science', edited by T.M. Goodeve and W. Merrifield, FRS, begun in the early 1870's, and, in 1873, 'adapted for the use of artisans and other students in public and other schools'. Longmans also published *Advanced Science Manuals*, 17 vols., 1890–1909; *Elementary Science Manuals*, 2 vols., 1890–1909; and H. Helmholtz's *Popular Lectures on Scientific Subjects*, with an introduction by John Tyndall.

[31] *The Academy*, **10** (24 July 1876) in a review of the SPCK's *Manuals of Elementary Science* (1875–); cf. the Clarendon Press Series of *School Books and Manuals* (1866), and Macmillans, *School Class Books*, 7 vols., 1870–1877, and *Manuals for Students* (1885), *Science Primers*, 10 vols, 1875–1880; and the *Nature* series, 1873–1880.

[32] For Romanes, see Jacob Gruber, *A Conscience in Conflict*, New York, 1960. There is unfortunately no recent life of Lankester.

[33] Madison (1966, p. 79). 'That volumes in a series like the "Progressive Science Series" should reach second and third editions is a welcome sign that we are beginning to recognise a good thing when it is offered to us. Certainly the three volumes before us are quite indispensable to students of the subjects dealt with, and they have also the rare advantage of being intelligible to the general readers.' These volumes were by J. Arthur Thomson, *Heredity*; T.G. Bonney, *Volcanoes*; Soddy, *Interpretation of Radium.*

[34] The intellectual and literary dimensions of this reaction are described in Ch. 10, 'The Revolt from Reason', Jerome Buckley, *The Victorian Temper*, Cambridge, Mass., 1951.

[35] Cf. F. Fernseed, 'The Future Martyrdom of Science', *Quarterly Journal of Science*, **III** (3rd Series), (1881), 201–208; Frances Power Cobbe, 'The Scientific Spirit of the Age', *Contemporary Review*, **54** (1880), 127–129.

[36] This series was published by the SPCK from 1881.

[37] Cf. the series of titles on *English Men of Science* (ed. J.R. Green), published by Dent from 1906–1919.

[38] *Athenaeum* (7 January 1893), 27: 'Should the present work come to a second edition', *Nature* suggested, 'we would advise Dr. Wright to get some scientific fund to assist him in its revision' (*Nature*, **47** (15 December 1892), 148–149.

[39] Stebbing described himself for *Who Was Who* as ' a serf to Natural History, principally employed about crustacea'.

[40] See, for example, the *Cambridge Natural Science Manual* (1892); and Dent's *Scientific Primers*, 1909; also the 'History of Science' series begun by Watts in 1911; and the series by Cassells, edited by H.E. Roscoe – 'Century Science Series'. One of the specialised textbook writers was Alfred Pearce, editor of the Manchester School Series, from 1893 onwards.

[41] See, *Nature*, **37** (9 February 1880), 340.

[42] The publisher's price for the entire series in 1900 was £21.10.0; *Nature* offered the series for £14.14.0 by instalments, or £13.0.0. by cash; cf. *Nature*, **61** (4 January 1900), ci.

[43] Cf. The contemporary comments on Kegan Paul's most successful ventures: *The Critic*, **III** (September 1883), 354. But see also Charles Kegan Paul, *Memories*, London, 1899. Because each national house kept its own records, of which most have been destroyed, it is difficult to estimate the financial success of the series as a whole. By using known details about the size of editions one can make approximate statements about the most successful titles, but for volumes which went through less than one edition it is difficult to calculate loss.

[44] This series is not to be confused with the *Bibliothèque des Sciences Humaines*, published by Gallimard, or *La Nouvelle Bibliothèque Scientifique* created by Flammarion in 1962. (There is also the *Bibliothèque de Philosophie Scientifique*, ed. Paul Gauthier, begun after 1945.) The original Flammarion series was begun by Gustave le Bon in 1905, and appeared alongside the Felix Alcan series for nearly half a century.

References

The place of publication is London, unless otherwise indicated.

Academy (1876): **10**, 386.
Allen, G. (1887): *Fortnightly Review*, **41**, 883.
Athenaeum (1890): 30 August, 293.
Athenaeum (1896): 8 August, 196.
Bain, A. (1904): *Autobiography*. p. 321.
Barrington, R. (1914): *Life of Walter Bagehot.*
Critic (1883): **3**, 354.
Duncan, D. (1908): *Life and Letters of Herbert Spencer.*
Dyer, H. (1893): *Science Teaching in Schools*, p. 17.
Ellis, H. (1967): *My Life*, pp. 170–1.
Eve, A.S. and Creasey, C.H. (1945): *Life and Work of John Tyndall.*
Fiske, J. (1894): *Life and Letters of Edward Livingston Youmans.*
Huxley, L. (1900): *Life and Letters of Thomas Henry Huxley*, Vol. I.
Huxley, L. (1918): *Life and Letters of Sir Joseph Dalton Hooker*, Vol. I.
Huxley Papers (1871): Huxley Papers (Imperial College), f. 263 Youmans to T.H. Huxley (August, 1871).
Jevons, H. (1886): *Letters and Journals of W. Stanley Jevons*, p. 324.
Kegan Paul, C. (1899): *Memories.*
Le Galienne, R. (1899): 'Grant Allen', *Fortnightly Review*, **66**, 1024.
Lubbock Papers (1880): British Museum, Add. MSS. 49, 642, f. 164, A.J. Mundella to Lubbock, 26 November 1880.
Lubbock Papers (1889): British Museum, Add. MSS. 49, 469, Tyndall to Lubbock, 18 April 1889.
Lubbock Papers (1891): British Museum, Add. MSS. 48, 652, f. 138, Kegan Paul to Lubbock, 19 November 1891.
Lubbock Papers (1893): British Museum, Add. MSS. 49, 654, f. 6, Kegan Paul to Lubbock, 18 December 1893.

MacLeod, R.M. (1969): 'The Genesis of Nature', *Nature*, **224**, 423–440.
MacLeod, R.M. (1970): 'The X-Club: A Scientific Network in Late-Victorian England', *Notes and Records of the Royal Society*, **24**, 305–322.
Madison, C. (1966): *Book Publishing in America*, New York.
Masson et Cie, (1954): *1804–1954: Un siècle et demi d'edition medicale et scientifique*, Paris, 1954, pp. 52–53.
Mumby, F.A. (1934): *The House of Routledge, 1834–1934*.
Nature (1874): **9**, 477.
Nature (1875): **12**, 105.
Nature (1876): **14**, 328.
Nature (1882): **26**, 121.
Nature (1888): **37**, 340.
Nature (1891): **45**, 195–196.
Nature (1893): **49**, 74–75.
Nature (1896): **53**, 145–146.
Nature (1896): **53**, 433.
Nature (1903): **67**, 337.
Newcomb, S. (1874): *North American Review*, **119**, 286.
Overton, G. (1925): *Portrait of a Publisher: The First Hundred Years of the House of Appleton, 1825–1925*, New York.
Publishers Weekly (1873): **3**, 275.
Saturday Review (1873): 29 November, 701.
Scribner's Monthly (1873): **6**, 608.
Spectator (1874): 23 May, 664.
Spencer, H. (1904): *An Autobiography*, Vol. II.
Strong, T.B. (1906): *The Method of Science*, Oxford.
Unwin, S. (1926): *The Truth About Publishing*.
Webb, B. (1926): *My Apprenticeship*, p. 123.

V

EDUCATION: SCIENTIFIC AND TECHNICAL

Commentary

Introduction

The history of scientific and technical education in Britain can be seen in at least three different levels, which in context coalesce. It is, first, the history of sustained social pressure, generated by the ideas of a few small groups of 'reformers' who could combine a sense of patriotism and a zealous belief in technological progress, and who were both sufficiently knowledgeable to discern a "public problem" and politically capable of pressing for parliamentary action; and, second, a study in the problematics of institutional change, in which new administrative apparatus could be conceived, first to prompt and then to satisfy the educational demands of self-defining managerial and middle-class interests. Third, it is a study in the development of educational objectives in an industrial society, the goals of which were changing in response to varying perceptions of economic needs and political necessities.

However much they may be divided on questions of emphasis, historians of social administration, of education and of economic change, will invariably confront three central premises: first, that there came into currency between the 1830s and the 1850s a widely held assumption that national strength, which had been traditionally defined in terms of military and naval strength, and which by extension depended materially on industry, commerce and trade, now depended no less on a deliberate cultivation of the nation's educational and scientific talent; that this in turn relied upon a diffusion of the means of acquiring scientific and technical education throughout society; and third, that this diffusion, made possible through the efforts of 'reformers', revealed considerable differences in objectives, and prompted repeated attempts to redefine the concepts of 'scientific' and 'technical' education to fit the changing intellectual and industrial circumstances of the country. These factors were, together, decisive in fashioning the attitudes and assumptions upon which government policies would rest. The three reports in this series, reprinted by the Irish University Press, are important benchmarks in this development; as such, they are central to the historian's understanding of the process by which externally derived social and economic concepts have been conceived and introduced into the educational assumptions of modern Britain.

The history of scientific and technical education in this country of course considerably pre-dates the developments with which the reports in this series are concerned, and the interested student may find the bibliography given below merely a convenient point of departure. But in a sense one may be excused for beginning a discussion of this series with the Great Exhibition of 1851,[1] an event of considerable intrinsic importance, which ushered in a period of intense and unprecedented activity in eduction and a vigorous review of Government policy towards science and industry.

Perhaps more than any other single event in Victoria's long reign, the success of the Great Exhibition represented the triumph of British industry and commerce, founded upon the principles of Liberalism and Free Trade, dedicated to the progress of art and industry through the encouragement of technical change. To contemporaries, the Exhibition appeared to vindicate the virtues of peaceful competition, enshrined in material accomplishment and industrial leadership. As J B Bury wrote, it provided 'a public recognition of the material progress of the age and the growing power of man over the physical world'.[2] In conceiving and directing the Exhibition, a Committee of men of supreme importance to the cause of education, including Henry Cole[3] and Lyon Playfair,[4] had, through the leadership of the Prince Consort[5] and the Society of Arts,[6] forged a new public interest in the progress wrought by technology, and a new faith in the inevitability of stability and prosperity through a cultivated combination of hard work and sound knowledge. Within the first few years following the Exhibition, reverberations of its educational lessons were felt throughout the kingdom. 'People's Colleges' opened; essays on industrial and technical education were commissioned; and proposals for converting Mechanics' Institutes into technical colleges were heard across the industrial midlands, from Nottingham to Manchester.

For some, however, the Exhibition was not a source of unalloyed optimism. No sooner had the Great Exhibition closed than the Royal Society of Arts, at Albert's request, took steps to organise a series of lectures on the results of the Exhibition.[7] In perspective, Britain's achievements would be measured, not only by its lion's share of the prizes, but by its ability to seize the commercial opportunities displayed by the Exhibition. There were twelve lectures in the series, led by William Whewell;[8] of the twelve, eight stressed the three compelling features upon which would depend Britain's future prosperity – the provision of scientific and technical education at the secondary level in all public and private schools; the introduction of systematic scientific education in the universities; and government provision for scientific research, both 'pure' and 'applied'. The second of these objectives was already within sight, owing in part to a series of Government enquiries into the ancient universities, and in part to reforms already underway. The first and third objectives were, however, less easily realised, and it is chiefly they that form the starting point of the volumes in this series.

The era inaugurated by the Great Exhibition was heralded by the Queen's Speech from the Throne in November 1852, in which Parliament was told that

> The advancement of the Fine Arts and Practical Science will be recognised by you as worthy of the attention of a great and enlightened nation. I have directed that a comprehensive scheme should be laid before you, having in view the promotion of these objects . . . .[9]

Following the efforts of Lyon Playfair and Henry Cole, a scheme emerged by which a new 'Science Division' was added to the Department of Art already existing within the Board of Trade. Within two years a new Department of Science and Art emerged, to which were transferred the Government School of Mines, and later, the Royal College of Chemistry.[10] The School of Mines was meant to provide for the metropolis a school of applied science of the highest class; while, for the country at large, it would serve as a focus for the education of a new force of scientific teachers who would diffuse scientific knowledge through the 'industrial classes'. At the same time, the Department was given the task of helping to establish Science Schools and classes, to regulate payments by results and to grant teacher's certificates.[11] The first Science Minute of the Department was issued in 1859; by 1860, nine 'science schools' were in operation, and 500 students were under instruction.

Throughout the 1860's, these classes made, however, a limited impression on science education, and did not in any case affect technical education at all. This had remained the province of the Society of Arts, which conducted a set of examinations which were still influential in many cities. But by the 1860's, the Society of Arts examinations declined, and the Mechanics' Institutes attracted fewer and fewer artisans.[12] There was widespread disagreement concerning the proper place of technical education and vocational training, coupled with a general assumption that technical training, insofar as this meant practical workshop techniques, was a private, commercial issue, and not a public responsibility. Indeed, the general absence of agreement on a satisfactory definition of 'technical education' seriously retarded a unified approach to the problem.

In the meantime, however, the importance of scientific and technical education was given fresh prominence in the national self-criticism following Britain's performance at the International Exhibition of 1862 in Paris. In the Great Exhibition of 1851, Britain had been at a distinct advantage. Not only was she displaying on home ground her best manufactures, but, having escaped the ravages of 1848 which had upturned much of Europe, she was enjoying a political stability which, however fragile it might seem to contemporaries, had put British industry in a formidably competitive position. In 1862, international circumstances remained favourable. While Britain was at peace, Italy was suffering the agonies of national unification, and Austria was locked in a struggle for national survival. France was anxiously attempting to speed her rate of industrialization, while Prussia was preoccupied with her struggle against Denmark.

Despite her advantages, however, Britain's performance in 1862 was notably less impressive than in 1851. The Society of Arts, ever watchful of the national 'image', issued a special report which expressed the pious hope that at least the Paris Exhibition demonstrated the good sense of Free Trade. But this defence could not long mask the truth. By 1867, when the next Exhibition was held, again in Paris, it was impossible to ignore the considerable improvements in the competitive position of France and particularly of Germany. There was no question that Britain's industrial leadership was severely challenged. Those appointed to investigate looked round again for reasons. Given the political context of *laissez-faire* and Free Trade, Government aid to industry was never a possibility. What did present itself, as within the realm of the 'possible', was the prospect of national self-improvement through technical education.

Beginning in 1861, a series of Royal Commissions had been launched to investigate the organization and content of British education.[13] The Newcastle Commission in 1861 had considered the state of 'popular' education, and three years later, the Clarendon Commission reported on the condition of the nine leading 'public schools'. In 1864, Lord Taunton was appointed to consider the education provided in schools which had thus far escaped parliamentary scrutiny, including the endowed schools, private schools and proprietary schools. It was in 1867, while the Taunton Commission (the 'Schools Enquiry Commission')[14] was sitting, that Lyon Playfair, then serving as a juror at the Paris Exhibition, wrote to Taunton, informing him of Britain's performance and drawing from that experience the view that 'England had shown little inventiveness and had made little progress in the peaceful arts of industry since 1862'[15]. Playfair urged Taunton, and through him, Lord Granville, Vice-President of the Council for Education, to institute an immediate enquiry into the country's need for technical education. This would necessarily include a survey of foreign attempts to generate technical skills and apply them to industry.

The Taunton Commission obligingly sent copies of Playfair's letter in the form of a circular to sixteen 'experts' in industrial manufacture, most of whom had sat as jurors to the 1867 Exhibition, asking their views. While most disagreed that the nation was in jeopardy, virtually all agreed that scientific education, for the great majority of people, was better on the Continent. The implication, as John Tyndall[16] put it, was clear. England could be industrially outstripped if her rivals were technically better equipped: 'As sure as knowledge is power this must be the result'.[17]

In his second report on technical education, Taunton recommended that an enquiry should be launched 'into the state and effects of technical education abroad'.[18] Throughout 1867 the Society of Arts lent its voice in support of the Commission's Report, and in January 1868 the Society held a Conference to discuss the future place of science in education, and the necessity for an efficient system of primary schools for children of the industrial classes.[19] In September 1867, the British Association, which for several years previously had urged Parliament to improve provision for science teaching in the schools, heard a report from its Parliamentary Committee which unequivocally insisted upon the importance of scientific education for all classes.[20] Thus prompted by a growing sense of international industrial rivalry, appeals for scientific and technical education soon acquired a momentum of their own. By 1870 they contributed significantly to demands for universal primary education, the cornerstone upon which it was believed, the edifice of secondary education could alone be securely created.

I The Samuelson Committee

Meanwhile, events moved rapidly in the direction of special parliamentary enquiry. Emerging from the turmoil of the Reform Act of 1867, Disraeli's Government, coming into power in February 1868, almost immediately appointed a Select Committee – the first of three in the present series – under the chairmanship of Bernhard (later Sir Bernhard) Samuelson, the famous German-born ironmaster who had served on the Acland Committee in 1867 and who was Secretary of the Society of Arts Committee on Technical Education.[21] The 'Select Committee on the Provision for Giving Instruction in Theoretical and Applied Science to the Industrial Classes' was instructed specifically to find answers to questions raised by Playfair's letter to the Taunton Committee.

The Samuelson Committee convened for the first time in April 1868, met twice a week throughout May and June, held a total of 22 meetings and submitted its short final report with thundering speed in July. The Committee consisted of eighteen members, notably including George Dixon, MP (Lib., Birmingham), who would soon work closely with Forster in drafting the Education Act of 1870,[22] Lord Robert Montagu, MP (Con., Huntingdonshire), then Vice-President of the Council[23] and a strong advocate of scientific education, and Thomas (later Sir Thomas) Dyke Acland, MP (Lib., North Devonshire)[24] who had already served on the Endowed Schools Commission. In the chairman, Bernhard Samuelson we find a central figure who was to be a common denominator in all three reports in the I.U.P. series.

The Samuelson Committee interviewed a total of 56 witnesses, drawn from government educational establishments (including the Science and Art Department, the Committee of Council, and the School of Naval Architects); from the Universities of Oxford and Cambridge; from the 'few secondary schools in which science has been for some time systematically taught'; from the 'managers and teachers of science classes and Mechanics' Institutes receiving aid from the state, or supported exclusively by voluntary efforts'; and from 'the population engaged in the present staple industries carried on in the principal manufacturing towns and districts'. Rather arbitrarily, the Committee divided the educational sphere into three 'classes' – foremen and workers; smaller manufacturers and managers; and the proprietors of large industries – and a further sector – the relation of industrial education to industrial progress. The evidence of witnesses demonstrated that the first 'class' was desperately ill-equipped in numbers of both schools and teachers. For manufacturers and managers, the best instruction available was no better than that given by the best elementary schools, while for the proprietors, college and university courses were few, brief, and often irrelevant.

As to the relation of industrial education to industrial progress, the Committee expressed itself puzzled. There was, perhaps surprisingly in view of the energetic work of the Society of Arts, a strong difference of opinion on this question. This was, of course, the central issue:— Was technical education (and by implication, scientific education) vital to industrial progress, or not? There were quite possibly important social and economic factors quite separate from the *general* level of scientific and technical knowledge among the workforce, which were in their net effect more significant to the economic prosperity of individual industries. But, unfortunately, the Committee's greatest weakness was its haste. Not surprisingly, it was unable to draw firm conclusions about British industry as a whole. In fact, the issue was (and is) extremely complex, and answers were neither easy nor easily come by. In an important sense, however, this was possibly immaterial; the Samuelson Report became principally an educationalists', not an industrialists' charter. A fuller analysis could have disaggregated industry, not by class of worker or manager, but by product or process sector, and could have proceeded sector by sector to gather written and oral evidence on the relevance of different kinds of educational skills – whether scientific, technical, 'technological' (to perpetrate an anachronism) or managerial – to specific industries. This was, however, an expensive and lengthy task, and one, of course, inconsistent with the terms of reference of the Committee, and incompatible with the pressures on the Committee for quick and simple answers for Parliament. Indeed, the Samuelson Committee of 1867 reflects many of the strengths and weaknesses of Select Committee enquiries.

In retrospect, the pressure of a strong lobby of 'zealots' gave the Committee's recommendations a force which its evidence scarcely warranted. Most of this evidence focused on the educational inadequacies of existing schools and government departments, rather than on industry itself, a fact which did not escape criticism. From what manufacturers' evidence was available, for example, it appeared that industry was by no means unanimous in the view that the absence of scientific instruction was in itself disastrous. As the Committee reported, economic factors were involved:

> Although the pressure of foreign competition, where it exists, is considered by some witnesses to be partly owing to the superior scientific attainments of foreign manufacturers, yet the general result of the evidence proves that it is to be attributed mainly to their artistic taste, to fashion, to lower wages, and to the absence of trade disputes abroad, and to the greater readiness with which handicraftsmen abroad in some trades, adapt themselves to new requirements.[25]

Furthermore, the Committee defended the British workman for possessing 'in a pre-eminent degree' requisite practical experience and manufacturing skill. Nonetheless, in their next paragraph, the Committee hurried on to register itself and its conclusions, in favour of greater educational provision. Given the composition of the Committee, a conclusion of this kind was perhaps predictable; but it was a measure of the gathering consensus that a basic system of scientific and technical education was a necessary, if not itself a sufficient, condition for industrial leadership.

Despite its inadequacies, the Samuelson Report had at least four important consequences. First, the Committee had agreed that existing provision for the early education of foremen and workers was too brief and too narrow to be of lasting value. If the working classes were ever to benefit from science, elementary instruction was essential for every child. This in itself led directly to pressure for universal primary education. Secondly, the Committee argued for the reorganization of educational provision at its higher levels, including the introduction of adult science classes, the reorganization of secondary instruction, the redesignation of certain endowed schools as 'science schools', the more equitable distribution of state grants among the provinces, and the encouragement of 'superior' schools of science, supported jointly by the State, local and private benefaction in centres of industry and commerce. These recommendations laid the foundation for many decisions subsequently taken by the Department of Science and Art. Moreover, the Committee recommended improved provision for science teachers, including 'some slight addition' to their pay, and greater attention to science in their training, which contributed to the system of payment by results, and to the case for extended provision of fellowships in science at Oxford and Cambridge. Finally, in urging greater liaison between government institutions for scientific instruction in London, the Committee argued for the creation of a 'central scientific school', proposed as early as the 1850s – a proposal which years later, was embodied in the Imperial College of Science and Technology at South Kensington.

In the context of the time, however, the Committee's recommendations, apart from those respecting science education, were largely shelved. Some would undoubtedly have required new legislation, which the Government were not prepared to undertake. Indeed, Disraeli's 1868 Government was a minority government – a 'caretaker administration', dependent upon the goodwill of the Opposition. The session was accordingly difficult, with the Irish and Scottish Reform Bills consuming a good deal of parliamentary time. Moreover, the Irish Church question anticipated a general election, and Parliament was prorogued on 31 July, just two weeks after the Samuelson Committee's report was tabled. The Liberals, returned in November 1868, outlined a careful strategy of reform in education, the army and the civil service, but Samuelson's other recommendations had to wait.

With Parliamentary action momentarily in abeyance, the field was left to the technical education "lobby" outside Westminster. According to Lyon Playfair, the nation had been "much moved" by the Samuelson Report, 'and public conferences were held in London, Edinburgh and in the leading provincial towns to consider a question so grave to our interest as an industrial country'.[26] At such conferences, the thorny question was this:— given that industrial training may be desirable, of what should it consist? Apprenticeship, and the imparting of manipulative skills and 'practical' knowledge, was obviously important and was generally urged by the manufacturing interests. But what was the role of the school? 'Instruction in manipulative skill', Playfair told an Edinburgh audience in 1870, 'is not education at all; and, such as it is, belongs to the workshop, not to the school'.[27] What would 'technical education' in fact, come to mean – technical skills or scientific knowledge? Playfair and some educationalists stressed that the teaching of scientific and technical principles was also important. But if so, where would the balance be struck between science and practice, school and workshop? In 1868, the Society of Arts accepted the principle that technical education should be taken to include school work as well as workshop work. Through the 1870's, moreover, at the suggestion of Col. John Donnelly,[28] of the Science and Art Department, the Society of Arts gradually moved from its limited examination of schemes for specific industries, to a system of technical examinations for artisans generally. With the aid of the Livery Companies, inspectors were sent abroad to study different trades, and a body of experience and slow progress continued, on an *ad hoc* experimental basis.

II The Devonshire Commission

The question of science education, as opposed to technical education, had rather been submerged in the questions set by the Samuelson Committee: It already deserved treatment in its own right. Throughout the 1860's, impelled by the new discoveries in geology, zoology, chemistry and physics, T H Huxley,[29] John Tyndall, John Lubbock,[30] James Wilson[31] and others sought for science a place in the school and university curriculum equal to that held by theology and classics. Through public lectures and journals dedicated to the critical spirit, they took their campaign to the country. By far the single most visible device for the diffusion of scientific attitudes in the 1860's was the British Association which had become in its thirty years of existence a respected British institution, with a political and intellectual importance that won its pronouncements a national hearing. In August, 1866, the British Association's Parliamentary Committee reported its regret that no steps had been taken by Government to promote the study of science in schools,[32] and submitted a memorandum to the Schools Inquiry Commission, supporting the Commission's recommendations.

In 1867 the Parliamentary Committee announced optimistically to the British Association meeting at Dundee that 'the attention of the public appears to have been awakened to the necessity for introducing scientific teaching into our Schools, if we are not willing to sink into a condition of inferiority as regards both intellectual culture and skill in art, when compared with foreign nations'.[33] In 1868 the Council of the British Association meeting at Norwich considered a resolution (provisioned with arguments by Huxley and Tyndall, urging the Government to remedy defects in secondary education) which in March 1868 was printed and sent to the Education Department.[34] Meanwhile, owing to the efforts of J N Lockyer,[35] and T H Huxley, occasionally abetted by Tyndall and Playfair, who consistently argued for the importance of applying the benefits of new knowledge in a systematic fashion, the question of government provision for scientific teaching and research was brought into the foreground.

In August 1868, Alexander Strange, FRS[36], delivered a paper on the 'Necessity for State Intervention to Secure the Progress of Physical Science'. The paper aroused immediate attention. Strange, with Playfair, had recently been a juror at the Paris Exhibition. When Playfair returned to begin his struggle for technical education, Strange looked instead to the support of fundamental research. In his paper, he appealed to the spirit of one of the British Association's earliest objectives: to remove any disadvantage of a public kind which impeded the progress of science.[37]

There was no doubt in Strange's mind that State participation was vital to research. He took for granted that any abstract question whether scientific investigation should be aided or even carried on by Government had already been settled by the irresistible verdict of circumstances. William Crookes, writing in the *Quarterly Journal of Science*, agreed that science instruction could not prosper at the universities until students wanted to learn science;

and students could not turn to science until there was hope of a career.[38] At Magdalen College, Oxford, a demyship in science once went begging because, as the best candidate asked, 'What am I to do for a living, after I have completed my studies?'[39]

Following the Norwich meeting, the British Association, in November, 1868, appointed a committee of twelve, which included John Tyndall, Edward Frankland, Thomas Hirst, T H Huxley – essentially the famous 'X-Club'[40], plus Playfair and J N Lockyer (soon to become the first editor of *Nature*). The committee was asked to determine whether there was sufficient provision for the vigorous prosecution of physical science in Britain. In March 1869, the committee issued a circular to several men of science, asking:

1. Whether any course could be adopted to improve their field, whether by Government or by private initiative.
2. Whether there were grounds for asking for such initiative, and
3. Whether the interests of the community were 'sufficiently involved in the more vigorous development of scientific knowledge' to warrant the employment of state aid, either
   (a) through the creation of new institutes for experimental and observational research,
   (b) the extension of old institutions, or
   (c) the enlargement of grants to individuals for apparatus and materials.

Several months later, the committee reported its conclusion that, judged by any responsible standards, there was insufficient national provision for research. Hitherto research had paid the community 'but it had not paid the man'.[41] The community advantage wrought by research implied an obligation of support on behalf of the community itself. This obligation would not be satisfied merely by the establishment of a scientific observatory, by sponsoring physical measurements, creating one or two museums and providing out-of-pocket expenses for talented amateurs, but by allocating reasonable sums of money to individuals prepared to pursue a full-time vocation in scientific research.

Many in the British Association shared the committee's view, but others feared State interference. *Nature*, for example, published a letter from Alfred Russell Wallace, protesting against what *Nature* called the 'Science Reform Movement' because 'Experience shows that public competition ensures a greater supply of the materials and a greater demand for the products of science and art, and is thus a greater stimulus to true and healthy progress than any Government patronage.'[42]

On the other hand, there seemed grounds for arguing that 'where the State has a finger there will be patronage, and the preference of inferior agents who have the support of friendly recommendations, to superior agents who are standing alone'.[43] Already, in any case, the State was spending £140,000 a year on Museums, Gardens and Surveys, while the Science and Art Department Estimate for 1869-70 was over £225,000.[44] More important, however, was the categorical belief, evident especially among university men, that research should not be divorced from instruction which sugested that research could be properly endowed only through the universities. Benjamin Jowett, for example, argued that there was no opposition between original research and a moderate amount of teaching.[45]

Even where reformers agreed that state support was needed, they differed about how this should be distributed. On balance, those most enthusiastic for state intervention and 'Science Reform' had little experience of university life and were not Oxford or Cambridge men. This was particularly evident among the 'X-Club' and its concentric networks. On the other hand, strong arguments in favour of linking teaching with research came chiefly from the universities. In his inaugural lecture at University College, London, in October, 1869, Professor A W Williamson[46] delivered 'A Plea for Pure Science' and insisted that research should not be allowed to develop outside the context of education. Henry Roscoe[47] whose spacious laboratories at Owens College, Manchester, were among the best in England, urged that scientific research, was, through the teaching of scientific method, the best training for everyday commercial life;[48] and James Clerk Maxwell, in his inaugural professorial lecture at Cambridge in October, 1871, in a laboratory made possible by private benefaction, stressed the advantages of doing research in the university with the resources available there.[49] At Oxford, Sir Benjamin Brodie wished to found new research institutions within the universities, with staff closely connected to university teaching. In London, on the other hand, Huxley and Lockyer were strongly in favour of independent support for research, however much this research might in practice contribute to university teaching.[50]

Whatever the best connection between university science research and Government research might be, the case for support was sufficiently clear to the British Association. On 18 August 1869, the British Association Council at Exeter recounted the three Government Commissions on secondary education, and its own attempts to bring science into the curriculum, and resolved that

> the study of Natural Science, whether as a means of disciplining the mind, or for providing knowledge useful for the purpose of life, is of essential importance to the youth of this country; and that it ought to form a part of education in all Secondary Schools.[51]

The Council accordingly presented a petition for Lord Lyttelton to present in the House of Lords; and for Sir William Tite to present in the House of

Commons. One week later the British Association took separately a decision to exert its influence to obtain a Royal Commission to consider:

1) the character and value of existing institutions and facilities for scientific investigation, and the amount of time and money devoted to such purposes.
2) what modifications or augmentations of the means and facilities that are at present available for the maintenance and extension of science are requisite and
3) in what manner these can be best supported.[52]

In effect, the British Association's twin pronged "attack" for the support of science in secondary education and in research was met by a single response. In February 1870, Gladstone's administration agreed to appoint a Royal Commission under William Cavendish, seventh Duke of Devonshire, to study national provision for 'scientific instruction and the advancement of science'.[53] The Duke, a disant relative of Henry Cavendish, the chemist, had taken the Tripos as revised by Babbage in1829. In 1861, he had succeeded the Prince Consort as Chancellor of Cambridge where, in October, he donated the cost of building the Cavendish Laboratory.

The Secretary of the Devonshire Commission was the ubiquitous Norman Lockyer. Two of its nine members (Sharpey and Miller) had been active in building research schools at University College. Two others (Huxley and Lubbock) were members of the 'X-Club'. One other, Bernhard Samuelson, was by now a familiar figure in educational circles.

The Commission sat for six years, met 85 times, interviewed over 150 witnesses and published eight reports and four massive volumes of evidence. Although instructed to look particularly at scientific instruction in schools, museums and universities, it soon raised searching questions about the state of British science as a whole and research in particular.[54] Some reports were more conclusive than others; some virtually assumed the truth of what they set out to prove. The first report, which appeared in March 1871, examined the institutions concentrated in South Kensington, and recommended that the Government combine the Royal School of Mines and the Royal College of Chemistry into a single science school with a single staff of professors in mathematics, physics, chemistry and biology.[55] The second report, issued in March 1872, was concerned with primary education and with endorsing improvements in the policies of the Science and Art Department which had long been sponsored by the London network. The third report, dealing with Oxford and Cambridge, appeared in August 1873; the fourth, on national museums, in January 1874; the fifth, on provincial university colleges, in August 1874; and the sixth, seventh and eighth, on secondary schools, the Scottish universities and Government provision for science, respectively, appeared in June 1875.

Over these four years, editorial comment in *Nature* and the literary quarterlies kept public interest in the Commission alive. The Commision revealed much evidence of large government expenditure on what could be considered 'scientific' research, justified on grounds of prospective naval or commercial rewards; indeed, as the history of the Arctic and *Challenger* expeditions demonstrated, this definition could be interpreted widely. Moreover, the state had increased its expenditure on science-related activities dramatically since the 1850's.[56] It was, in fact, doing a good deal, but it was not *seen* to be doing so. Much science that was supported was hampered by lack of coordination and by narrow notions of departmental control. This absence of strategic thinking, and the implied waste of scientific resources, was the chief criticism of the 'Science Reformers' who repeated in public views which they had expressed officially to the Commission. In 1870, 1871[57] and again in 1872, Alexander Strange demanded that scientists should have rewards and standing comparable to the practitioners of any other learned vocation.

> It has hitherto been too much of the custom, [he said] to treat men of science as exceptions to all other professions; to assume that whilst it is quite proper to enrich and ennoble soldiers who fight for pay, lawyers who evade or apply the law according to circumstances, physicians who kill or cure as seemeth best to them, and even divines, whose missions to save souls might be deemed a sufficient privilege ... [the man of science] should work for love, and die, ... in poverty.[58]

It seemed plausible that, in the interpretation of nature, no less than in the interpretation of religion, the labourer should be worthy of his hire. W B Carpenter, the physiologist, dwelt on England's peculiar need:

> If England was behind Germany in original investigations, it was not, as sometimes is said, because Englishmen are inferior to Germans in ideal power, but because the German universities are so arranged as to afford a career to men who choose to devote their lives to study. In England such men, having no means of making a livelihood by the pursuit of science, are obliged to turn their attention to a 'practical profession'.[59]

Through the early 1870s, the distressed state of scientific men was recounted again and again. 'It is known to all the world', lamented *Nature*, 'that science is all but dead in England'. By this was meant not the 'Jury' science of exhibitions and practical technology, but 'that searching after knowledge which is its own reward'.[60] 'Whether we confess it or not', Lockyer wrote, 'England so far as the advancement of knowledge goes, is but a third or fourth rate power'.[61] From the start, the solution was evident. 'It is the question of "scientific careers" that is the pressing one, and the one most difficult to settle', one correspondent agreed.[62]

The fifth report showed that Owens College (Manchester) and University College (London), as well as other university colleges united by the examination system of London University, were unable to expand their science faculties in the absence of outside help. Industry, with few exceptions, had shown little interest. City and merchant guilds were often generous, but unpredictable, sponsors. Private philanthropy in support of scientific research was rarely to be seen. Government aid, both in capital and maintenance grants, was the only real alternative. In June 1873, the sixth, seventh and eighth reports of the Devonshire Commission appeared. In the sixth report, the Commissioners found that, despite repeated attempts, little progress had been made in the introduction of science in the public or endowed schools of Britain since the Clarendon Commission of 1861 and the Taunton Commission of 1864. Readers of the IUP series will see the range of arguments presented once again. Here, perhaps it will suffice to draw attention to the Commissioners' verdict, delivered in measured tones of deliberate rebuke:

> The omission from a Liberal Education of a great branch of Intellectual Culture is of itself a matter for serious regret, and, considering the increasing importance of Science to the Material Interests of the Country, we cannot but regard its almost total exclusion from the training of the upper and middle classes as little less than a national misfortune.[63]

Perhaps the most original aspect of the sixth report was the appendix especially prepared by Norman Lockyer, after reviewing the evidence of the Endowed Schools Commission. Between September 1872 and the Spring of 1875, Lockyer himself conducted a personal survey of secondary schools, and assembled a wealth of information concerning the 'mechanics' of science teaching, including the provision of appropriate textbooks and apparatus. The experience of this piece of educational 'research' for Lockyer was, as we shall see, extremely significant to his development as an educational reformer. In this study, Lockyer found that there were 205 schools in Britain with annual endowments of over £200. Of the 128 schools which replied to Lockyer's circular, only 63 had science courses (affecting perhaps 2,400 boys, or 25% of the total registered) averaging 3-4 hours of a pupils's time per week. The position among the schools which did not reply was presumably even worse. The greatest weakness of all, he found time and again, was the sheer dearth of suitable staff. This, in turn, he reported to the Commission, was due to the time-honoured difficulty — 'the utter absence of an assured career for any student of science'.[64] In 1874, the Devonshire Commission's seventh report appeared, testifying that only two universities in Britain (Edinburgh and London), actually offered degrees in science, [65] and demonstrating widespread discrepancies in faculty salaries. It was this same question of 'pay and prospects' which guided the Commission in its eighth report, and made the development of a 'profession' of scientific graduates no less important for teaching than for scientific research.

V

With their eighth Report, on the 'endowment of science', the Devonshire Commission drew its work to a close. The Government, was in fact, doing a good deal to aid science. According to the Civil Estimates of 1869-70, the Government spent about £400,000 on scientific museums, expeditions, learned societies and the scientific work of central departments. But the Commission accepted that the 'progress of Scientific Research must in a great degree depend upon the aid of Government. As a nation we ought to take our share of the current scientific work of the world.'[66] To ensure adequate consideration of both science and research, they followed Strange and *Nature* in recommending that the Government create a Ministry of Science and Education, with a council of scientific advisers; that the scientific facilities of schools and civil departments be augmented; and that a national technical laboratory and physical laboratory be built.[67] The Commission, in fact, embraced the recommendations of the reformers 'it would be their own fault', J. S. Cotton observed, 'if they let the issues remain any longer in the domain of theory.'[68]

Unfortunately, the Devonshire Commission's recommendations proved too sweeping for the Government to implement. After repeated prodding by Playfair,[69] Disraeli's Government did accept certain specific recommendations — notably, by increasing the Government Grant for research, and creating a Solar Physics Committee — but the constitutional and administrative difficulties posed by the creation of a Ministry of Education and Science, which would embrace not only scientific research but also secondary, technical and higher education could not be quickly resolved.[70] Over the next fifty years, many of the Devonshire recommendations were realised in some form, each highly significant for the development of central policy towards science and education. But none was of greater significance than the proposal that science should form an *indispensable* part of the country's educational apparatus.

III The Samuelson Commission

While the campaign for the recognition of science was still in its infancy, the movement for technical education gathered speed. In the late 1860's, at the time of the Samuelson Committee, 'technical education' was a generally unfamiliar concept, associated with the 'useful knowledge' of the Mechanics' Institutes, but distinct (in both social and pedagogical forms) from the 'scientific instruction' meant to inform the schools and universities.[71] From the point of view of government, this distinction was not merely of academic interest. Scientific education could arguably be presumed to have some claim on public educational expenditure. 'Trade subjects', however, were in a different category altogether. According to generally accepted principles, the subsidy of such subjects would be very near the subsidy of industry itself, and this dangerously close to policies of Protection. Moreover, through the 1870's and 1880's, strong differences of opinion about the sort of technical education desired festered within the educationalists' ranks. Huxley's view, expressed clearly to a London audience in 1877, was that technical education was 'simply a good education, with more attention to physical science, to drawing and to modern languages than is common and there is nothing especially technical about it.'[72] Others, however, including Playfair, felt that it should be rather more 'technical' in content, without sacrificing scientific principles; while others held that it should be wholly technical or commercial work, or the pragmatic business of applying techniques to industry. All agreed, however, that there should be technical education at both secondary and 'higher' levels.

Over the next forty years, the redefinition of technical education to include both practical, vocational experience and applied science gradually took hold. The movement received a great push in 1877 when a group of men including Sir W. G. Armstrong, T. H. Huxley, John Donnelly, Douglas Galton and H. Trueman Wood, (Assistant Secretary of the Royal Society of Arts) were invited by a Committee of the Livery Companies of London to advise them concerning the prospect of encouraging technical instruction.[73] In 1878, several reports were published. Huxley's views, which commanded great weight, recommended the 'diffusion of such instruction' among the industrial population; adequate training for suitable numbers of teachers; systematic arrangements for apprenticeship in industry; and provision for the best to continue their studies beyond the ordinary school-leaving age and towards careers as teachers or as 'original workers in applied science'[75] . In 1880, the plans of the 'Livery Companies Committee' were given substance by the incorporation of the 'City Guilds of London Institute for the Advancement of Technical Education'. Its first organising Director and Secretary was Philip (later Sir Philip) Magnus.[76] In 1866, the Institute opened a technical college in Finsbury,[77] and between 1882 and 1884 a Central Institute for the training of teachers was built on the Gore Estate in South Kensington. This was to become the predecessor of the City and Guilds of London Institute, later part of Imperial College.

But these developments did not fully answer the demand for technical education, particularly at the secondary level. During the 1870's, new voices were heard in support of the educationalists, notably that of A. J. Mundella, who was credited with being 'one of the earliest to mention the words "technical education" in this country'.[78] Before coming into office as Vice-President of the Committe of Council, with the Liberal Government of 1880, Mundella had travelled widely to Germany and had recognised the enormous strides made by German industry since 1870.

V

In April 1881, George Anderson, Liberal MP for Glasgow,[79] requested Gladstone's Government to set up a Royal Commission to visit the continent to study different schemes of technical education. Mundella was not greatly in favour, believing that 'to appoint a roving commission to travel all over Europe would be a very expensive, and, I think, a needlessly tedious process'.[80] Mundella suggested instead that Bernhard Samuelson, with experience of both the Select Committee of 1868 and the Devonshire Commission of 1870-1875, might make an unofficial study himself. 'What Englishmen do for themselves', he remarked, 'is better done than what a government does for them.'[81] In July, the issue was again raised. Mundella announced that he had spoken with four, interested men two of whom, Samuelson and John Slagg,[82] had consented to conduct such an investigation at their own expense. In fact, the Commission, when finally announced in August 1881, consisted of men who were by definition already deeply committed advocates of technical education, including Samuelson, Magnus, Roscoe, Swire (later Sir Swire) Smith,[83] John Slagg and William Woodall, MP,[84] together with Gilbert Redgrave, the Secretary appointed by the Education Department. Of the six only three were then MP's, although two of the others eventually entered Parliament.[85] Four were progressive industrialists, two were educationalists and one (Roscoe) was a practising scientist. In this extraordinary fashion, a self-confessed 'lobby' acquired the status of a traditionally 'neutral' Royal Commission.[86]

Given these unusual origins, the product of the Commission was remarkable. At a public cost of £2,600 it travelled throughout Europe to the heart of industry in Germany, Belgium, Switzerland, Holland and France. Its terms of reference — 'to inquire into the Instruction of the Industrial Classes of Certain Foreign Countries in technical and other subjects for the purpose of comparison with that of the corresponding classes in this country . . .' — gave it a virtually free hand. Sitting for over three years, the Commission produced in May 1884, five huge volumes of data and memoranda on economic, industrial and social conditions. The first volume presented reports on technical education, with recommendations, while the second contained reports by H. M. Jenkins, FGS (Secretary of the Royal Agricultural Society) on agricultural education in Europe, and by William (later Sir William) Mather (a mechanical engineer from Salford) on technical education in the United States and Canada. The remaining volumes included reports by Thomas Wardle on the silk industry, and by Professor Sullivan on technical education in Ireland.

The Commission's findings were, in a word, pessimistic. Britain still led in invention, they argued, but was falling steadily behind in the application of new technical skills. Britain had perhaps a good system of evening instruction, but European elementary and secondary education surpassed their

British equivalents. The Commission viewed with apprehension progress made in Europe in the technical training of foremen and managers, and argued that this could not have been achieved without a high degree of technical instruction and a general appreciation of learning and research.[87]

Given this, their recommendations were sweeping. In the 16 years since the Samuelson Committee, British secondary schools had still not begun to offer technical instruction at a sufficient level and scale. Accordingly, the Commission suggested the transfer of all secondary education to public expense; recommended that the new teacher training courses be set up at the Normal School of Science, complemented by additional training colleges to be created by School Boards, and urged that Britain develop institutions following the model of the Zurich Federal Polytechnic.[88] When the report appeared, *The Times* put on a brave face:

> It must be regarded as satisfactory that the Commissioners have been able to terminate their labours . . . with the conviction that, whatever may be the progress of other nations in technical education and in manufactures, our own industries also are full of vigour; that we already possess considerable opportunities for theoretical instruction in the technical sciences and in art applied to industry; that these opportunities are capable of increase on their present lines; that the value of such instruction and the necessity for its further development are felt by those most directly interested; and that this development is making sure, though gradual, and perhaps somewhat tardy, progress.

But *The Times* went on to admit, in a leading article, that

> With the whole world competing for the supply of the universal market, it is vital to Great Britain that its artisans should understand how to march straight to their industrial goal by the shortest and most certain routes. Beyond a doubt the ordinary public appliances of education have not hitherto aided them as a body very efficiently in the race . . . At last, however, a serious commencement has been made of a national edifice of technical education.[89]

The Samuelson Report, according to Magnus, was 'widely circulated . . . and was for many years the recognised work of reference on all questions of technical education'.[90] Its belief that good secondary schools held the key to healthy industry became a cornerstone upon which later Education Acts were based. From the Government of the day, however, the Report met with a stony silence. The events of 1884-1885, witnessing Gladstone's leadership of the Liberal Party shaken by the driving wedge of Home Rule, left little parliamentary time for educational reorganization on the scale Samuelson required. Between July 1885 and August 1886, there were no less than three successive governments, and the issue, accordingly, went back into the

V

public domain, where Huxley and others, long veterans of such campaigns, prepared themselves for the next phase of combat.

IV The Aftermath

The sequence of events in the wake of the Samuelson Commission are important to the history of technical education and readers of this IUP series may wish to follow further developments in greater detail. The following paragraphs will give merely an outline of the chief characteristics of this development.[91]

Following the report of the Commission in 1884, its members toured the country trying to persuade manufacturers that if Britain's commercial supremacy was undermined, continental and particularly German military supremacy would certainly result. Two years after the Samuelson Commission, a new Royal Commission on elementary education was set up under R. A. Cross.[92] The Cross Commission agreed with Samuelson that despite common knowledge of glaring deficiencies in technical education, very little had been done, and 'still less in that scientific education which is the foundation of such technical instruction as can be given in the school'[93]. Despite the parliamentary work of Roscoe, Acland, and Playfair between 1884 and 1889, during which time there were seven successive attempts at legislation, these arguments failed to make any headway. Moreover, there remained a problem of the definition of 'technical education' which hampered attempts to introduce subjects into existing curricula at Board Schools. For example, Rosamund Davenport-Hill referred in the *Contemporary Review* to 'two distinct meanings — one the teaching of a specific art or trade; the other instruction in elementary science bearing on all arts or trades and the training of hand and eye'.[94] This was a familiar difficulty, dating at least from the 1860's, but the Cross Commission in 1888 attempted to resolve the question by arguing that the one would be incomplete without the other:—

> By technical instruction we understand instruction in those scientific and artistic principles which underlie the industrial occupations of the people . . . as well as instruction in the manual practice involved in the application of these principles.[95]

In the meantime, through the initiative of Henry Roscoe and A. H. D. Acland, the National Association for the Promotion of Technical Education was created in 1887.[96] For twenty years, this body provided a forum for systematic attempts to put technical education firmly in the hands of local School Boards and firmly on the rates. Their efforts had their reward. In July 1889, after several unsuccessful attempts, Salisbury's Government (through Sir William Hart-Dyke, Vice-President of the Council) introduced and carried over bitter opposition, a long awaited Technical Instruction Act. A short amending act followed in 1891. This legislation defined 'technical instruction' and 'instruction in the scientific and artistic principles applicable to industry and special trades, but not the practice of these trades'. It gave powers to county councils and urban sanitary authorities ('local authorities' under the Act) to levy a rate not exceeding one penny in the pound for the promotion of technical education. In operation, the Science and Art Department, with Donnelly as its chief, helped local authorities by providing examinations on both scientific principles and their applications. The newly created London County Council (LCC) inaugurated several 'maintained' technical schools, and many provincial cities, including Birmingham, Huddersfield, Leicester, Manchester and Salford, created new or expanded technical schools. Meanwhile, several new polytechnics were proposed. By the mid 1890's, 'technical education' had taken on an even wider meaning, embracing both scientific, technological and general education, both at the secondary level and at the more advanced level of the polytechnic.[97]

In 1889, the Association, given a new lease of life, added 'Secondary Education' to its title, and began to press for means to increase the funds available to local authorities for the support of technical education. The situation was transformed in the wake of the Local Taxation (Customs and Excise) Act, 1890.[98] The Act was intended to direct to the local authorities, newly created under the Local Government Act, 1888, certain surplus funds created for compensating publicans whose licenses were not renewed. During its passage, Acland tried to insert an amendment which would give local authorities the power to use these sums for technical instruction; the amendment failed, but three years later, G. J. Goschen, then Liberal Chancellor — a man of strong German associations, one feels, and a willing convert to technical education — moved that this 'whisky money' be made over to County Councils for either technical education or the relief of rates. As one MP quipped, local authorities were thus enabled to 'distil wisdom out of whisky, genius out of sin, and capacity for business out of beer'.[99] Many local authorities snapped at the chance of 'whisky money'. In 1892-93, £472,560 was used for technical education in this way; by 1900-01 the figure had risen to £836,847 (rather more than the total income received by all the schools from their endowments).[100] This contrasted with the sums of £12,762 and £106,209, raised in the respective years by the penny rate provided under the Technical Instruction Act, 1889. By all accounts, local authorities used their funds rigorously, and technical schools became a source of local prestige.

The battle for education won, educational reformers turned themselves to the even more formidable task of rationalizing the structure of education. By the end of the century what Lord Goschen called 'the chaos of local government' was reflected in the multiplicity of educational authorities up and down the land. It was clear that the School Boards, created under the 1870 Act to administer elementary education, had also created higher-grade schools where scientific education was often tailored to local circumstances.[101] Moreover, the Science and Art Department, beginning with science classes in elementary schools, had gradually extended their grants to secondary schools. Many of these (92 up to 1895) were recognised as 'organised science schools', and over 3,000 students received some science instruction. But the Science and Art Department, in expanding in this direction, without legislative provision, made for innumerable difficulties of confusion and overlapping. Finally, the Technical Instruction Act, which gave powers to County Councils and County Boroughs, extended the range of authorities eligible to be called 'educational' even further.

In the meantime, many of the old campaigners had left the field of battle. Huxley died in 1895, and Mundella in 1897. Only Magnus and Lockyer survived well into the present century, and lived to write the history of the movement they helped to launch.[102] Nevertheless, their campaign continued. Stimulated by international rivalry and the threat of German competition — vividly captured in Ernest Williams's *Made in Germany*[103] — the educationalists won new ground in 1895 with the appointment of the Royal Commission on Secondary Education (the Bryce Commission). The Bryce Commission summarised several of the recommendations which had emerged over the years in the wake of the Devonshire and Samuelson Commissions. It proposed a strong central Ministry of Education, and recommended placing secondary education into the hands of reconstituted local education authorities. It also recommended that

> ample provision must be made in schools for scientific teaching . . . [and] the chief tongues of modern Europe . . . and that full opportunities to boys and girls to prepare themselves for the particular occupation which they intend to follow in after life, whether industrial or commercial, ought to be supplied by the teaching of the practical arts . . .[104]

The Bryce Commission's Report gained greater urgency from a second visit of Magnus, Woodall, Swire Smith and Redgrave to Germany in 1896.[105] Their report was widely read and reviewed, and in Magnus's view, hastened progress towards reorganising educational administration. This was completed by the Education Act of 1902, which swept away the School Boards and constituted 'local authorities' as 'local educational authorities' in their place. In the meantime, by the Board of Education Act, 1899, the old Department of Science and Art was dissolved, and replaced by a single Board of Education with a unified secretariat. The century opened with the beginnings of a coherent structure for primary, secondary and technical education for all of England and Wales.

In 1906, the National Association for the Promotion of Technical and Secondary Education was wound up, its duty done. An important era appeared to be ending. In 1910, Philip Magnus felt confident enough to write:

> The necessity of the higher scientific and technical instruction and of its specialisation to industry and commerce is generally realised. No one doubts it. The questions now engaging attention are the kind of training that should lead up to it, the best means of providing avenues through which the children of our elementary schools may pass to the higher technical institutes, the character of the teaching most suitable for children after the elementary school age, and the relation between university and technological instruction.[106]

After years of work, the way had been cleared, administratively and politically for a 'silent social revolution' in British scientific education that was to become a central part of the history of Britain in the twentieth century. The principles stated, it remained for a new generation, made next to familiar with the threat of international industrial and technological competition, to attempt their implementation.

Footnotes

1 On the Great Exhibition itself, see C. R. Fay, *Palace of Industry,* 1851, (Cambridge, 1951); and 'The Great Exhibition (i) How it all began, (ii) Frustration and Fruition', in K. W. Luckhurst, *The Story of Exhibitions,* (London, 1951), pp. 85-116. On the aftermath of the Great Exhibition, see W. L. Burn, *The Age of Equipoise,* (London, 1964). See also D. S. L. Cardwell, *The Organization of Science in England: A Retrospect,* (rev. ed. London, 1972), pp. 76-81. Cardwell ably summarizes the chief points at issue in the wake of the Great Exhibition and should not be ignored by the student. This section of his book does, however, oversimplify many of the issues and neglects the industrial contribution to the structure of the educational debate.

2 J. B. Bury, *The Idea of Progress* (London, 1920), p.329.

3 Henry (later Sir Henry) Cole, (1808-1882); civil servant; assistant keeper of the Record Office, 1838; served on managing committee of exhibitions of 1851, 1862 and 1871-4; British commissioner at the Paris Exhibitions of 1855 and 1867; joint secretary of the Science and Art Department, 1853, and sole secretary, 1858-73; KCB, 1875. Cf. H. Cole, *Fifty Years of Public Life* (London, 1884).

4 Lyon (later Sir Lyon) Playfair, (1818-1898); educated at St Andrews, Glasgow and Giessen; assistant to Thomas Graham at University College, London; honorary professor of Chemistry at the Royal Institution, Manchester, 1842-45; professor in the School of Mines, Jermyn Street, 1845; FRS, 1848; President of the Chemical Society, 1857-9; helped organize the Great Exhibition; CB, 1851; Secretary of Science to Department of Science and Art, 1853, and Secretary for Science and Art, 1855-8; professor of chemistry at Edinburgh, 1858-69; Liberal MP for universities of Edinburgh and St Andrews, 1868-85; Postmaster General, 1873; chairman and deputy speaker of the House of Commons, 1880-83; Liberal MP for South Leeds, 1885-92; Vice-President of Council, 1886; raised to peerage, 1892: GCB, 1895. Cf. Wemyss Reid, *Memoirs and Correspondence of Lyon Playfair* (London, 1899).

5 Albert, Prince Consort of England (1819-1861), son of Duke of Saxe-Coburg-Gotha; educated privately, then at Brussels and Bonn; married Queen Victoria, 1840; placed by Peel at the head of the Royal Commission on the rebuilding of the Houses of Parliament, 1841; Chancellor of Cambridge University, 1847, where he successfully

V

advocated alterations in the curriculum; projected the idea of the Great Exhibition; died of typhoid fever in December, 1861.

6 Cf. D. Hudson and K. W. Luckhurst, *The Royal Society of Arts, 1754-1954,* (London, 1954). With the strong support of the Prince Consort, the Society of Arts received its Royal Charter in 1847.

7 *Second Report of the Commissioners for the Exhibition of 1851*, HC. 1852, pp.72-74.

8 Whewell, 'The General Bearing of the Great Exhibition on the Progress of Art and Science', *Lectures on the Results of the Exhibition delivered before the Society of Arts, Manufacturers and Commerce, at the suggestion of HRH Prince Albert, President of the Society* (London, 1852). William Whewell (1794-1866) was Master of Trinity College, Cambridge, 1841-66; he supported the election of the Prince Consort as Chancellor in 1847, and assisted in the introduction of the Moral Sciences Tripos and the Natural Sciences Tripos, he published distinguished works in the natural and mathematical sciences, philosophy and theology.

9 'The Queen's Address to both Houses of Parliament', *Hansard,* 3rd series, 123, c. 19. For comparative comments on the development of art education, see Celina Fox, *Critical Commentary: Fine Arts and Design, supra,* pp.68-93.

10 The Government School of Mines and Science Applied to the Arts was founded in 1851 in association with the Geological Museum in Jermyn Street. The Royal College of Chemistry was created in 1845 with the aid of the Prince Consort. It thrived briefly under the direction of A. W. Hoffman, but fell into financial loss until its 'rescue' in 1853. See G. Roberts, *The Royal College of Chemistry,* (1845-1853): A Social History of Chemistry in Early Victorian England (unpublished Ph.D. thesis, Johns Hopkins University, 1973).

11 H. Butterworth, *The Science and Art Department,* (Unpublished Ph.D. thesis, University of London, 1951).

12 Cf. Thomas Kelly, *A History of Adult Education,* (Liverpool, 1970).

13 J. W. Adamson, *English Education, 1789-1902,* (Cambridge, 1930), chaps VIII-X. For general reviews of these developments, cf Richard Johnson, *Critical Commentary: Education, supra,* pp.5-29, and Gillian Sutherland, *supra,* pp.137-159.

14 *Report of Commissioners on Education in Schools in England not comprised within Her Majesty's two recent Commissions on Popular and Public Schools,* (the "Schools Enquiry Commission"), HC 1867-68, (3966), XXVIII (parts 1-17).

15 *Report of the Schools Enquiry Commission Relative to Technical Education,* p.261. HC 1867, XXVI. Letter from Playfair to Taunton, 15 May 1867, 266-70.

16 John Tyndall, (1820-1863), natural philosopher; teacher at Queenwood College, Hampshire; studied at Marburg with Bunsen, 1848-1850; FRS, 1852; professor of natural philosophy at the Royal Institution, 1853; succeeded Faraday as Superintendent, 1867-87; made important investigations on heat, sound and light, succeeded Faraday as scientific adviser to Trinity House and the Board of Trade, 1866. Well-known for popularising science. Cf. A. S. Eve and C. H. Creasey, *Life and Work of John Tyndall,* (London, 1945).

17 *Report of the Schools Enquiry Commission relative to Technical Education,* p.10, HC 1867, XXVI, 261.

18 *Ibid.,* p.264.

19 For an account of the Society of Arts Conference in January, 1868, see the *Journal of the Royal Society of Arts, XVI,* 31 January 1868, p.183; and for the report arising from the conference, see *J. Roy. Soc Arts., XVI* (1868), 627.

20 Report of the Parliamentary Committee, *Report of the British Association for the Advancement of Science,* [Dundee], (London, 1867).

21 Bernhard (later Sir Bernhard) Samuelson (1820-1905); manager of Manchester Engineering firm, 1842-6; began famous Britannia ironworks at Middlesbrough. Liberal MP for Banbury, 1859 and 1865-85, and for North Oxfordshire, 1885-95; baronnet, 1884; FRS, 1881; presented technical institute to Banbury, 1884. Samuelson also served on the Royal Commission on Elementary Education in 1887, and was chairman of Select Committees on the patent laws, (1871-2), and the railways (1873).

22 George Dixon (1820-1898), merchant of Edgbaston. In 1868, helped found the National Education League; attempted to amend Education Act of 1870 in accordance with 'advanced' Liberal views; member of first Birmingham School Board, 1870; MP for Birmingham, 1867-76; 1885-98.

23 Lord Robert Montagu, (1825-1902), liberal Conservative MP, Huntingdonshire, 1859-74; advocated legislation on sewage disposal, cattle plague, and prevention, and vaccination; helped found School of Naval Architecture.

24 Thomas (later Sir Thomas) Dyke Acland, (1809-98); scholar and educational reformer; Liberal MP for North Devonshire, 1865-85; served actively on the Endowed Schools Commission, 1864-73 and in the debates in Committee on the Education Bill of 1870.

25 *Report from the Select Committee on Scientific Instruction,* HC. 1868 (432), I, vii.

26 *Hansard,* 3rd series, 197 c. 204.

27 Adamson, *English Education,* p.398.

28 John (later Sir John) Donnelly (1834-1902); Royal Engineers, retired with honorary rank of major-general, 1887; assisted Sir Henry Cole in reorganizing Science and Art Department, 1858; Inspector of Science, 1859; arranged for payment of teachers by results of examinations of pupils, 1859; became 'director of science', 1874; supervised science schools and institutions throughout the country; secretary and permanent head of Science and Art Department, 1884-1889; CB, 1886; KCB, 1893.

29 Thomas Henry Huxley (1825-1895); M.B. London, 1845; assistant surgeon on HMS *Rattlesnake,* 1846-50; FRS, 1850; lecturer on natural history at Royal School of Mines, 1854; naturalist to Geological Survey, 1855; served on many royal commissions, including that on the sea fisheries of the United Kingdom, 1864-5, the Royal College of Science for Ireland, 1866, on Scientific Instruction and Advancement of Science, 1870-75, on vivisection, 1876, and on Scottish Universities, 1876-8; Hunterian professor at Royal College of Surgeons, 1863-69; Fullerian Professor at the Royal Institution, 1863-7; President of the Royal Society, 1883-5; Privy Councillor, 1892. Cf. Cyril Bibby, *T. H. Huxley: Scientist, Humanist and Educator,* (London, 1959).

30 Sir John Lubbock (1834-1913), fourth baronet and first Baron Avebury; banker, scientist and author; educated at Eton; Liberal MP for Maidstone, 1870 and 1874, London University, 1880-1900; promoted the Bank Holidays Act (1871), and the Act for Preservation of Ancient Monuments (1882); Privy Councillor, 1890; baron, 1900; FRS, 1858; author of numerous scientific and ethical works.

31 James Wilson (1836-1931), schoolmaster and antiquary; educated at St John's College, Cambridge; senior wrangler, 1859; Fellow, 1860; mathematical master in Rugby, 1859-70; ordained, 1879; headmaster of Clifton, 1879-90; archdeacon of Manchester, 1890-1905; Canon of Worcester, 1905-26.

32 Report of Parliamentary Committee, (22 August, 1866), *Report of the British Association,* [Nottingham], (London, 1866).

33 Report of Parliamentary Committee, (September, 1864), *Report of the British Association,* [Dundee], (London, 1867).

34 *Report of the British Association,* [Norwich], (London 1868).

35 Norman (later Sir Norman) Lockyer (1856-1920); astronomer; clerk in the War Office, 1857; pioneered observations of sun-spot spectra, 1866 and of solar promi-

V

nences, 1868; secretary to Royal Commission on Scientific Instruction and Advancement of Science, 1870; transferred to Science and Art Department, 1875; director of Solar Physics Observatory and professor of astronomical physics, Royal College of Science, 1890-1913; editor of *Nature*, 1869-1920; FRS, 1869; CB, 1894; KCB, 1897. See A. J. Meadows, *Science and Controversy: A Biography of Sir Norman Lockyer*, (London, 1972).

36 Lieutenant-Colonel Alexander Strange (1818-1876), educated at Harrow; Madras Light Cavalry; trigonometrical survey of India; Inspector of scientific instruments for use in India, 1861; FRGS, and FRAS, 1861; FRS, 1864. Cf. J. G. Crowther, *Statesmen of science*, (London, 1965), pp.237-269.

37 'On the Necessity for State Intervention to Secure the Progress of Physical Science', Address to the Mathematics and Physics Section, *Report of the British Association*, [Norwich], (London, 1868).

38 [W. Crookes], 'On the Teaching of Natural Science at the Universities', *Q. J. Sci.*, VI, (October, 1869), 500.

39 *Ibid.*

40 Roy MacLeod, 'The X-Club: A Social Network of Science in Late-Victorian England', *Notes and Records of the Royal Society, 24*, (1970), 305-322.

41 [W. Crookes], 'On National Institutions for Practical Scientific Research', *Q. J. Sci.*, VI, (January, 1869), 48.

42 'Government Aid to Science' *Nature*, I, (1870), 279-88.

43 W. Crookes, *Q. J. Sci.*, VI, (1869), 48-9.

44 See *Nature, I*, (1870), 589-90.

45 Benjamin Jowett, 'Suggestions for University Reform, 1874' in Lewis Campbell, *The Nationalization of the Old English Universities*, (London, 1901), pp.183-208.

46 Alexander Williamson (1824-1904), chemist; educated at Dijon, Heidelberg, Giessen and Paris; professor of practical chemistry, University College, London, 1849-87; President of Chemical Society, 1863-5, 1869-71; President of the British Association, 1873; FRS, 1855; helped introduce science degrees in London University.

47 Henry (later Sir Henry) Roscoe, (1833-1915), educated at University College, London, and Heidelberg; researched with Bunsen on the measurement of chemical action of light; professor of chemistry, Owens College, Manchester, 1857-85; knighted, 1884; Liberal MP, South Manchester, 1885; PC, 1909.

48 H. Roscoe, 'Original Research as a Means of Education', Address at the opening of Owen's College, *Nature, VIII*, (1873), 538 ff.

49 J. Clerk Maxwell, *Introductory Lecture on Experimental Physics*, (London, 1871).

50 *Nature* pointed out that almost all continental researchers taught as well, and that even 'in this country the greater number of our most distinguished men of science are professors and teachers', 'The Royal Commission on Science', *Nature, I*, (1870) 375. See also, *The Academy*, (1 December 1872), 460.

51 *Council Minutes*, (18 August, 1869), *Report of the British Association*, [Exeter], (London, 1869).

52 *Council Minutes*, (25 August 1869), *Report of the British Association*, [Exeter], (London, 1869).

53 The Commission (hereafter the 'Devonshire Commission') included the Marquis of Lansdowne; Sir John Lubbock; James Kay-Shuttleworth, MD; Bernhard Samuelson, MP; W. A. Miller (London); H. J. S. Smith (Oxford); William Sharpey (London); George Stokes (Cambridge); T. H. Huxley (London). Historical commentaries on the Commission can be found in D. S. L. Cardwell, *The Organization of Science in*

England: A Retrospect, (London, 1972), pp.119-126, and in Crowther, *Statesmen of Science* (London, 1965), pp.222-33.

54 *Nature,* IV (1871), 302.

55 The first brief report was printed in full in *Nature,* III (1871), 421.

56 Cf. 'Government and Scientific Investigation', *Spectator,* XLIV (1871) 882-4; Cf. also Anon., *Correspondence between the Royal Geographical Society and the Government* (London, 1873).

57 C.f. *J. Roy. Soc. Arts,* XVIII (1870); reported in 'The Relation of the State to Science', *Nature,* I (1870), 589-91.

58 'On the Necessity for A Permanent Commission on State Scientific Questions', *J. Royal United Services Institution,* XV (1870), 537-66, reprinted in *Nature,* IV (1871), 133, 'On Government Action in Science', *Report to the Mathematics and Physics Section of the British Association* (1871), p.56.

59 *The Academy,* (1 December 1872), 460.

60 'A Voice from Cambridge', *Nature,* VIII (1873), 21.

61 'Our National Industries', *Nature,* VI (1872), 97.

62 *Nature,* II (1870), 25.

63 *Sixth Report of the Devonshire Commission,* HC. 1575 [C. 1279], XXVIII, p.10, para. 46.

64 *Appendix to the Sixth Report of the Devonshire Commission,* p.60.

65 *Seventh Report of the Devonshire Commission,* HC. 1875, [C. 1297], XXVIII, pp.3, 6.

66 *Eighth Report of the Devonshire Commission,* H.C. 1875 [C. 1298], XXVIII, p.2. Both reports were abstracted at length in *Nature,* XII (1875), 285-8, 305-8, 361-3, 389-92, 469-70. The seventh report, dealing with the universities of London, Scotland, Trinity College, Dublin and Queen's University, Ireland, was discussed in *Nature,* XIII (1875), 21-2.

67 *Eighth Report of the Devonshire Commission,* p.47. See Strange's letter to *The Times* (6 February 1874); 'A Minister for Science', *Nature,* IX, (1874), 277. A Science Council of thirty members were proposed anonymously by Strange in *Nature,* XII (1875), 431.

68 *The Academy* (4 September, 1875), 252.

69 Cf. *Hansard,* 3rd Series, 227 c. 551.

70 Lewis Gunn, 'Organizing Science in Britain, Some Relevant Questions', *Minerva,* 5, (1967), 167-97.

71 Cf. 'What is true technical education?', *The Economist,* (25 January 1868), 87-8.

72 T. H. Huxley, 'Technical Education', (1877), *Collected Essays,* III, (London, 1893) pp.411-12.

73 See G. T. Millis, *Technical Education: Its Development and Aims* (London, 1925), p.56. A. Albu, 'Introduction', to E. E. Williams, *Made in Germany* (London: 1896, 1973), pp. xxvii-xxviii.

74 Philip Magnus, *Educational Aims and Efforts* (Longmans, 1910), pp.86-7.

75 Livery Companies' Committee, *Report on Technical Education,* (London: privately published, 1878). A volume of essays prepared by Sir William Armstrong, G. J. C. Bartley, Col. John Donnelly, Capt. Douglas Galton, T. H. Huxley and H. Trueman Wood.

76 Philip (later Sir Philip) Magnus (1842-1933), Secretary and Organizing Director of the City and Guilds of London Institute, 1880-1913; MP for London University,

V

1906-1922; a consistent advocate of education — particularly technical education — he sat on several commissions and was a staunch supporter of Lockyer's British Science Guild. See Frank Foden, *Philip Magnus: Victorian Educational Pioneer* (London, 1970), esp. pp.128-129.

77 On Finsbury Technical College see E. G. Walker, 'Finsbury Technical College', *Central, 30* (1933), pp.35-48, H. E. Armstrong, 'The Beginning of Finsbury', *Central, 31* (1934), pp.1-14.

78 *Hansard,* 3rd series, 260. c. 537; cf. W. H. G. Armytage, *A. J. Mundella, 1825-1897: The Liberal Background to the Labour Movement* (London, 1951).

79 George Anderson (1881-95); retired merchant; published pamphlets on National Education; (Dod's *Parliamentary Companion,* 1881).

80 *Hansard,* 3rd series, 260. c. 542.

81 *Ibid.,* c. 543.

82 John Slagg (1841-1889); Liberal MP for Manchester (1880-85); MP for Burnley (1887-89); merchant at Manchester Chamber of Commerce.

83 Swire Smith (1842-1918), knighted, 1898; Liberal MP for Keighley Division of Yorkshire, 1915-1918: Member of the Royal Commission on Technical Instruction, 1881-1884; Vice-Chairman of the Royal Commission on Technical Instruction, 1909; LL.D., University of London, 1912.

84 William Woodall (1832- ?), Liberal MP for Stoke on Trent; partner in china works; magistrate for Staffordshire; 'in favour of the equalisation of the borough county franchise, a simplification of its land laws'. (Dod's, *Parliamentary Companion,* 1881).

85 Swire Smith in 1915 and Roscoe in 1885.

86 Given Mundella's surprising reluctance, it has been argued that the Commission was in fact engineered by Samuelson, with Magnus and Mundella acting as accomplices, to by-pass the Government's reservations towards a full-scale enquiry. Cf. Michael Argles, 'The Royal Commission on Technical Instruction, 1881-1884: Its Inception and Composition', *Vocational Aspects,* II, (1959), 98.

87 *Second Report of the Royal Commission on Technical Instruction,* 1884, [C. 3981] I, pp.508-510. In England, on the contrary, opportunities for applied research were rare and advanced technical instruction virtually non-existent.

88 The development of polytechnics could be fostered in London under the provision of the City of London Parochial Charities Act, 1883.

89 *The Times,* 16 May, 1884, quoted in M. Argles, 'The Royal Commission on Technical Instruction', *Vocational Aspects,* II, (1959), 103.

90 Magnus, *Educational Aims and Efforts,* p.94. One version was published by Francis Montague, *Technical Education: A Summary of the Report of the Royal Commission Appointed to Inquire into the State of Technical Instruction,* (London, 1887).

91 For further reading, see the well chosen excerpts from reports and legislation of the period in J. Stuart MacLure, *Educational Documents – England and Wales, 1816-1963* (London, 1969). For legislative details of the 'Technical Instruction Movement', Cf. Philip Magnus, 'Technical Education', *Encyclopedia Britannica,* 11th ed (1910-11).

92 *Royal Commission on Elementary Education,* first Report, HC 1886 (4565), XXV; second Report, HC. 1887 (5056), XXIX; third Report, HC 187 (5158), XXX; Final Report, HC 1888 (5485), XXV.

93 Adamson, *English Education,* p.407.

94 R. Davenport-Hill, 'Technical Education in Board Schools', *Contemporary Review, 53,* (1888), 672-685.

95 *Final Report of the Royal Commission on Elementary Education,* 1888, HC. (5485),

XXV, p.217. The celebrated debate between Armstrong and Playfair on the nature of technical education, in terms of its 'civilizing' and conceptual satisfactions and its vocational objectives, is extremely instructive. See W. G. Armstrong (Lord Armstrong), 'The Vague Cry for Technical Education', *Nineteenth Century, 24* (July 1888), 45-52; 'Lord Armstrong on Technical Education', *Nature,* XXXII (1888), 313-14; Lyon Playfair, 'Lord Armstrong and Technical Education', *Nineteenth Century, 24* (September 1888), 325-333; W. G. Armstrong 'The Cry for Useless Knowledge', *Nineteenth Century, 24* (November 1888), 653-668.

96 See the Annual Reports of the National Association for the Promotion of Technical (and Secondary) Education, 1888-1906; and the *Record of Technical and Secondary Education,* published bi-monthly by the Association from November 1891 to December 1906.

97 For changing definitions and their significance, see P. W. Musgrave, 'The Definition of Technical Education', *Vocational Aspects, 16,* (1904), 105-111.

98 Cf. Adamson, *English Education,* p.409.

99 G. A. N. Lowndes, *The Silent Social Revolution* (Oxford, 1937), p.40. In fact, the disposition of the 'whisky money' was influenced strongly by the temperance party during a sessional crisis which endangered the safety of several Liberal Unionist seats. Cf. the allusive evidence of Arthur Elliot, *The Life of George Joachim Goschen, first Viscount Goschen, 1831-1907* (London, 1911), p.167. P. R. Sharp, 'Whisky Money and the Development of Technical and Secondary Education in the 1890's', J. of Educational History and Administration, 4 (1971), 31-36.

100 *Ibid.,* p.40. For contemporary criticism of the school, see Lowndes, *ibid.,* chap. IV, pp.47-63.

101 *Ibid.,* p.40.

102 Philip Magnus, *Industrial Education* (London, 1888); *Educational Aims and Efforts* (London, 1910); R. D. Roberts, *Education in the 19th Century* (Cambridge, 1901). For an account of Lockyer's advocacy of technical education see Norman Lockyer, *Education and National Progress: Essays and Addresses, 1870-1905,* (London, 1906).

103 Ernest Edwin Williams, *Made in Germany* (London, 1896; reprinted, Harvester Press, 1973). Cf. also, G. R. Searle, *The Quest for National Efficiency* (Oxford, 1971).

104 *The Royal Commission on Secondary Education* HC. 1895 [C. 7862], p.284.

105 Cf. *Report on a Visit to Germany with a View to Ascertaining the Recent Progress of Technical Education* (HMSO, 1896): Magnus, *Educational Aims and Efforts,* p.121.

106 Philip Magnus, *Educational Aims and Efforts* (London, 1910), pp.126-7.

Bibliography

The reader interested in pursuing the subject of technical education will find many suggestive references in the text and footnotes to primary and secondary sources of particular importance. The following list is therefore intended as only a brief supplementary guide to further reading, and as such includes many items of a contextual and general nature. In many cases, bibliographies to specific themes form part of these treatments.

John William Adamson, *English Education, 1789-1902* (Cambridge: Cambridge University Press, 1930).

Michael Argles, "The Royal Commission on Technical Instruction, 1881-1884: Its Inception and Composition", *The Vocational Aspect of Secondary and Further Education,* II (1959), 97-104.

V

Michael Argles, *South Kensington to Robbins* (London: Longmans, 1964).

W. H. G. Armytage, *The Rise of the Technocrats* (London: Routledge and Kegan Paul, 1965).

Eric Ashby, *Technology and the Academics* (London: Macmillan, 1959).

Cyril Bibby, *T. H. Huxley: Scientist, Humanist and Educator* (London: Watts, 1959).

E. J. T. Brennan, (ed), *Education for National Efficiency: the Contribution of Sidney and Beatrice Webb* (London, 1975).

W. H. Brock, (ed), *H. E. Armstrong and the Teaching of Science, 1880-1930* (Cambridge, 1973).

W. L. Burn, *The Age of Equipoise* (London: Allen Unwin, 1964.

H. Butterworth, "The Development of Technical Education in Middlesbrough, 1844-1903", *Durham Research Review,* III (September 1960), 27-34.

D. S. L. Cardwell, *The Organisation of Science in England: A Retrospect,* (London: Heinemann, 1957, 1972).

J. G. Crowther, *Statesmen of Science* (London: Cresset Press, 1965).

Frank Foden, "Technical Education in England", *Pedagogica Historica,* VI (1966) 68-97.

Frank Foden, *Philip Magnus: Victorian Educational Pioneer* (London: Vallentine, Mitchell, 1970).

D. Hudson and K. W. Luckhurst, *The Royal Society of Arts, 1754-1954* (London: John Murray, 1954).

T. Kelly, *A History of Adult Education* (Liverpool: Liverpool University Press, 1970).

David S. Landes, *The Unbound Prometheus: Technological Change and Industrial Development in Western Europe from 1750 to the Present* (Cambridge: Cambridge University Press, 1970).

David Layton, *Science for the People: The Origins of the School Science Curriculum in England* (London: George Allen and Unwin, 1973).

G. A. N. Lowndes, *The Silent Social Revolution* (Oxford: Oxford University Press, 1937).

A. J. Meadows, *Science and Controversy: A Biography of Sir Norman Lockyer* (London: Macmillan, 1972).

Stuart MacLure, *Educational Documents: England and Wales, 1816-1963* (Londón: Chapmune Hall, 1965).

G. T. Millis, *Technical Education: Its Development and Aims* (London: Edward Arnold, 1925).

J. Monagham, "Some Views on Education and Industrial Progress a Hundred Years Ago", *The Vocational Aspect of Education,* XX (1968), 187-194.

P. W. Musgrave, "The Definition of Technical Education, 1860-1910", *The Vocational Aspect of Education,* XVI (1964), 105-111; "Constant Factors in the Demand for Technical Secondary and Further Education, 1860-1900", *British Journal of Educational Studies,* XIV, 1966; reprinted in Musgrave, P. W. (ed.), *Sociology, History and Education: A Reader* (London: Macmillan, 1970), pp.65-74; 143-157.

K. O. Roberts, "The Separation of Secondary from Technical Education", *The Vocational Aspect of Secondary and Further Education,* XXI (1969), 101-105.

G. W. Roderick and M. D. Stephens, "Scientific Studies in the Public Schools and the Endowed Grammar Schools in the 19th Century: The Evidence of the Royal Commissions", *The Vocational Aspect of Secondary and Further Education*, XXIII (1971), 97-105.

G. W. Roderick and M. D. Stephens, *Scientific and Technical Education in Nineteenth Century England* (Newton Abbot: David and Charles, 1972).

Michael Sanderson (ed), *The Universities in the Nineteenth Century* (London: Routledge and Kegan Paul, 1975).

Michael Sanderson, *The Universities and British Industry, 1850-1970* (London, 1972).

Geoffrey R. Searle, *The Quest for National Efficiency* (Oxford: Blackwell, 1971).

Mabel Tylecote, *The Mechanics Institutes of Lancashire and Yorkshire* (Manchester: Manchester University Press, 1957).

P. F. R. Venables, *Technical Education: Its Aims, Organization and Future Development* (London: G. Bell and Sons, 1956).

Ernest Edwin Williams, *Made in Germany* (London: Heinemann, 1896; repr. Harvester Press, 1973, with introduction by Austen Albu).

VI

Fathers and Daughters: Reflections on Women, Science and Victorian Cambridge†

Roy MacLeod and Russell Moseley

Introduction

In England, as elsewhere, the last few years have witnessed growing interest in the role of women in modern science, and in the effect of scientific developments upon women's domestic and professional lives.[1] However, it is remarkable how little historical attention has been paid to the way in which women have viewed their own participation in the scientific enterprise.[2] There were many heroic women who, from at least the late eighteenth century, filled the first ranks of science and mathematics.[3] But, as is well known, the profession of science with all its ascriptive connotations has offered few prospects to women. Indeed, so pervasive was the 'masculinization' of science, that a sociologist of science can describe historically the 'scientist's role' without reference to women at all.[4]

We know little about the first English women who studied science. From where did they come? What were the prospects open to them? And what, in relation to science, did they achieve? There is no consensus as to whether science acted as an avenue, as an opportunity, or as a barrier lying in the path of women's general educational progress. Historians have generally acknowledged that in science women had a place, but that no matter what their individual achievements that place was at the side of their scientific husbands, or fathers—precisely where one finds individuals as remarkable as Caroline

† The research on which this note is based has been supported by a grant from the Social Science Research Council. We are grateful for the helpful comments of Dr Brian Harrison and Dr Negley Harte on several points raised in this paper.

1 Cf. K. Lonsdale, 'Women in science', *Impact*, XX (1970), 45–59. This entire issue was devoted to 'Women in the age of science and technology'. Dr Lonsdale was one of the first women FRS's. See Note 70 below. See also M. Oliver, 'Women in Chemistry', *Journal of the Royal Institute of Chemistry*, 79 (1955), 413–420; Alison Kelly, 'Women in Science: A Bibliographic Review', *Durham Research Review*, No. 36 (Spring 1976), 1092–1108; and Audrey B. Davis, *Bibliography on Women with Special Emphasis on their Roles in Science and Society* (New York, 1974).

2 An important exception lies in the work of Margaret Rossiter. Cf. M. Rossiter, 'Women scientists in America before 1920', *American Scientist*, 62 (1974), 312–323, and her forthcoming book. See also Deborah Hean Warner, 'Science education for Women in Antebellum America', *Isis*, 69, No. 246 (March 1978), 58–67.

3 Mrs Marcet's *Conversations on Chemistry*, it was said, 'opened Faraday's eyes to Science'. Cf. Martha Somerville, *Personal Recollections from Early Life to Old Age of Mary Somerville* (London, 1874), 114. For Mrs Somerville, see E. L. Patterson, 'The case of Mary Somerville: an aspect of nineteenth century science', *American Philosophical Society Proceedings*, 118 (1974), 269–275; George Basalla, 'Mary Somerville', *New Scientist* (7 March 1963), 531; M. Sanderson, 'Mary Somerville', *Geog. Rev.*, 64 (1974), 410–420; A. Rebiéne, *Les Femmes dans la Science* (Paris, 1897). Cf. also L. Buccharelli's current work (at M.I.T.) on Sophie Kovalewska.

4 Cf. Joseph Ben-David, *The Scientist's Role in Society: A Comparative Study* (Englewood Cliffs, N.J. 1971).

Herschel, Louisa Tyndall, and Mary Lockyer.[5] Occasionally, women could 'ornament' scientific gatherings, as did Ellen Lubbock, Georgina Busk and Henrietta Huxley.[6] They attended classes at the Ladies Educational Society, Friday Lectures at the Royal Institution, and soirées at the Royal Society. They could botanize or 'galvanize' in their leisure; and one day they will command a place in the great history that remains to be written of the English scientific 'amateur'.[7] They could even attend, in their hundreds, the annual meetings of the British Association. But in these activities it is assumed that women were ancillary, merely consorts or companions.[8] Ironically, when women sought to move into the world of science, outside the world of the family and their husbands' work, they often became a force to be reckoned with. Unfortunately, in most cases they have since been ignored.[9]

This preliminary note will outline some considerations which may be involved in their 'rediscovery'. These considerations form part of a much larger work now in progress on the development of science in Victorian Cambridge.[10] Here, we confine ourselves to an outline of the general prospects open to women in science during the late nineteenth century, and to the particular position of those women who read for the Natural Sciences Tripos at Cambridge between 1880 and 1916.

Women, science and education

In three decades after 1880, in Britain as elsewhere, the movement for women's professional education achieved its first victories.[11] Beginning in the late 1870s, small but increasing numbers of Englishwomen, chiefly from the professional and managerial middle-classes, began to read scientific subjects. Institutionally, as is well known, there were enormous obstacles in their way. The claims of class and marriage, of biological and social convention;[12] the absence of any conception of scholarship as a recognized 'calling' for women; the reservations, generally at Oxbridge, of Church and College; the doubtful wisdom of providing expensive instruction to a section of the population

5 Cf. M. B. Ogilvie, 'Caroline Herschel's Contribution to Astronomy', *Annals of Science*, 32 (1970), 149–161; J. Friday, C. Gregor and R. MacLeod, *John Tyndall, Natural Philosopher, 1820–1893* (London, 1974); T. Mary Lockyer and Winifred Lockyer, *Life and Work of Sir Norman Lockyer* (London, 1928). Lord Rosse, the distinguished astronomer, it was remarked, 'married wisely, a lady of congenial taste, the daughter of a civil engineer, and, 'tis said, a better mathematician than himself'. Rev. Richard Owen, *The Life of Richard Owen* (London, 1894), 220. This example could doubtless be multiplied.

6 Cf. W. H. Brock and R. MacLeod (eds.), *The Life and Journals of Thomas Archer Hirst, FRS, 1830–1892* (London, in press).

7 Women, remarkably, do not figure in J. B. Morrell's otherwise excellent 'Individualism and the structure of British science in 1830', in R. McCormmach (ed.), *Historical Studies in the Physical Sciences*, 3 (1971), 183–204.

8 Cf. Gerald D. Meyer, *Science for English Women, 1650–1760: The Telescope, The Microscope and the Feminine Mind* (Ph.D. thesis, Columbia University, 1951).

9 In some cases, to be later recovered by anxious biographers; cf. Evelyn Sharp, *Hertha Ayrton, 1854–1923* (London, 1926); Lady Frances Balfour, *Dr Elsie Inglis. A Biography* (London, 1918).

10 R. MacLeod and R. Moseley, *Science with Honours: Science and Victorian Cambridge, 1850–1916* (in preparation).

11 Cf. 'Women in the professions', in Ishbel Maria Gordon (Countess of Aberdeen) (ed.), *International Council of Women, Report, 1899* (London, 2 vols., 1900), II, 157–187; Edith Morley, *Women Workers in Seven Professions* (London, 1914); Elisabeth Adams, *Women Professional Workers* (London, 1921); Nellie Franz, *English Women Enter the Professions* (Cincinatti, 1965); S. Barbara Kanner, Bibliography in M. Vicinus (ed.), *Suffer and be Still: Women in the Victorian Age* (Bloomington, 1972); Sara A. Burstall, *Retrospect and Prospect: Sixty Years of Women's Education* (London, 1933); and O. R. McGregor, 'The social position of women in England, 1850–1914: a bibliography', *British Journal of Sociology*, 6 (1955), 48–60.

12 Cf. Joan N. Burstyn, 'Education and sex: the medical case against higher education for women in England, 1870–1900', *American Philosophical Society Proceedings*, 117, No. 2 (April 1973), 79–83.

which was unlikely ever to gain materially from its use; the possible dangers of thereby encouraging 'inattentive' wives and mothers—such difficulties were raised often and applied to female education generally. When applied to scientific education, they acquired a new force. As James Stuart, Professor of Mechanism at Cambridge, himself a firm advocate of women's education, wrote in 1869:

> There is scarcely any subject on which there is among men so much of dogmatic assertion based on the imaginary results of a hypothetical experiment, and sometimes having not even so much basis as this. It has been affirmed, and not in this age and country alone, that women do not *need* to be taught Science, an immediate and direct utility being all that has been kept in view, and the true end lost sight of.[13]

Indeed, there were other arguments as well; for example that women did not need to learn science to make them 'better women', or 'abler to help and bless the world'; and that women were, in any case, generally incapable of participating in the advancement of scientific knowledge. In the event, the first argument relied on the prior assumption that the study of science was a 'harsh, masculine' activity, one opposed to 'woman's grace and woman's simplicity' and to her 'noble character'.[14] The second held up only so long (but for quite long, as it turned out) as poor elementary preparation in mathematics restricted women's opportunities for higher education in the physical sciences. But however dubious their logic, these arguments persisted. Science for women was at best useless, at most brutalizing. What could be said in its favour?

From the mid nineteenth century, there were perhaps three common arguments for encouraging the study of science by women: the 'better wives and mothers' argument; the 'equal opportunity' argument; and the conviction, Darwinist in overtones, that women were particularly 'fitted' to pursuing certain kinds of knowledge.[15] The third argument was heard frequently among those who claimed that women's skills were best demonstrated by a 'capacity for noting details—patience and delicacy'.[16] The second argument was to gain currency towards the end of the century. But, in context, the first argument was possibly the most compelling. As *Nature* pointed out in 1870,

> If, as we are constantly being told, the "sphere of woman" is at home, what duty can be more clearly incumbent upon us than that of giving her the opportunity of acquiring a knowledge of the laws which ought to guide her in the rule of her house. Every woman on whom the management of a household devolves may profit by such knowledge.[17]

This prospect was not confined to the middle-class 'wife and mother'. In his opening lecture to the Ladies' Course of Physiology at the University of Edinburgh in 1871, Professor John Bennett stressed the need for better knowledge among women to safeguard the family: 'Among the working classes it is too frequently the improvidence and ignorance of women which lead to the intemperance and brutality of the men, from which originate half the vice and crime known to our police officers and courts of justice...'. In his view, 'science [physiology] contains for them, more than any other, all the elements of real and useful knowledge.'[18] Indeed, women's education in science was widely understood in terms of biology, physiology, and botany,[19] as the examinations of the Science and Art Department largely testified.

13 James Stuart, 'The teaching of science', in Josephine Butler (ed.), *Woman's Work and Woman's Culture* (London, 1869), 134.

14 *Ibid.*, 136. Cf. L. E. Becker, 'On the study of science by women', *Contemporary Review*, 10 (1869), 386–404.

15 Cf. letter from Mary Somerville to Josephine Butler (10 May 1869) quoted in Stuart, *op. cit.*, 138–140.

16 'Women's work in science', *Research*, I, No. 8 (February 1889), 130.

17 'The scientific education of women', *Nature*, 2 (1870), 117.

18 John Bennett, 'Physiology for women', *Nature*, 5 (1871), 73–74.

19 The London Society of Apothecaries, for example, announced an open competition in 1878 for women students of botany, based upon Hooker's *Science Primer-Botany*, and Oliver's *Lessons in Elementary Botany* (*Nature*, 17 [1878], 314).

Whatever the force of these arguments, from the 1850s onwards growing numbers of women demonstrated their firm intention of reading science. One route led through medicine. The long struggle for the admission of women into professional medicine has been well documented.[20] Some women found that medical schools offered a convenient pathway to an education in science. For example, the opening of the London School of Medicine for women in 1874 produced a variety of courses in science, most of which could be followed without a medical career in view, and *Nature* pointed out that 'there is no reason why women should not receive a scientific reputation and earn a fair competency by engaging in these studies'.[21] Opportunities to develop a 'reputation' were not, however, very great. The Armstrong College of Science at Newcastle (a constituent college of the University of Durham) allowed women to attend its lectures from its foundation in 1871 and although the first women did not register until 1880, by 1893 there were about 30 women taking courses in science.[22] At Manchester, however, women were admitted to the University in 1883, with the 'utmost reluctance'.[23] Indeed, according to Mabel Tylecote, 'it required much courage for a woman to enter [Manchester] and much effort to retain her place...'.[24] Between 1899 and 1914 only in Zoology (nine) and in Botany (30) did women graduates predictably outnumber men, and 'it was to the laboratories and advanced classes in science that there was the greatest delay in admitting women'.[25] Thus, in the same period, only four women took degrees in physics, and only 14 in chemistry.

In London, the situation was not much better. When London University opened its degrees to women in 1878, the Ladies Educational Association transferred all its interests to University College (including over 200 full-time students in all fields) which became the country's first co-educational establishment (even though admission to the College's Chemical and Physical [and other] Societies was denied to women).[26] In 1880, Bedford College for Women opened a physical laboratory under the direction of Oliver Lodge; this joined a chemical laboratory which had been in existence for some time, where biology courses were taught to women by the staff of St. Thomas' Hospital Medical School.[27] In South Kensington, the Royal College of Science attracted an appreciable number of women students (66 attending courses between 1881 and 1916), three quarters of whom qualified for the College's Associateship.[28]

Elsewhere in Britain, the picture was even less promising. Even where science was taught, few women took part. This was certainly well-known at the time, and was reflected in secondary level examinations. As *Nature* diffidently observed in 1878, the Cambridge Higher Local examination results showed that 'scientific study does not yet appear to have attracted a superfluity of women'. That year, only 30 of 500 women candidates took science subjects.[29]

20 On women and medicine see K. C. Mead, *A History of Women in Medicine from the Earliest Time to the Beginning of the Nineteenth Century* (Haddam, 1938); Enid Moberly Bell, *Storming the Citadel* (London, 1953); A. MacGehee Harvey, *Adventures in Medical Research* (Baltimore, 1976); E. G. Anderson, 'History of a movement', *Fortnightly Review*, 53 (1898), 404.

21 *Nature*, 10 (1874), 529.

22 C. E. Whiting, *The University of Durham, 1832–1932* (London, 1932), 153. Durham's Convocation voted in favour of admitting women to degrees in 1894.

23 Edward Fiddes, *Chapters in the History of Owens College and of Manchester University, 1851–1914* (Manchester, 1937), 39.

24 Mabel Tylecote, *The Education of Women at Manchester University 1833 to 1893* (Manchester, 1941), 55.

25 *Ibid.*

26 H. Hale Bellot, *University College London, 1826–1926* (London, 1929).

27 *Nature*, 22 (1880–1881), 181.

28 From the *Register of Old Students of the Royal College of Science* (London, various editions).

29 *Nature*, 18 (1878), 423. 21 candidates took botany, 26 physical geography and geology, seven geology, and nine chemistry.

To those wishing university degrees, London offered the greatest opportunity. Between 1879 and 1911 just over 600 women received the London B.Sc. degrees, over two-thirds of whom were 'internal' students of Bedford College, University College, Royal Holloway, the Royal College of Science, or of King's College, Birkbeck College, the South-West Polytechnic, or the Northern Polytechnic Institute.[30] In the same period, Manchester had fewer than 70 women science graduates. Given the opportunities of London, it is perhaps remarkable that women preferred Cambridge at all. Nevertheless, over 400 women between 1881–1916 did pass honours examinations in science at Cambridge.[31] *Why* they came, remains secondary in importance only to the fact that in going there, they opened a new chapter in the history of women's education in England.

The natural sciences at Cambridge

From its creation in 1848, and its first 'class' in 1851, graduates of the Natural Sciences Tripos (NST) came gradually to occupy important positions in English professional life. Created at a time of close parliamentary scrutiny into the contribution of the ancient universities to scientific education and research, the NST became a landmark in the history of 'reform' at Cambridge. As such, it played an important part in winning acceptance for the scientific movement in Britain.

Beginning with only six students in 1851, the 'naturals' as W. Aldis Wright called them, could not take a full degree in natural science until 1860. After that, their numbers grew slowly, partly because of the challenge of satisfying examiners in an enormous range of disciplines; and partly because of the colleges' ambivalent attitude towards the creation of lectureships and fellowships in the sciences. As the *University Reporter* commented in 1870, the colleges felt themselves 'guilty of extravagance' when they appointed scientific dons.[32] Much, moreover, remained of the traditional distrust of laboratory science, attributed (among others) to Isaac Todhunter, according to whom, it was unnecessary for students to see experiments performed, when the results could be vouched for by their teachers, all of them men of the highest character, and many of them clergymen of the Church of England.[33]

From the 1870s, after interminable battles, the number of science fellowships began to increase and laboratory facilities expanded. Undergraduate numbers steadily grew, until by 1914, an average of over 300 men and women were annually reading for the NST. In 1881, to appease both those demanding greater specialization, and those supporting a wide-ranging liberal education in science, the Tripos was divided into two parts. Thereafter, Part I retained a general nature, while Part II enabled a candidate to specialize.[34]

The division of the Tripos in 1881 coincided with the appearance of the names of women in the class lists. From the foundation of their colleges in the 1870s, students from both Girton and Newnham were able to attend University lectures, but their

30 Between 1878 and 1911 these women accounted for 16% of all London B.Sc. degrees. Taking only 'internal' degrees, women accounted for over 30% between 1903 and 1911.

31 Between 1881 and 1916, 299 women took only Part I of the NST, and 104 took Part II as well. In all, they accounted for 18% of the women who took a Tripos. K. T. Butler and H. I. McMorran, *Girton College Register 1869–1946* (Cambridge, 1948) and *Newnham College Register 1871–1950* (Cambridge, 1965).

32 *Cambridge University Reporter* (19 October 1870).

33 Cf. Isaac Todhunter, *The Conflict of Studies and Other Essays* (London, 1873), pp. 8–9, 16–21.

34 R. MacLeod and R. Moseley, 'Breadth, depth and excellence: sources and problems in the history of university science education in England, 1850–1914', *Studies in Science Education*, 5 (1978), 85–106.

existence was not officially recognized by the University.[35] Women could, however, ask to take a Tripos, and many did so during the 1870s. In science, their progress began in 1865, when it was announced that 'teachers and governesses' could be examined in science and given special certificates during an experimental period of three years.[36] Hitchin College, forerunner of Girton, formed a natural science class in 1871. On the advice of George Liveing, University Professor of Chemistry, emphasis was placed on chemical studies and classes were taken by J. W. Hicks,[37] and Mrs Whelpdale, 'a lady who has had experience in teaching the subject'.[38]

There was opposition, however, and not only from established (especially, and notoriously, medical) quarters.[39] Emily Davies, the founder of Girton, argued that the Moral Sciences Tripos and the NST were less demanding than either the Mathematical or Classical Triposes and discouraged her students from taking them. Nevertheless, Girton women sat the examination for the NST in 1874. Mary Kingsland,[40] passed with marks equivalent to a second class honours degree, and Jane Frances Dove,[41] passed at ordinary degree level. The first student from Newnham to enter for the Tripos, Amy Ogle, received first class honours (though not, of course, a degree) in 1876. Kingsland, Dove and Ogle were followed by eight other women, five from Girton and three from Newnham, before the Senate voted in 1880 to allow their names to appear in the Tripos lists.[42]

During the 1880s the numbers of Cambridge women reading for the NST grew steadily, reaching an average of 14 each year in the 1890s. After 1900, as table 1 illustrates, the NST failed to match in numbers of women the expansion in mathematics, history and medieval and modern languages. The women who took the NST before the First World War formed an impressive group,[43] but as our research has shown, there is little archival material from which one can infer their motivations or expectations. Partly, of course, this fact reflects the position then occupied by women in the University. Sources which are rich in information about male undergraduates (for example, the University *Reporter* and the *Cambridge Review*) provide little about

35 Women were not awarded degrees at Cambridge until 1948. See Rita McWilliams-Tullberg, 'Women and degrees at Cambridge University, 1862–1897' in M. Vicinus (ed.), *A Widening Sphere: Changing Roles of Victorian Women* (Bloomington, Indiana, 1977), pp. 117–145.

36 Examinable subjects were divided into six groups, one of which (Group E) comprised botany, zoology, geology and physical geography and chemistry. One subject in this group had to be taken by each student. (*Athenaeum* [31 October 1868], 571.) It was expected that these certificates would be used as evidence of competence to teach. London University, Queen's University, Belfast, and the University of Dublin all held similar examinations. *The Times* noted that 'there are rare cases in which no test or stimulus is needed to call out a pupil's powers. But, taking the mass of scholars, nothing, so to speak, brings them up to the mark except the certainty of an independent examination.' (*The Times*, [19 March 1873]).

37 Liveing, a staunch advocate of higher education for women, had intended to take the classes himself, but was prevented from doing so by ill health. Hicks was one of the first NST graduates.

38 *Nature*, 5 (1871), 49.

39 In 1920, *The Times Educational Supplement* noted that 'Medical men in all races fear women, but it is remarkable that this particular superstition should survive amongst their descendants at Cambridge.' Quoted in Rita McWilliams-Tullberg, *Women at Cambridge* (London, 1975), 179.

40 Kingsland was appointed assistant lecturer in natural science and mathematics at Girton, and later became a school teacher and social worker.

41 Dove became an assistant mistress at Cheltenham Ladies College, with special responsibility for teaching physiology.

42 The Senate vote on this issue was 398 in favour to 32 against, a 'very unexpected majority'. (Barbara Stephen, *Girton College, 1869–1932* [Cambridge, 1933], 75).

43 Undoubtedly a large number completed their Tripos studies but were unsuccessful in the examination; unfortunately we know nothing of these women. In the case of men, we know that between 1881 and 1904, 546 (25%) of a total of 2144 candidates took Part I of the NST and failed, J. A. Venn, *Alumni Cantabrigienses* (Cambridge, 1927–1953).

Table 1. Numbers of students taking triposes in sample years between 1882 and 1916

Tripos†	1882			1890			1900			1910			1916		
	Women	Men	% women of total	Women	Men	% women of total	Women	Men	% women of total	Women	Men	% women of total	Women	Men	% women of total
Mathematical	5	208	2·3	17	119	12·5	20	71	22·0	39	157	19·9	28	44	38·9
Classical	14	81	15·1	9	129	6·5	15	122	11·0	19	112	15·7	15	10	38·9
Historical	3	11	21·4	5	25	16·7	27	61	30·1	41	178	18·7	41	19	68·3
Medieval and modern languages				5	4	55·6	14	9	60·9	33	29	53·2	33	7	82·5
Moral sciences	1	3	25·0	2	6	25·0	3	12	20·0	4	13	13·5	3	4	42·9
Natural sciences	7	58	10·8	14	91	13·3	16	145	9·9	17	234	6·8	18	49	26·9
Total	30	361		52	374		95	420		153	723		138	133	

† Other Triposes attracted negligible numbers of women.

Source: The Historical Register of the University of Cambridge to the year 1910 and *Supplement, 1911—1920.*

women. Nor are the several Royal Commission reports on Oxford and Cambridge of great use in this respect. It was not until 1922 that the position of women was discussed in a parliamentary report, and in that instance attention was focused solely on the vexed issue of women's membership of the University and the consequences of university membership for the status of Girton and Newnham.[44]

College histories are likewise of little assistance. The growth of Girton was first documented by Barbara Stephen in her *Girton College 1869–1932* and *Emily Davies and Girton College*,[45] and has been discussed more recently by Megson and Lindsay[46] and by Bradbrook[47] in 1969. Newnham's history has been less frequently explored: the only full-length work remains Gardner's *Short History*[48] in 1921. All of these accounts address themselves briefly to a description of the struggle to establish the claims of women to a higher education. They are not concerned with the motivation and intellectual preferences of women students, nor with their roles once they had left Cambridge. Unfortunately, these issues were also neglected by Rita McWilliams-Tullberg, who has produced the most recent work on women at Cambridge.[49] Although her book includes a number of interesting incidental references to women and science, her chief concern lies in the story of the University's changing attitudes toward women from the 1860s.

To date, historians of education have scarcely begun to ask important questions about women in science, let alone produce conclusive answers. In view of this, a different approach may be required. Perhaps the most promising alternative involves focusing on the students themselves. Although few left diaries or autobiographies (and the *Life and Letters* style of Victorian biography appears to have been a predominantly male phenomenon) there are at least comprehensive registers of Newnham and Girton students, which contain a wealth of valuable information on the social backgrounds, education and careers of women who had studied at Cambridge.[50] With these, it becomes possible to use the techniques of collective biography[51] in order to examine more closely questions concerning the 'why', the 'how', and the 'results' of women's education both in science and in other areas. What follows outlines some of the progress we have made in this direction.

From their regional and social origins, their secondary education, their marital patterns and eventual occupations, there emerges a remarkable profile and a host of interpretative questions. Performing a 'census' on the women in question reveals that most, like their male contemporaries, came either from Lancashire and Yorkshire, the South East or London.[52] The Midlands, East Anglia, the North West, Scotland, and the West Country are significantly under-represented, as is generally true of the Cambridge population in the period. Until the turn of the century 'northerners' were dominant, although after 1900, numbers from London and the Home Counties increased considerably. Precisely why is not yet clear. Possibly part of the answer rests in the distribution of fathers' occupations.

44 *Royal Commission on Oxford and Cambridge* 1922, (Cmd 1588), x.

45 Barbara Stephen, *Girton College 1869–1932*, *op. cit.*, Note 42, and *Emily Davies and Girton College* (London, 1927).

46 B. Megson and J. Lindsay, *Girton College 1869–1959. An Informal History* (Cambridge, 1960).

47 M. C. Bradbrook, *That Infidel Place* (London, 1969).

48 A. Gardner, *A Short History of Newnham College* (Cambridge, 1921).

49 Rita McWilliams-Tullberg, *op. cit.*, Note 39.

50 *Girton and Newnham Registers*, *op. cit.*, Note 31.

51 Cf. Lawrence Stone, 'Prosopography'. *Daedalus*, 100 (Winter 1971), 46–79.

52 That is 75, 72, and 88 respectively. *Girton and Newnham Registers*, *op. cit.*, Note 31.

Table 2. Fathers' occupations of men and women students reading natural sciences, 1881–1916

Background	Women	Men
Church	24 (6·0%)	312 (11·6%)
Medicine	26 (6·5%)	375 (13·9%)
School teaching	20 (5·0%)	62 (2·3%)
University teaching	20 (5·0%)	24 (0·9%)
Industrial/commercial	99 (24·6%)	393 (14·6%)
Law	24 (6·0%)	95 (3·5%)
Civil service	8 (2·0%)	43 (1·6%)
Others	32 (7·9%)	54 2·0%)
Not known	150 (37·2%)	1334 (49·6%)
Total	403 (100·2%)	2692 (100·0%)

Source: Girton and Newnham Biographical Registers, and J. A. Venn, *Alumni Cantabrigienses*.

As table 2 reveals, both men and women reading for the NST tended to come first from industrial or commercial, and second from medical or clerical families. However, 'trade' families were about twice as prominent among the women. This pattern does not vary very substantially over the period 1881–1916. Why was this so?

There is some evidence that northern manufacturers approved of the NST for their sons;[53] they also seem to have approved of the possibility that science would be a sensible 'modern' preparation for young women.[54] To demonstrate this conjecture, however, it would be necessary to know more about both family attitudes and the preparation available in northern girls' schools. This knowledge must, in turn, be set in the context of College and University admissions. At a time when science had hardly secured a firm place in schools generally, Cambridge could expect few girls to arrive with a sound basic education in science.[55] Almost all girls reading science were educated either at private schools, or, eventually at Girls' Public Day School Trust (GPDST) high schools.[56] Women who succeeded in Part II of the Tripos were more likely to have been educated at a GPDST school than a private school.[57] Grammar schools, state schools, or private tuition accounted for only a small number of Cambridge women scientists.[58]

When they arrived, it was to Newnham, rather than Girton that most prospective science students came. Of the 403 women NST students, Newnham accounted for 247, compared with Giron's 156 and this proportion was the same for both Part I and Part II.[59] Newnham also achieved better Tripos results in science, possibly reflecting Emily Davies's preference for classics and mathematics.[60] At first both Girton and Newnham

53 Particularly as a route to a career in medicine—this accounts for no fewer than 45% of male NST graduates between 1851 and 1916. Venn, *op. cit.*, Note 43.

54 Family connections and wealth were essential factors in determining the social background from which students were drawn. High fees at Oxford and Cambridge were cited among reasons advanced for the foundation of Aberdare College for Women in Cardiff in 1885. *Nature*, 38 (1888), 257.

55 McWilliams-Tullberg, *op. cit.*, Note 39, 124.

56 The numbers were 157 and 177 respectively. *Girton and Newnham Registers, op. cit.*, Note 31. For the GPDST, see Josephine Kamm, *Indicative Past* (London, 1971).

57 50 and 37 respectively, out of a total of 104. *Girton and Newnham Registers, op. cit.*, Note 31.

58 That is, 10, 4 and 6 respectively. *Ibid.*

59 Girton produced 120 successful Part I and 36 Part II students; Newnham, 179 and 68. *Ibid.*

60 Among those who took Part I only, Girton students obtained the equivalent of 14% first class degrees, 48·7% second class degrees and 37·3% third class; while for Newnham the corresponding figures were 18%, 54·5% and 27·3%. In Part II, 24% of Girtonians obtained the equivalent of first class degrees, 47·7% seconds, 28·9% thirds; for Newnham, the figures were 28%, 47·8% and 24·6% respectively. *Ibid.*

Table 3. Subsequent occupations of women 'naturals' graduating between 1881 and 1916

Occupation	Year of graduation						Total
	1881–1886	1887–1892	1893–1898	1899–1904	1905–1910	1910–1916	
School teaching	14	34	39	45	53	34	219 (54·3%)
Medicine	5	4	5	3	11	19	47 (11·7%)
University teaching	4	8	6	9	8	12	47 (11·7%)
Social committee work		2		3	8	6	19 (4·7%)
Government research	1	2		3	2	6	14 (3·5%)
Others	7	7	11	8	9	15	57 (14·1%)
Total	31	57	61	71	91	92	403 (100·0%)

Source: Girton and Newnham Biographical Registers.

had to rely for their science on the goodwill of such enlightened dons as George Liveing, John Wale Hicks and Michael Foster. Between the 1880s and the turn of the century, however, a few women were offered teaching posts in the natural sciences. All (seven at Girton and nine at Newnham) were, with one exception, products of the NST.

The future opening to these 400 women was problematic. In retrospect, however, three occupations accounted for almost 80% of the total—medicine, school teaching and university teaching (see table 3). At first sight, the proportion of women who later did medicine, among those who followed any career at all, seems remarkably high. It is well known that many men in the medical community at Cambridge (and elsewhere) were hostile to women. This opposition derived from a host of contemporary attitudes towards the 'masculinization' of women who took up any professional career; medicine was perhaps merely more 'profane' than most. As Walter Rivington wrote in 1879:

> many of the most estimable members of our profession perceive in the medical education and destination of women a horrible and vicious attempt deliberately to unsex themselves—in the acquisition of anatomical and physiological knowledge the gratification of a prurient and morbid curiosity and thirst after forbidden information—and in the performance of routine medical and surgical duties the assumption of offices which Nature intended entirely for the sterner sex.[61]

As late as 1920, the defeat of a proposal to award degrees to women was seen by contemporaries to be due to 'science men in general and the medical profession in particular'.[62] Women were not allowed to attend lectures and dissections in the school of anatomy in Cambridge until 1911, and were not permitted to sit for the first and second (i.e. pre-clinical) MB examinations until 1916. Intending women doctors were therefore obliged to take the examinations of the London examining bodies, entailing both delay and expense.[63] Indeed, Elizabeth Garrett Anderson, England's first qualified woman medical practitioner, wanted her daughter to read for the NST, but advised her to enrol at London University, where she could go directly to a medical degree.[64] When the anatomy restrictions at Cambridge were lifted in 1911, the number of women taking the NST and going on to medicine, grew quickly; in the four years after 1912, 17 women took Part I and did so, compared with only 22 in the previous 30 years.[65]

61 W. Rivington, *The Medical Profession* (London, 1879), quoted in J. Martin, *Elizabeth Garrett Anderson* (London, 1965), 66.

62 McWilliams-Tullberg, *op. cit.*, Note 39, 179.

63 *Ibid.*, 126.

64 *Ibid.*, 146.

65 Few women who intended to enter medicine went on to take Part II—only eight between 1881 and 1916. *Girton and Newnham Registers, op. cit.*, Note 31.

It accords with common experience to expect Cambridge women to become teachers, in science no less than in other fields. Of the total, 54·3% (as against 12·6% of men) took up school teaching. Most of these 200-odd graduates were quickly employed, but the market seems to have saturated quickly. In 1916, the Board of Education's Consultative Committee on Scholarships for Higher Education found that of the 72 girls' schools investigated, only 39 had two full-time science mistresses.[66] Possibly these poor career prospects ensured that fewer Cambridge women science students were becoming teachers by the outbreak of the 1914–18 War. Certainly, many were beginning to find employment in government, in industry and in a range of occupations (such as social work or law) where their science degrees were of no direct relevance.

Overall, 117 (31·5%) became professional scientists. It can be argued that science, as distinct from medicine, offered a more 'pure', less profane and presumably therefore more 'acceptable' occupation. Yet, the 'purest' sciences attracted the smallest numbers. Whether because of, or in spite of, popular expectations and previous schooling, most 'scientific women' showed a marked preference for biology and physiology. Of the 117, the medical sciences accounted for 36·2%, other biological sciences for 29·1%; chemistry attracted 16·5%, geology attracted only 7·1% and physics only 6·3%. Altogether, these were among the first waves of women graduates to break against the wall of established professional attitudes. Revealingly, standard sources such as *Who Was Who* or the *Dictionary of National Biography* do not include many women in this group. Some of these, it is true, received posthumous praise in the obituaries of learned journals,[67] but few were elected to the leading scientific societies. There, again, they found considerable opposition. As late as 1889 a proposal put before the Geological Society to admit women as Fellows was rejected by 79 votes to 23. It was also decided that women would not be admitted to meetings even as visitors.[68] In 1908, confronted by a similar question, the Chemical Society decided to exclude women from the fellowship, but to admit them as subscribers, the appellation to be printed with a capital 'S'. This remarkable decision prompted one of the better poems to appear in *Nature:*

> Daughters of Eve! So zealous to pursue
> The work in Life by which you seek to live!
> When F.C.S. you claim, as is your rightful due—
> The "S" alone is what they, grudging, give!
>
> Be patient! Time is on your side.
> Reason and justice will your course defend
> Ignoble spite and arrogance of pride
> Shall meet their retribution in the end![69]

66 *Report of the Consultative Committee on Scholarships for Higher Education*, 1916 (Cd. 8291), para. 72.
67 For example 11 women (two who read Part I only and nine Part II) were recognized in *Nature*.
68 *Research*, I, No. 11 (May 1889), 197.
69 Letter, signed 'T', in *Nature*, 79 (1908–1909), 221.

Ultimately, some of these early pioneers did gain access to the leading scientific societies and two became FRSs.[70] But such worldly honours aside, Cambridge women of science built reputations in their fields. They scattered across the globe, often as doctors (like Norah Larwood who spent most of her working life in China, and Marguerite Bell, a missionary in India between 1919 and 1953). Violet Wainewright abandoned a scientific career to farm in Kenya; Beatrice Lindsay became the resourceful editor of the *Dietetic Reformer;* Hanna Hilton became a designer and craftswoman; Ethel Skeat became the first woman to be admitted to Munich University; Alice Taylor went first to the National Physical Laboratory and then left to become a potter; Eva Rayner became Mayoress of Cambridge. There were also perhaps 20 or so who actively supported the women's movement, including Muriel Brickworth, Secretary of the Bath Suffrage Society, and Mabel Hardie,a doctor recalled by her college register as a 'militant suffragist', who spent a post-graduate period in Holloway Prison. Women who became academics were not always confined to Newnham or Girton. May Holmer became Professor of Physiology in Delhi, and Elinor Phillips taught for 39 years at the Japanese Women's University in Tokyo.

Above all, however, one gains from the most cursory acquaintance with Cambridge women a picture of formidable personal determination. The will to succeed in medicine was important to some; great force of character was needed by others. There were equally formidable obstacles. As late as 1895, a university education was for women by no means a passport to a future profession.[71] For many of those a profession was an alternative to marriage. It is striking that over 60% of the women reading the NST never married and also that most of these, whatever their family economic circumstances, chose to follow professional careers until late in life. There is clearly no causal relationship between marriage and career, but most of the 40% who married, disappeared, to all intents and purposes, from the masculine world of professions and occupations. Given the competition for posts and widespread contemporary prejudices against professional women (often based on popular psychological *obiter dicta*), this is perhaps not surprising. Under the circumstances, it is remarkable that so many women persevered. Whether they did so in part because they reflected certain ascribed values that we can take as representative of 'Cambridge women', remains a large and open question.

Of those who did marry, perhaps the most visible were those who married Cambridge dons. The removal of celibacy restrictions on fellowships in 1882 was followed by a spate of marriages. Those women who read natural science and who

70 The attitude of the learned societies to the admission of women varied considerably. Women were admitted to the British Association for the Advancement of Science by purchased ticket after 1843. There was some early opposition on the grounds that the proceedings would degenerate into a meeting of dilettanti. Women were in theory eligible for election as officers and to serve on committees, although it was not until 1913 that Ethel Sargeant became the first woman president of a section (botany). (O. J. R. Howarth, *The BAAS: A Retrospect* [(London, 1931), 16, 98–99.] Both the Zoological and Royal Entomological Societies admitted women from their foundation in 1829 and 1833 respectively; the Linnaean Society from 1905; the Royal Microscopical Society from 1909; and the Geological Society from 1919. Women were first elected to the Royal Society in 1945. In that year Marjory Stephenson, an NST student from 1903–1906, and Kathleen Lonsdale, educated at Bedford College, London, became the first women Fellows. Stephenson, a student of Newnham, was a distinguished biochemist and had held a number of university posts at Cambridge. The second Cambridge woman to be elected to the Royal Society was Agnes Robertson (Mrs Arber). A botanist, she had been a Research Fellow at Newnham and had undertaken research in both botany and philosophy. She was elected in 1946.

71 Cf. Alice Gordon, 'The after-careers of university-educated women', *Nineteenth Century*, 37 (1895), 955–960. We are indebted to Dr Brian Harrison for this reference.

remained in Cambridge to occupy the new houses on the Madingley Road and Storey's Way included Edith Whitlock, who married J. S. Gardiner, Professor of Zoology; Anne Fletcher, wife of S. J. Hickson, University Lecturer in Zoology; and Laura Howell, wife of S. F. Harmer, Superintendent of the Museum of Zoology. These alumnae of Newnham and Girton were certainly an important and influential group, and their presence within the Cambridge academic community undoubtedly helped to advance the cause of women's education.

Conclusion

This brief sketch has outlined some of the characteristics of women 'naturals' and some of the consequences of their reading science at Cambridge. Other questions remain, and require closer attention. What part, for example, did this development play in the 'revolution of the dons'?[72] What can be said about the new 'student culture' among women (a subject remarkably neglected by recent accounts of 'male' Oxbridge)? What became of those who 'rowed', relatively speaking, as against those who 'read'? And what factors seem to have influenced the particular character of science education among the different disciplines? Certainly, the 'woman question' featured among the issues dividing the progressive and conservative factions in the medical community. The ostensibly 'liberal' caucus of science and medicine—men who held strong views on religious tests, on life fellowships and on celibacy—was not unequivocal when the issue of 'science for women' arose. Even as late as 1921, a letter to *The Times* complained of the 'stamping and bad manners to which some science men treated the girl students who joined them for lectures'.[73] It would be instructive to know more about the role of the 'reformers' among Cambridge physiologists and anatomists in altering this picture.[74]

It will also be important to disentangle the influences operating upon the young women who accommodated themselves to science. We recognize the enormous influence of particular academic families in encouraging their daughters both in going to Cambridge and in reading science. Thus, Vernon Harcourt, Karl Pearson, Arthur Schuster, J. S. Gardiner, William Bateson and A. C. Haddon, all Fellows of the Royal Society and distinguished scientists, had daughters who chose science at Cambridge. But what can be said of other non-academic family influence? There has been much speculation about the factors impelling fathers to acquire the ideals of sociability and service in 'reformed' Cambridge.[75] Much more remains to be said about the parallel personal influences acting upon women. Clearly, once the earliest generations of Cambridge women had opened the way, others more easily followed, and a familiar pattern of school-family-collegiate relationships developed. To understand the nature of these relationships it will be necessary to know more about the extent of science teaching available before this process began, and about the influence of Cambridge in modifying its direction. At the very least, we must expect to find that the scientific interests of Cambridge women began to crystallize long before they arrived at the Porter's Lodges in Sidgwick Avenue or Huntingdon Road.

72 Cf. Sheldon Rothblatt, *The Revolution of the Dons: Cambridge and Society in Victorian England* (London, 1968).

73 *The Times*, 25 October 1921.

74 See Gerald Geison, *Michael Foster and the Cambridge School of Physiology: The Scientific Enterprise in Late-Victorian Society* (Princeton, 1978).

75 Rothblatt, *op. cit.*, Note 72; Rothblatt, 'The student sub-culture and the examination system in early nineteenth century Oxbridge', in L. Stone (ed.), *The University in Society* (Princeton, 1975), I, 247–303.

VII

The 'Naturals' and Victorian Cambridge: reflections on the anatomy of an elite, 1851–1914

ROY MACLEOD & RUSSELL MOSELEY

> Other universities may set out to rear proconsuls, and succeed therein, but Cambridge leaves her actions to define her ambitions. If she prepares for nothing, there is nothing which she is not prepared for, if only opportunity offers. (W. Napier Shaw, *Nature*, 24 July 1913, p. 527)

The Natural Sciences Tripos (NST), one of the most successful and least studied innovations in the history of university science education in England, was introduced at Cambridge in 1848. From its first 'class' in 1851, the 'naturals', as W. Aldis Wright called them, began to occupy important positions in British public life. Created at a time of close parliamentary scrutiny into the English universities, the Natural Sciences Tripos became instrumental in winning acceptance for the scientific movement in Britain. In its inauguration and operation, the NST also raised a number of questions important both to the history of education and the history of science. These questions centre upon the adaptation of the English universities to the special imperatives of science, and upon the nature of English university science ever since. That these questions are worth considering bears directly upon the assumption that

> the university's own built-in conservatism and the pressures upon it to adapt to new external conditions is one of the most potentially illuminating, but most practically obscure, aspects of the process of historical change. [1]

Elsewhere, we have described the importance of the NST to contemporary educational debates on specialisation and the merits of examinations [2]. In a later study, we hope to show the significance of the NST to the history of Englsih liberal education and university life [3]. Here, we confine ourselves to an outline of the important prosopographical features of the 4223 men who took the examination between 1851 and 1916 [4].

THE NATURAL SCIENCES AND THE 'UNIVERSITY QUESTION'

There are perhaps four interconnected reasons why the introduction of the natural sciences through the NST had an important bearing on the social history of Cambridge. [5] These may be summarised briefly as dealing, respectively, with: (1) institutional influence, (2) recruitment and social access, (3) educational philosophy, and (4) professional interests. By 1848, Oxford and Cambridge had endured for nearly two decades the repeated fire of public criticism. In parallel with pressures for constitutional reform in the learned societies, in parliament and in the Civil Service, critics of the universities had focused upon their antiquated modes of government, the injustice of their religious tests, the shortcomings of college teaching and the misuse of college endowments [6]. As wealthy private institutions, closed to all but Anglicans, closely linked to the interests of Church and Crown, and with examinations set by traditional allegiances to mathematics and classics, they did not escape attack. Such attacks were not new, but the economic and political circumstances of the period gave them added force.

In this period of political turbulence, attention inevitably turned to the universities. Their role in English society was central and, in government, essential. The majority of Disraeli's readers, it was said, "could as easily fancy a county member not being a freeholder, as a university not being Oxford or Cambridge" [7]. Until at least the mid-nineteenth century, the pyramid of Oxbridge graduates was sharpening, while their influence increased [8]. Recruitment by the colleges was traditionally disposed towards leading families, and particular regional affiliations. The chief function of college teaching, and university examination, had become the training of men for the Anglican clergy. For how long could they and their universities remain unrepresentative of new middle-class, manufacturing, often Dissenting interests; how could they resist behind centuries-old statutes any attempt to refashion them as fully 'national' institutions? If Cambridge was to establish its influence among new middle-class economic interest, and maintain its influence over a reformed Parliament and civil service, new accommodations must be made. One route lay through a sequence of collegiate and university reforms. Another, through the admission of new studies, particularly in the natural and moral (including economic) sciences, and through the incorporation of these into the university's examination structure [9].

These accommodations must be seen in the light of quite dramatic changes in the size and complexity of the university. Between 1780 and 1900 there were two important surges in Cambridge undergraduate numbers, which nearly trebled the size of the university. The first began about 1800, and ended about 1820. The second began about 1860 and continued until the turn of the century [10]. Matriculations leapt from 450 per annum between 1850 and 1859 to nearly 1000 pa between 1890 and 1899 [11]. Many factors may account for these increases, including the end of the French wars, the desire of colleges to increase the economic multipliers working upon their Cambridge rents; the changing position of the Anglican Church and its demands for clergy; and the University's efforts to secure the loyalty of greater numbers of professional and manufacturing families lying outside the traditional landholding elite. The introduction of new studies in the 1840s was as much a consequence as a cause of the University's changing relationship with the country. There are, however, reasons to believe that the natural sciences, requiring disciplined hard work, professing a species of useful knowledge, and taught at first in a spirit of reverence for 'religion and sound learning', had special attractions for parents who, profiting from a second or third generation of capital accumulation, wished to have their sons acquire the social éclat of a Cambridge education. Thereafter, the steady development of the NST had a powerful self-promoting effect; men attracted to the Tripos gave it a good name and encouraged others to follow.

In this context, it would be naive to suppose that intellectual arguments for the introduction of new studies could be unaffected by larger political considerations. Science had by the 1830s become popular throughout Britain, and popular scientific education boasted a wide artisan and working class following. Baden Powell, writing from Oxford in 1843, saw the likely political result:

> ["if] scientific knowledge is rapidly spreading amongst all classes except the higher . . . [then] the consequence must be that that class will not long remain the higher" [12].

Accordingly, university reform in science teaching involved not only questions of educational principle, but also questions of power and the distribution of wealth. The new demands of science—for lecture rooms, laboratories, museums, and for the appointment of new staff—not only transformed the physical landscape of Cambridge, but also altered the balance of power between the colleges and the universities, and the respective roles in education and examination. This affected college life quite directly. As Arthur Engel has observed, college teaching in Georgian and early Victorian Oxbridge was viewed by dons not as a career in itself, but as a rung on the eccesiastical ladder [13]. If new studies were introduced, it was reasonable to believe that patterns of undergraduate recruitment would change, and that the social function of dons and colleges might alter accordingly.

These issues of recruitment, wealth and access cannot be separated from related questions, familiar to economic historians, of social composition and mobility. Could reforms in college recruitment and teaching, prompted by the introduction of new studies, open the University's gates to a wider section of English society? Indeed, could the 'reformed' University attract the sons of second and third generation industrial and manufacturing entrepreneurs, whose reproduction was vital to Britain's economic performance? Alternatively, would Cambridge, in its introduction of new studies, become accountable for the diversion of the country's intellectual elite towards modern professional careers, and away from more economically productive activities? [14]

The introduction of modern studies, and the natural sciences in particular, compelled fresh discussion of the University's educational objectives. Those objectives had been traditionally defined by the reproduction of an educated elite, largely directed towards the Church; and based on the ability of Cambridge to induce a sense of 'social fusion' among this elite, building character upon a common foundation in classics and mathematics. Given the advent of new studies, a diversification of objectives seemed likely; in the end they contributed to a quiet revolution in both the appearance and the reality of English liberal education. In the process, professional motivation and academic preferment began to occupy an important place in university history. In the natural sciences, and especially in medicine, professional interests acquired special importance. And with the expansion of the natural sciences came the diffusion of a 'scientific ethos' so pervasive that no academic subject could escape its influence. Historians of science conventionally agree upon the importance of the disciplinary 'research revolution' which gathered momentum within physics and physiology in Victorian Cambridge [15]. But the influence of this revolution was felt in all subjects. Where the introduction of science and scientific methods were defined as synonymous with the highest ideals of service and scholarship, the position of their advocates in the debates of the Regent House seemed almost invincible.

EXAMINATIONS AND ELITES

Upon these issues of influence, recruitment, access, educational philosophy and professional preferment, the history of university examinations offers an important perspective. Within scarcely fifty years, the NST both altered the organisation of science at Cambridge, and defined new, self-exemplifying characteristics among a substantial section of the student population. For all its obvious importance, the relation between the introduction of examinations and the production of elites has been little studied [16]. On the one hand, the rate of progress within scientific knowledge is rarely considered a function of the examination system obtaining at a given time; on the other hand, any possible relationships existing between early education and later eminence have been much less interesting to historians of science than to social or educational historians. Certainly, it is commonly observed that the study of scientific elites is much less well developed than the study of political and social elites [17]; and for no university is this more true than for Cambridge. In 1950, as part of a series of studies on social class, H. Jenkins and D. Caradog Jones attempted a broad historical survey of Cambridge graduates, 1752–1900. This study emphasised both the resistance of Cambridge to any change in its social structure, and the very gradual diminution of clerical and landed class dominance in its affairs [18]. Shortly afterwards, an American study concerned with social mobility through university access concluded that the English universities were successful in making political accommodations with the country without undertaking substantial alterations in their patterns of recruitment [19]. Both studies suggested that changes did take place, but since both took university populations in aggregate, neither was concerned with the relevance of educational innovations to the course of institutional development, or to the production of new elites within the given social pyramid.

In the last few years, particularly following the publication of Lawrence Stone's Princeton seminars on 'The University and Society' [20], it has become clear that university history cannot be complete if it ignores the nature of the different elites they nourish. It is also clear, however, that few university (or Oxbridge college) histories have adopted this approach. In the context of Cambridge, it is both difficult and imprudent to discuss the connotations of class, mobility and occupation without a clear grasp of the structure of Cambridge university life. This is obviously true of prevailing social categories, of sizars, fellow commoners and noblemen; it is arguably also true of the academic categories of 'honours' men and 'poll' men. The examination structure was the mechanism by which intellectual and eventually professional distinctions between men were made and maintained.

In introducing the NST, a new set of elites was produced, which, in the fullness of time, began to reproduce itself, much as the Anglican clergy had done since the Reformation. This possibly unintended consequence was eventually to provide what C. P. Snow later called the "new man" of English science [21]. But did this 'new man' differ from his contemporaries in some Arnoldian sense? What social characteristics define this 'new undergraduate'? We can agree with Lawrence Stone that

> it is by asking these kinds of questions, rather than by thinking up macroscopic theory or by studying the internal administrative history of the individual institutions, that new understanding of the shifting relationships of the university to society is most likely to emerge. [22]

The model for the NST was inevitably the Mathematical Tripos, which, dating from the middle of the eighteenth century, was the pivot of Cambridge education [23]. The NST was to be 'natural' in the sense of complementing the Newtonian sciences (notably mechanics and optics) which were already represented in the Mathematical Tripos. In the language of Whewell, an honours degree in the natural sciences, would demonstrate a candidate's ability to grasp the 'progressive sciences', and in so doing complete his liberal education [24]. As such, honours in the NST (and in the Moral Sciences Tripos, begun in the same year) were at first available only to graduates already grounded in the 'permanent' studies of mathematics and the classics, who had previously taken the BA degree [25]. In the same idiom, the NST was founded upon the assumption that different kinds of knowledge could be examined for specific purposes—principally, to discipline the mind and to develop the faculties of memory and reasoning. Beyond these, there was a higher moral purpose. At a time when a large proportion of men took only poll degrees, or failed to take degrees at all [26], the incorporation of the natural sciences into written university examinations would not only encourage the instruction of new subjects; they would also encourge undergraduate effort and achievement. Academic preparation was to be coupled to a new ethic of work and a new ideal of service.

The Tripos began with only six students in 1851, and numbers grew slowly, partly because of the unprecedented challenge of satisfying examiners in a Whewellian range of subjects. At first, there were five subjects in what we have called the 'Old Exam' (see Table I). By 1880, there were 12 papers in eight subjects [27]. As Alex Hill, later a

TABLE I. *Graduates of the Natural Sciences Tripos* 1851–1916

Examination	*Period*	*Men*	*Women*
'Old Exam'	1851–1881	405	–
Part I only	1881–1916	2954	299
Part II	1881–1916	864	104
	Total	4223	403

Source: *The Histological Register of the University of Cambridge to the Year* 1910, *and Supplement* 1911–1920.

university lecturer in medicine, recalled, "A candidate's place depended upon his aggregate of marks . . . and to secure a first class he must show that his knowledge, like that of the Master of Trinity, ranged from Zoology to Mineralogy" [28]. Such a range of subjects inevitably led to charges of superficial learning. Moreover, as the *University Reporter* commented in 1870, some colleges would consider themselves "guilty of extravagance" in appointing scientific dons. Much remained of the traditional disdain of laboratory science epitomised by Isaac Todhunter, according to whom it was unnecessary for students to see experiments performed, since the results could be vouched for by their teachers, all of them men of the highest character, and many of them clergymen of the Church of England [29].

From the 1870s, however, the number of fellowships in the natural sciences began to increase, laboratory facilities expanded, and the number of men reading for the Tripos steadily grew; in the first five years of the decade there were 79; in the next five, 125. In 1881, in an attempt to reconcile the competing demands of 'breadth' and 'depth', the NST was divided into two 'Parts'. Part I consisted of three days' written work (six papers on eight subjects), and two days, practical examinations. From 1899 to 1916 the examination was lengthened to seven days with two papers in each of the eight subjects. Part II (for which success in either Part I of the NST or Part I of the Mathematical Tripos was a prerequisite) presented a wider choice of advanced questions and encouraged a higher degree of specialisation [30]. By 1881, when the Tripos was divided, 405 men had graduated with honours in the natural sciences. Whereas less than 2% of Cambridge undergraduates read for the NST in 1861, the proportion had doubled by 1880, and had reached 18% by 1910. From 1883 the NST had annually more graduates than the Mathematical Tripos. The now familiar division of labour (mathematical physics to the Mathematical Tripos; experimental physics to the NST) helped focus the University's demand for physics teaching, which in turn greatly stengthened the institutional position of the Cavendish. In 1910 a pre-war highwater mark was reached when 216 'naturals' accounted for almost one-third of all Cambridge honours graduates that year [31].

Perhaps equally important, by the last quarter of the century, the NST had provided a successful disciplinary model, which was adopted with little modification by other subjects [32]. Largely through its influence Cambridge regained its lost position in the history of science education [33]. Academically both demanding and far-reaching, the NST acquired an intellectual reputation possibly more distinctive than the London BSc, and an image far more charismatic.

THE 'NEW UNDERGRADUATES'

Our analysis suggests that the 'naturals' predictably shared in many ways the same student culture as much of the Cambridge student population, or at least that proportion of it reading for honours. Accepting social continuities in college admissions throughout this period, this was unavoidable. However, there are two senses in which it is useful to refer to the 'naturals' as 'new undergraduates'. First, their social and school backgrounds differed in certain important respects from their contemporaries. Secondly, especially from the 1880s, they imported qualities relatively new to Cambridge. By the end of the century, it is evident that 'naturals' were less likely than their contemporaries to be sons of clergymen, lawyers or landed gentry; they were less likely to have been educated at a Clarendon school; they were more likely to be the sons of medical men, or have fathers in industry or commerce [34]. If, as is likely, the NST was attracting, by the 1880s, students who might not otherwise have gone to Cambridge, the NST may well have influenced the overall composition of the Cambridge 'student body'.

Between 1851 and 1916, the regional origins of NST students remained relatively constant. The only major change occurred in the 1870s when the numbers of sons coming from families resident in the north of England rose appreciably. For the NST population as a whole, students from London comprised a large percentage (slightly under one-fifth) of the total, while the south-east of England contributed a similar proportion. Regional differences between graduates taking Parts I and II between 1881 and 1916 were not particularly marked, although the latter attracted a higher percentage of northern, midland and London students; the former a greater proportion from the south-east [35]. After the 1880s, medics' sons and men with fathers in industry, business and commerce were more dominant [36], while the proportion of sons of clergymen (while still numerically significant) continued to decline steadily. Certain social groups remained almost entirely unrepresented. Very few 'naturals', for example, were drawn from farming families; George Liveing complained in 1879 that despite the undoubted benefits of a Cambridge education "farmers cannot be bribed by the scholarships, freely offered by many county councils to take advantage of it" [37].

During the 1860s and 1870s, Lyon Playfair, Henry Roscoe, T. H. Huxley and other educational reformers argued that Oxbridge made no attempt to adapt itself to the interests of 'middle-class' parents—particularly professionals and those in manufacturing and commerce [38]. Trinity, for example, stood apart in taking far fewer NST men from industrial or commercial families by comparison with other colleges. The possibility of attracting larger numbers of middle-class students to both universities was discussed throughout the period. C. S. Roundell, a Fellow of Merton, admitted to the Select Committee on the Oxford and Cambridge Universities Bill in 1867 that

> "We fail to reach the commercial and mercantile class", and attributed this failing to the predominantly classical nature of the studies and "the exclusiveness of the place . . . the University is in the hands of a privileged class, and a privileged church." [39]

William Bateson of St John's claimed that the problem of attracting more sons of manufacturers lay in "the desire to put young men into business early in life" [40].

The introduction of the natural sciences to Cambridge did not serve necessarily as a 'drawing card' for parents wishing a professional career for their sons. On the contrary, through the 1840s it was conventional wisdom that

> should a father observe in his gifted son a tendency to physical philosophy, he anxiously endeavours to dissuade him from this career, knowing that not

> only will it tend to no wordly aggrandizement, but that it will have the inevitable effect of lowering his position in what is called, and justly called, good society . . . [41]

However, by the 1870s, with the growing recognition of science as an occupation, and with new professional careers opening in scientific medicine, it could be argued that the NST was perceived as a route to a 'profitable' degree. Pressures for a more 'useful' education were felt in many quarters. Writing in 1877, Henry Latham argued that

> In our time the attention of parents and some of the students is pretty closely directed to the question of profit. This arises partly from the increasing severity of the struggle to "get on," partly from the fact that a much larger proportion of our students is now drawn from a class used to take "business" views of matters than was nomally the case, and more than all, from the competition for scholarships, when at school, having brought pecuniary rewards much under the notice of young people. [42]

From the 1860s, middle-class parents concerned with 'getting on' took increasing advantage of University College, London, and later the provincial university colleges. Sir Henry Roscoe, speaking of industrialists from the north of England, claimed that there

> were few engaged in that district in any large way of business in which chemistry plays a part who did not show their appreciation of the value of scientific education by sending their sons or their managers to learn chemistry at Owen's College. [43]

Oxford's Professor of Natural Philosophy, Bartholomew Price, agreed. When asked where the "sons of rich manufacturers in the north" were educated he replied that "A great many of them, I think, go to Owen's College, Manchester" [44]. At Cambridge, George Liveing was aware of this. In 1868 he told the Select Committee on Scientific Instruction:

> We want a different class of students here besides those who now come; it is not likely, or desirable, that the actual artisan class should assemble here in numbers, but the master manufacturers might well be educated here. At present, the tradition and habits of the place are such that few who are engaged in trade think of sending their sons whom they intend for like pursuits. I want to see these traditions broken through, and many of that class coming here. [45]

Liveing's remarks were aimed at least in part at the Tests, which had systematically deterred Dissenters, Catholics and Jews, and which almost certainly had, at the very least, discouraged boys from the industrial cities. Following the Cambridge University Act of 1856, Dissenters could take degrees (except degrees in theology) without subscribing to the Thirty-Nine Articles. However, they could not hold college fellowships until 1873, and this bar to academic preferment provoked widespread concern [46]. Although it is difficult to establish, outside such celebrated individuals as Neville Goodman, whether Dissenting parents actually preferred Cambridge over, say, Owen's College or UCL, the abolition of the Tests clearly helped, in Goldwin Smith's phrase, to "restore the unity of the nation". Almost by definition, the existence of the NST thereafter helped, as Liveing put it, to "open the door [of Cambridge] as wide as possible" in social terms.

However, Liveing also admitted, in his evidence to the Devonshire Commission in 1870, that the success of Cambridge in attracting an audience of 'modern students' depended largely on the ability of schools to produce suitable men. Indeed, through the 1870s, the number of undergraduates reading natural science remained low, and it was clear that not only were such studies failing to prove attractive to those classes which traditionally sent their sons to Cambridge in the expectation of a liberal education, but that many men made a poor showing in the examinations themselves. Michael Foster claimed that students' poor performance was not due so much to bad college teaching or idleness, but to the fact that they

> for the most part come up to the University, not only wholly ignorant of even the elementary facts and ideas of chemistry and physics, but with their minds unprepared by any adequate discipline for the reception of the truths of experimental science. [47]

His criticisms found their targets among the public schools which traditionally sent boys to Cambridge. The 23 'greater' public schools [48] produced only about one-quarter of all the 'naturals' graduating between 1851 and 1881. But over one-third of those who failed Part I between 1881 and 1904 (some 546 candidates) had been educated at major public schools.

It can be inferred that the introduction of natural sciences at Cambridge exposed one of the major weaknesses in the University's customary feeder schools. The high failure rate among public school boys suggests either that the schools selected academically weaker boys for science, or that school science teaching gave a poor preparation. In all probability both explanations are valid. The schools' lamentable provision in science teaching, and some of the reasons for it, are well known [49]. Indeed, as late as 1901, six of the most prestigious public schools won only a handful of certificates in science (see Table II).

TABLE II. *Oxford and Cambridge Schools Examination Board Certificates awarded to six public schools*

	Total number issued	Total in science
Eton	45	0
Winchester	48	4
Rugby	58	6
Cheltenham	32	7
Clifton	23	8
Marlborough	18	4

Source: *Nature*, 66, p. 459, 1902.

From the early 1880s, however, there were signs that far more boys wishing to read science were coming from 'lesser' public schools, minor private schools and middle class grammar schools [50]. Between 1881 and 1904, slightly more than 30% of successful candidates who read only for Part I came from the greater public schools (compared with just over 25% between 1851 and 1881); but less than 20% of the men who went on to read for Part II came from the greater public schools. Grammar schools provided only 8% of men who stopped at Part I, but nearly 16% of those who went on to Part II.

This evidence suggests that if success in the demanding Part II of the Tripos was in any way attributable to previous teaching in science then it is clear that a distinction between the 'two cultures' began at school and was thence reinforced at university. Other Triposes continued to recruit most of their candidates from the greater public schools. Jenkins & Jones showed that about one-half of all Cambridge undergraduates between 1850 and 1899 attended one or more of these schools compared with less than one-third of all NST candidates over the same period. Conversely, proportionally more NST men had been educated at grammar and at lesser public and private schools [51]. In view of this it is possible that Liveing's call to open Cambridge to a wider middle-class intake was as much a response to the perceived intellectual demands of the sciences, as much as a wish to cultivate the sons of the bourgeoisie.

Although there was a striking improvement in the quantity, quality and backgrounds of science teachers between 1850 and 1914, it is clear that their recruitment remained a difficult problem [52]. Eventually however, NST graduates played a major role in staffing schools. Science teachers in the public schools during the 1850s and 1860s tended to be Cambridge Wranglers who were employed to teach mathematics, but whose experience of natural philosophy in the Mathematical Tripos gave them a grasp of physics [53]. By the 1870s and 1880s, however schools were employing men with degrees in the natural sciences; and NST graduates began to turn increasingly to school science teaching.

The vicious circle of low recruitment and poor preparation could only be resolved by fundamental reforms in both the schools and the colleges. Financial incentives in the form of exhibitions and scholarships were not in themselves sufficient inducements. In 1872, for example, scholarships in natural science were offered by Clare and Emmanuel but no elections were made. It was reported that "at neither of the colleges did candidates present themselves whose attainments, in the opinion of the examiners, entitled them to receive the distinction. The number of competitors was but small in each case, in one only three" [55]. However, it was argued that not until the colleges showed still greater commitment to the new studies could the situation improve. When T. W. Danby became the first 'natural' to win a fellowship in science (at Downing in 1867) an editorial in the *Lancet* cynically observed that it was the first such award in seventeen years:

> No wonder that the numbers who go in for the examination are few; no wonder that they do not show equal merit with the candidates for the other triposes if it be so. The wonder is that the Tripos keeps on its legs at all, and that any good men are found to devote themselves to natural science. Those who seek reward as well as honour, and they constitute the mass of hard-working students, see that the path of Natural Science will lead to little or nothing, and turn in the better-paid ways of Classics or Mathematics. [56]

By the early 1880s, this position had begun to change. Forming 'bridges' between the colleges and the University, several dons began to crystallise a new professional and educational image for the natural sciences. Of these, three deserve particular mention—Coutts Trotter and Michael Foster of Trinity, and Donald MacAlister of St John's.

Trotter graduated as 37th Wrangler in 1859 and became a Fellow of Trinity two years later. After studying experimental physics in Germany under Helmholz and Kirchhoff, he was appointed a lecturer in physical science at Trinity in 1864. For the next 10 years, first as tutor and later as Vice-Master, Trotter worked for collegiate and university reform and the expansion of teaching in the natural sciences. Writing at the turn of the century, John Willis Clark believed that "no one has been so completely identified with what may be termed modern Cambridge", while Michael Foster claimed that "the development of natural science at Cambridge is coincident in time with, and in great measure due to, Trotter's academic activity." [57].

Trotter and Foster were both members of the group Winstanley has called the 'Trinity Reformers.' Foster was not a Cambridge graduate, but from his arrival in 1870 as Trinity Praelector in Physiology he threw himself behind the reform movement. An entire generation of new biologists, including Walter Gaskell, J. N. Langley, Sheridan Lea and F. M. Balfour, were among his students. His close colleague, Donald MacAlister, appointed to St John's as a lecturer in medical practice in 1880, and admissions tutor between

1893 and 1905, was also a staunch advocate of the NST, and by 1914 his college had no fewer than ten dons in the natural sciences—two each teaching in chemistry and geology, and one each in physics, botany, zoology, physiology, minerology and human anatomy. Outside Cambridge's largest and richest colleges the claims of natural science were less well acknowledged, with striking exception of Caius. Indeed, between 1851 and 1916 we find Cambridge science (in teaching strength and undergraduate numbers) dominated by Trinity, Caius and St John's. These three colleges produced 18·7, 16·8 and 13·7%, respectively, of all NST graduates. The only other colleges which challenged this overall hegemony were Christ's and Emmanuel, with 7·8 and 7% [58]. No other college claimed more than 5% of the whole, Corpus, St. Catharine's and Magdalene each having less than 1%.

SUCCESS BEYOND THE CAM

The NST provided many different avenues to professional advancement, and these differences became more pronounced after the Tripos was divided. Given the falling rate of overall Cambridge graduate entry to the clergy, the 'naturals' featured prominently in the 'secularisation' of the student body. While about 20% of 'naturals' graduating in the period 1851–1881 entered the Church (a figure slightly below that associated with the Mathematical and Classical Triposes), the proportion of clergymen among Part I and Part II 'naturals' had fallen to 2% by 1905. Instead, among those who read for Part I alone, there came such men as Will Spens, later Master of Corpus Christi College; Edward Wilson who accompanied Scott to the Antarctic; E. G. Browne who became a professor of Arabic; and R. A. S. McFie, a world authority on gypsy matters. The range of employments reflect the wide world in which Cambridge was to play a part, if we consider H. E. I. Taylor, a former railway worker who entered Cambridge at the age of 30; D. Dewar, later a judge in India and Vice-President of the Evolution Protest Movement; and P. A. Cobbold, who spent years investigating ways of fighting forest conflagrations only to meet his end in an unexpected fire. Only the broad range of occupations, not their individual character, seemed predictable. But this was true of Cambridge men in general.

In another important respect, the contribution of the 'naturals' was both singular and unexpected. Perhaps more by accident than design, the NST became an important avenue for men entering the medical profession. This occurred at a turning point in the history of English medical education and practice [59]. In 1850, only 157 of the 1717 university-educated doctors in Britain held Oxbridge degrees; and Scotland and London dominated the scene. Hospital medical schools taught and examined, but did not license; the medical corporations examined and licensed, but did little teaching; while only Oxford and Cambridge could teach, examine and license. The NST consolidated the strengthening connection between Cambridge and the British medical profession. Between 1851 and 1881, 32% of NST graduates became medical men; between 1881 and 1904, this proportion rose to 57% of those taking Part I and 29% of those taking Part II. Altogether, over 50% of all undergraduates who took the NST between 1851 and 1916 did so à cheval on the way to medical practice. It was more than a 'survival tactic'. In fact, medicine offered the first, and throughout the period, the most profitable, if unintended, professional objective for the natural science student. Certain colleges thrived on the medical connection. Over 62% of the NST graduates of Caius between 1851 and 1881 became medics, as did 77% of its Part I men between 1881 and 1904. Trinity and St John's were some way behind, but the number of medical men each produced was large—

24% of men reading honours in both colleges between 1851 and 1881; and 55% of St John's and 53% of Trinity's men reading only Part I between 1881 and 1904. Cambridge medicine had firmly established itself as a 'family concern'. Between 1851 and 1904, 14% of fathers who sent sons to Cambridge to read natural sciences were doctors; and 80% of their sons themselves became doctors. In the hands of Michael Foster, George Humphrey, Donald MacAlister and G. D. Liveing, these graduates became an instrument in the vanguard of medical reform; for them, the NST provided a sound basis for future research (17 of those who became doctors also became Fellows of the Royal Society) and social recognition (14% were mentioned in *Who's Who*). At the same time it became increasingly clear that Cambridge could not ignore either the development of experimental medicine, or its interpenetrations with professional practice.

Outside medicine, the greatest importance attached to the new 'scientific class' [60]. Arguably, as career prospects improved, it appeared reasonable to stay on for Part II. Considerable numbers taking Part II also took up university posts (26%) and school teaching (16%). Government attracted a fair number (8·6%) but industry, the church and law, far fewer (each less than 4%). The important new category was that of the university scientists. Of those men taking Part II and going on to practise science, 10% became chemists; 8% became physicists; 7% geologists; and 6%, botanists, with physiology, zoology and biology each accounting for another 9%. Among the physicists were Charles Chree, A. S. Eve, T. C. Fitzpatrick, W. C. D. Whetham, C. T. R. Wilson, R. J. Strutt (the fourth Lord Rayleigh), J. A. Crowther, W. L. Bragg, George Paget Thomson and E. V. Appleton. Men who later restored the fortunes of British physiology (including H. K. Anderson, Joseph Barcroft, John Mellanby and C. S. Sherrington) passed through the same portal.

It is possible to capture the 'essence' of these new elites by considering four profiles, representing computer-based statistical averages drawn for men who took the NST.

Graduates between 1851–1881

Graduate A was the son of a medical practitioner from the north of England who attended a lesser public school and who went up to Trinity. After taking a second class degree, he would spend some time at a London teaching hospital before entering the medical profession. He stood a poor chance of achieving election to the Royal Society or of inclusion in the *Dictionary of National Biography*.

Graduates taking only Part I between 1881–1904

Graduate B was, similarly, the son of a medical practitioner from the north of England. He, too, attended a lesser public school and entered Trinity College, where he obtained a second (or equally likely, a third) class degree. Like his predecessor he would spend some time at a London teaching hospital before entering the medical profession. His chances of inclusion in either the *Dictionary of National Biography* or *Who Was Who* were extremely slight, while the possibility of his election to the Royal Society was negligible.

Undergraduates taking only Part I, but failing, between 1881–1904

Graduate C shared the same social background as his contemporaries, but was as likely to have attended one of the greater public schools as one of the lesser. Upon failing his Tripos examinations, he was awarded an ordinary degree, but this would not prove a barrier to his subsequent entry into the medical profession. His chances of inclusion in either the *Dictionary of National Biography* or *Who Was Who* are only marginally more flattering than the negligible odds against his election to the Royal Society.

Graduates successfully taking Part II between 1881–1904

Graduate D came, in this period, from the north of England and was only slightly more likely to be the son of a clergyman than of a commercial or business family. He attended a lesser public school, and either Trinity or St John's, obtaining first class honours in both Parts I and II. His subsequent career would be academic, at a university or university college. He stood a 3-to-1 chance of being remembered in *Who Was Who;* and less chance (9-to-1) of inclusion in the *Dictionary of National Biography;* one in seventeen of his contemporaries would be elected to the Royal Society.

By the end of the century, the 'naturals' were a major factor both in Cambridge life [61] and in its wider social influence. Certainly, their presence was a feature of the University's strenuous efforts to increase its endowments for science, and to win jobs for its graduates in new domains [62]. The Cambridge University Appointments Association (later Board) was established in 1899 with a view to "facilitating the employment of Graduates of the University in the various professions and occupations for which they are fitted by their University training" [63]. The reports of the Board included a section which enabled Cambridge graduates to advertise their skills and it is striking to see the frequency with which Part I graduates sought for industrial posts [64]. Certainly, within another generation, the University captured a larger share of the manufacturing classes. By 1936–37, 46% of Cambridge matriculations in all fields were from business backgrounds; only 2% were from landowning families [65].

In this context, the NST had preserved the appearance, if not the reality, of a liberal education. But by the outbreak of war in 1914 [66], the career expectations of undergraduates were undergoing a drastic change. For many, science had become of economic and professional significance, and Cambridge saw the emergence of a class of junior scientists seeking a future, and a future for science, quite different from that contemplated by the 'reformers' of 1849, 1870 and 1881 [67]. Consider William Bate Hardy, an NST graduate of 1884, who became the first Chairman of the DSIR's Food Investigation Board, Superintendent of the Low Temperature Research Station at Cambridge, and "an advocate of the need for more science in industry and . . . apostle of cooperation in research between the members of the British Commonwealth of Nations" [68]. The career of A. E. Shipley, who graduated with Hardy, brings the story full circle. With his appointment as Master of Christ's in 1910, and as Vice Chancellor in 1917, the 'revolution' seemed complete. The 'naturals' had arrived.

CONCLUSION

"The university", Lawrence Stone has written,

> has not been a Parsonian functionalist institution responding slavishly to social needs. Nor has it been a Marxist superstructure, automatically providing the ideological props for the group which currently controls the means of production. [69]

Cambridge in the late nineteenth century witnessed important shifts in its academic character and in its student population. The teaching of these natural sciences both prompted and profited from wider transformations in the social functions of the University. As we have elsewhere shown, the advent of the 'progressive sciences', introduced for reasons both academic and political, did not in itself divide the education of cultured gentlemen from the training of specialised experts. That came increasingly as the consequence of specialised examinations, of which the sciences merely provided an elaborate (or, as contemporaries described it, an 'aggressive') extension [70]. However, acting through the NST, the sciences did bear upon Cambridge in three important ways, quite aside from their implications for the progress of knowledge. First, they either led, stimulated or actually accounted for the greater part of any changes in social and occupational mobility observed in Cambridge during this period. Secondly, they created a new form of 'sponsored mobility'—especially within the medical and the new scientific professions—and established patterns of social reproduction within a powerful and growing section of the country's intellectual elite. As such, to borrow the phrase of John Roach, the NST added a new dimension to the role enjoyed by Cambridge in shaping the 'national intelligentsia' [71].

Thirdly, following the division of the Tripos in 1881, the NST contributed to a climate in which undergraduate education became closely associated with advanced study. As Michael Foster observed in 1888, the most remarkable difference in the life of Cambridge between 1868 and 1888 was the number of young Fellows doing research [72]. Research had not yet ceased to be a 'personal idiosyncrasy' and become instead "the normal completion of a University training" [73]. But by 1895, the creation of a category of 'Advanced Students', and the development of a research tradition at Cambridge, built in large measure upon students completing the NST, tended to strengthen the links between dons, demonstrators and undergraduates. By the 1890s, the popular image of the 'natural' as a 'liberally educated scientist' was not an academic contradiction, but a perceived fact. Indeed, at least as late as 1914, there remained a greater social distance between the Cambridge scientist, and his contemporary reading science at Manchester or UCL, than between him and his college contemporary reading, say, history or law at Cambridge. Writing in 1930, in his chiefly critical account of English university science, Abraham Flexner conceded of Cambridge that "there is no sounder or saner type than this unique combination of scholar, scientist, and man of the world . . . that the university produced, without having the organisation to produce the type in large numbers" [74]. The creation of this 'new man', whose social allegiances were uniform with the expectations of the country's governing elites, was among the most important social consequences of the NST. Hopes, or fears, that the growth of natural science would threaten the universities' traditional patterns of recruitment, for the pre-war generation at least, proved groundless. Cambridge, through its college admissions, accommodated the radical threat, or promise, posed by experimental natural science, and absorbed, until well into the new century, the sciences into a pattern of education predicated on the principles of 'religion and sound learning' and the traditions of the liberal arts.

This is not to deny that the teaching of the natural sciences profoundly influenced the intellectual character of Cambridge; eventually, the success of the NST and its student culture helped to undermine the dominance of revealed religion. Commitments to the Church were altered perceptibly by the demands of this new 'clerisy'. But it is in social terms that the 'naturals' may prove most interesting. Their history arguably demonstrates a form of 'lateral mobility' within the socio-economic groups that sent honours men to Cambridge. This lateral movement-away from the clergy towards medicine, away from landowning and towards manufacturing and business families—was well underway by the late nineteenth century. While it would be impossible to demonstrate cause, the movement was almost certainly assisted by the new accommodations between Cambridge and research, and between the NST and medical practice. By the 1890s, parents, headmasters and colleges increasingly saw the NST as an acceptable route to a professional career. At the same time Cambridge natural scientists offered leading examples of those dons, who, as Sheldon Rothblatt has eloquently described, demonstrated a new 'ideal of service', reflecting greater responsibility in teaching and in the reward of scholarship [75]. As

William Bateson agreed, the 'revolution in science' and the 'ideal of service' were closely connected [76]; blessed by the colleges, they were wedded in the NST.

The question however remained: whether the intellectual and social cohesion produced by the colleges would survive the particular specialised efforts required by the Tripos and the particular demands made through the NST upon the schools. Moreover, the implications of specialisation were moral and social, as well as intellectual; indeed, by the turn of the century, the 'naturals' had acquired an imprecise but distinctly 'different' image. Thus, Oscar Wilde in 1891 had Dorian Gray turn naturally to an 'expert' in his hour of distress:

> Alan Campbell was an extremely clever man, though he had no real appreciation of the visible arts, and whatever little sense of the beauty of poetry he possessed he had gained entirely from Dorian. His dominant intellectual passion was for science. At Cambridge he had spent a great deal of his time working in the laboratory, and had taken a good class in the Natural Science Tripos of his year. Indeed, he was still devoted to the study of Chemistry, and had a laboratory of his own, in which he used to shut himself up all day long, greatly to the annoyance of his mother, who had set her heart on his standing for Parliament and had a vague idea that a chemist was a person who made up prescriptions . . . He was strangely melancholy . . . but an excellent musician, who played both the violin and the piano better than most amateurs. [77]

Inadvertently, perhaps, Wilde summarised a host of convenient impressions. By the 1890s the 'naturals' had become a separate 'estate', intellectually brilliant, but spiritually distinct. Uniformity in school and class still pivoted upon the uniform expectations of the 'Little-go', including the requirement of compulsory Latin and Greek. But once the pressures of specialisation began to operate, working downwards through Part II, the 'common context' shared by the 'natural' and his arts contemporaries was destined to break. At least two 'cultures' developed within Cambridge, and (between Parts I and II) within Cambridge science, distinguishing general knowledge from academic knowledge. From the 1880s, this fragmentation in intellectual outlook was to be of increasing significance to the whole of Victorian science, religion and letters; in Cambridge, this process was quickened by the NST.

Looking beyond its contribution to this new elite, the NST, representing the combined interests of Cambridge science, had another far-reaching implication. By the 1890s, the financial demands of the Tripos were creating new institutional pressures, and unavoidably enhancing arguments for a stronger, more accessible university, sustained increasingly by public funds. These pressures worked selectively and slowly; but, on balance, the University by 1914 had become both strengthened internally, and increasingly vulnerable to outside opinion. This was of lasting importance. Beyond questions of access and recruitment, of social influence and mobility, the NST and its graduates helped begin and legitimate the process which historians would later describe, in the secular vocabulary of academic politics, as the "nationalisation of the universities" [78].

ACKNOWLEDGEMENTS

Research on which this essay is based was supported by a grant from the SSRC. Our computer program was prepared by Mr David Hitchen, of the Computer Centre at Sussex University; a full account of the project, and a full statistical guide, may be found in our report to the SSRC, *The Anatomy of an Elite: the Natural Sciences Tripos and its Graduates*, 1850–1914, two volumes (London, 1977).

We would like to acknowledge the assistance of Dr Elizabeth Leedham-Green of the Cambridge University Archives, and Professor Peter Mathias, at whose seminar our project was first discussed. We would also like to thank Mr A. G. L. Haig of the Australian National University for his helpful comments.

NOTES AND REFERENCES

[1] STONE, L. (Ed.) (1975) *The University in Society*, Vol. 1, p. v (Princeton University Press).

[2] MACLEOD, ROY & MOSELEY, RUSSELL (1980) Breaking the circle of the sciences: the Natural Sciences Tripos and the 'examination revolution,' in: MACLEOD, ROY (Ed.) *Days of Judgement* (Driffield, Yorkshire).

[3] MACLEOD, ROY & MOSELEY, RUSSELL *Science with Honours: Cambridge and the Natural Sciences Tripos* (in preparation). Case studies in this volume will include contributions by Drs R. Porter, G. Roberts, D. Wilson, W. Bynum, W. H. Brock and B. Norton.

[4] Those who read the natural sciences for an 'ordinary' degree—the 'poll' (or 'pass') men—would require a separate study. Their numbers were at first comparable: in 1869, there were seven poll men compared with 13 NST men; in 1870, 25 compared with seventeen; and in 1878, 22 compared with 23. After 1880, however, the number of poll men reading science fell sharply, while the number of NST men continued to rise. For a discussion of the *women* who took the NST in the period 1881 to 1916 see MACLEOD, ROY & MOSELEY, RUSSELL (1979) Father and daughters: reflections on women, science and Victorian Cambridge, *History of Education*, 8, pp. 321–333.

[5] For a convenient summary of the principal issues, see GREEN, V. H. H. (1969) *The Universities*, chapter 4 (Harmondsworth, Penguin), Reaction and reform at Oxford and Cambridge, and chapter 11, The universities and science. See also Macleod & Moseley, *op. cit.*, note [3].

[6] See WINSTANLEY, D. A. (1947) *Later Victorian Cambridge* (Cambridge University Press), and *idem*, (1940) *Early Victorian Cambridge*, (Cambridge University Press). For the relationship between the universities and the learned societies, see MACLEOD, ROY (1980) Whigs and savants: the Royal Society of London and reform, 1830–1848, in: INKSTER, I. & MORRELL, J. B. (Eds) *Metropolis and Province* (London).

[7] DISRAELI, BENJAMIN, *Vivian Grey*, quoted in BIBBY, C. L. (1959) *T. H. Huxley*, p. 215 (London).

[8] In the early century, there had been 200 graduates per 5 million population; this fell to 65 in the decade 1855–1865 (see Charles Daubeny, Report of the Economic Sciences and Statistics Section of the British Association, Nottingham, 1866). The proportion of Oxbridge matriculants per million men aged 15–25 fell from 732 in 1821 to 541 in 1861 (see GREENWOOD, M. (1935) University education, its recent history and function, *Journal of the Royal Statistical Society*, 98, pp. 1–23).

[9] Cf. ROTHBLATT, SHELDON (1975) The student sub-culture and the examination system in early 19th century Oxbridge, in: STONE, L. (Ed.) *op. cit.*, note [1], pp. 247–303.

[10] Cf. comparative statistics for Oxford and Cambridge in STONE, L. (1975) The size and composition of the Oxford student body, 1850–1910, in: STONE, L. (Ed.), *op. cit.*, note [1] p. 6.

[11] *Ibid.* cf. VENN, J. A. (1908) Matriculations at Oxford and Cambridge, 1544–1906, *Oxford and Cambridge Review*, 3, pp. 48–66.

[12] POWELL, BADEN (1832), *The Present State and Future Prospects of Mathematical and Physical Studies in the University of Oxford*, p. 27 (Oxford).

[13] ENGEL, A. (1975) From *clergyman to don: the rise of the academic profession in nineteenth century Oxford*, p. 5 *unpublished PhD dissertation*, Princeton University. Cf. Engel (1975) The emerging concept of the academic profession at Oxford, 1800–1854, in Stone, *op. cit.*, note [1].

[14] Cf. WILSON, CHARLES (1965) Economy and society in late Victorian Britain, *Economic History Review*, 18, pp. 194–195, COLEMAN, D. C. (1973) Gentlemen and Players, *Economic History Review*, 26, pp. 94–116.

[15] Cf. CROWTHER, J. G. (1974) *The Cavendish Laboratory* 1874–1974 (London), and GEISON, G. (1978) *Sir Michael Foster and the Cambridge School of English Physiology* (Princeton University Press).

[16] See HOSKIN, KEITH (1980) in MACLEOD, R. *op. cit.*, note [2].

[17] Cf. SHAPIN, S. & THACKRAY, A. (1974) Prosopography as a research tool in history of science: the British scientific community, 1700 to 1900, *History of Science*, 12, pp. 1–28; and PYENSON, L. (1977) Who the guys were, *History of Science*, 15, pp. 155–188. Of 93 men serving in British Cabinets between 1868 and 1955, 26 had been to Cambridge, 49 to Oxford, and 18 to no university. Of the Cambridge men, only one (Moulton) took the NST.

[18] JENKINS, HESTER & JONES, D. CARADOG (1950) Social class of Cambridge University alumni of the 18th and 19th centuries, *British Journal of Sociology*, pp. 93–116. The shortcomings of this study have been ably criticised in HONEY, J. R. de. S (1977), *Tom Brown's Universe: The Development of the Victorian Public School*, pp. 280–1 (London).

[19] ANDERSON, C. ARNOLD & SCHNAPER, MIRIAM (1952) *School and Society in England. Social Backgrounds of Oxford and Cambridge Students* (Washington).

[20] STONE, L. *op. cit.*, note [1].

[21] Cf. SNOW, C. P. (1961) *The New Men* (London).

[22] STONE, L. *op. cit.*, note [1] p. vi.

[23] The standard account of the Mathematical Tripos is that of BALL, W. W. R. (1918) The Mathematical Tripos, reprinted in his *Cambridge Papers*, pp. 252–316 (London). The first new Tripos was the Classical, established in 1827.

[24] Whewell's views are outlined in his *On the Principles of English Liberal Education* (London, 1837), and expanded (with a description of his proposed relationship between the Moral and Progressive Sciences) in his *Of a Liberal Education in General, and with Particular Reference to the Leading Studies of the University of Cambridge* (London, 1845, 2nd edn., 1850).

[25] For administrative details concerning the NST, see HOLMES, R. F. (1965) An historical note on the Natural Sciences Tripos, *Cambridge Review*, pp. 199–241, 23 January.

[26] Cf. ROTHBLATT, *op. cit.*, note [9], p. 296.

[27] The scientific subjects were chemistry, geology, zoology and comparative anatomy, mineralogy and botany, to which were added, physiology, physics and human anatomy.

[28] HILL, ALEX (1919) Aspects of science and the universities, *Nature*, 104, p. 256.

[29] TODHUNTER, ISAAC (1873) *The Conflict of Studies and Other Essays*, p. 17 (London).

[30] MACLEOD, ROY & MOSELEY, RUSSELL (1973) Breadth, depth and excellence: sources

and problems in the history of university science education in England, 1850–1914, *Studies in Science Education*, 5, pp. 85–106.

[31] Between 1851 and 1916, fully one-quarter of Cambridge men failed to take a degree at all, and of the remainder, one-half took the Ordinary (or 'poll') degree. Thus, the proportion of 'naturals' among those leaving Cambridge with an honours degree was, by 1910, particularly notable.

[32] Following the Natural Sciences and Moral Sciences Triposes, came Triposes in Theology, Law and History, History, Semitic Languages, Medieval and Modern Languages, Indian Languages, Oriental Languages, Mechanical Sciences and Economics. (We use, arbitrarily, the Latin rather than the Greek plural of Tripos.) Cf. MACLACHLAN, J. O. (1947–49) The origin and early development of the Cambridge Historical Tripos, *Cambridge Historical Journal* 9; STODDART, D. R. (1975) The RGS and the foundation of geography at Cambridge, *The Geographical Journal* 141, pp. 1–24.

[33] In fact, the number of men (and, after 1874, women) who read for the Natural Sciences Tripos does not exhaust the full extent of scientific instruction in the university. Until the 1880s, the Mathematical Tripos continued to attract most men reading physics, and there were considerable numbers who took the Special Examination in natural sciences for the Ordinary Degree. Finally, there were many who read for the MB, and later for honours in mechanical sciences, all of whom swelled the proportion of Cambridge 'scientists'.

[34] Between 1850 and 1899, 38% of Cambridge graduates (including 'poll men') became clergymen; but only 12% of NST graduates. In the same period, 10% of all Cambridge graduates, but 14% of NST graduates, were sons of medical men; 12% of all Cambridge undergraduates, but 14% of NST graduates, were from manufacturing or business families. As these comparisons are between the large pool of both honours and non-honours graduates, and the much smaller pool of NST graduates, the *relative* importance of the NST vis-à-vis other honours graduates is much increased. See note [31], *supra*. See also MACLEOD, R. M. & MOSELEY, RUSSELL (1976) *The Natural Sciences Tripos: Data and Commentary* (Brighton, typescript).

[35] Regional origins do not feature in the studies of Anderson & Schnaper or Jenkins & Jones, *op. cit.*, notes [18] and [19]. We were therefore unable to make comparisons within the Cambridge undergraduate body.

[36] Each group accounted for 14% of NST men between 1881 and 1904.

[37] LIVEING, G. D. (1897) The University of Cambridge, *British Medical Journal*, 19 June, 8.

[38] ARMYTAGE W. H. O. (1955) *Civic Universities: Aspects of a British Tradition*, p. 225 (London); MACLEOD, R. (1972) The resources of science in Victorian England, in: MATHIAS, P. (Ed.) *Science and Society, 1600–1900*, pp. 111–166 (Cambridge University Press).

[39] *Special Report from the Select Committee on the Oxford and Cambridge Universities Education Bill; together with the Proceedings of the Committee*, 1867, pp. 15–16.

[40] *Ibid.*, p. 38.

[41] GROVE, WILLIAM (1843) Physical sciences in England, *Blackwood's Magazine*, 54, p. 516.

[42] LATHAM, HENRY (1877) *On the action of examination considered as a means of selection*, p. 142 (London).

[43] Quoted in SANDERSON, MICHAEL (1972) *The universities and British Industry*, p. 84 (London).

[44] *Select Committee, op. cit.*, note [39], p. 120.

[45] *Select Committee on Scientific Instruction* 1868, (432), Appendix 20, p. 475. See also *Third Report of the Devonshire Commission* 1873, (c. 868) Q. 4636.

[46] See SKEATS, H. S. and MIALL, C. S. (1891), *History of the Free Churches of England*, p. 625n (London); FAWCETT, H. (1861) On the Exclusion of those who are not members of the Established Church from Fellowships and other Privileges of the English Universities, *MacMillan's Magazine*, 3, pp. 411–16.

[47] *Cambridge University Reporter*, 24 November 1885, p. 189.

[48] Namely, Charterhouse, Cheltenham, Clifton, Eton, Fettes, Haileybury, Harrow, Loretto, Malvern, Marlborough, Merchant Taylors, Oundle, Radley, Repton, Rossall, Rugby, St Paul's, Sedburgh, Sherborne, Shrewsbury, Uppingham, Westminster and Winchester. For some analytical limitations of this listing, see HONEY, *op. cit.*, note [18]

[49] See, for example, MEADOWS, A. J. & BROCK, W. H. (1975) Topics fit for gentlemen: the problem of science in the public school curriculum, in: SIMON, BRIAN & BRADLEY, IAN (Eds) *The Victorian Public School*, pp. 95–114 (London); cf. THOMAS, BERNARD (Ed.) (1957) *Repton, 1557–1957*, pp. 63–64 (London). Cf. WARD, DAVID (1967) The public schools and industry in Britain after 1870, *Contemporary History*, 2, 37–52.

[50] Between 1851–1881 less than 5% of NST graduates had been educated at grammar schools; 46·2% educated at 'lesser' public or private schools; 25·9% at one of the 'greater' public schools; and 4·9% at a grammar school. The education of the remainder is not known.

[51] Cf. JENKINS & JONES, *op. cit.*, note [18]. Of the entire undergraduate population between 1851 and 1881, 7% had attended a grammar school compared with 10% of NST men; and 42% had been to a lesser public or private school compared with one-half of NST men.

[52] See *Report of the Committee on the Position of Natural Science in the Educational System of Great Britain*, pp. 86–89 (HMSO, 1918). Indeed, by 1914 the majority of science graduates in Great Britain found employment in the schools.

[53] See CANNON, W. (1964) History in depth: the early Victorian period, *History of Science*, 3, p. 25.

[54] 153 Part I, and 78 Part II graduates between 1881 and 1904 became school teachers.

[55] *Nature*, 5, p. 429, 1872. At the turn of the century, the 'allied' colleges of Pembroke, Caius, King's, Jesus, Christ's, St. John's and Emmanuel awarded each year 59 scholarships and exhibitions with a value of £3040 but of these, only nine, worth £390, were for natural science. Trinity awarded five of its 21 scholarships and exhibitions for natural science (worth £235 of a total £1030) and Clare two out of ten (£120 out of £450). The last two colleges were held by *Nature* to have treated natural science 'quite fairly'. But for the rest: "£3,040 in literature and mathematics, and only £390 for science is a very unsatisfactory distribution of prizes, and does not encourage education in our schools and colleges" (*Nature*, 59, p. 116, 1898–99).

[56] Quoted in *The Cambridge Chronicle*, 9 November 1867.

[57] FOSTER, M. *et al.* (1888) *Coutts Trotter: in Memoriam*, p. 11 (Cambridge University Press).

[58] Christ's commitment to the sciences was primarily in medicine. Of 237 who read natural sciences between 1851 and 1904, 119 became medical men and five became physiologists. See also QUIGGEN, A. H. (1942) *Haddon the Head Hunter*, p. 34 (Cambridge University Press).

[59] NEWMAN, CHARLES (1957) *The Evolution of Medical Education in the Nineteenth Century*, pp. 284–91 (London).

[60] No fewer than 81 (16·5% of those who took Part II between 1881 and 1904) became Fellows of the Royal Society. This compares with only 1% of those who read Part I alone. The former figure strongly suggests that the Royal Society's close association with Cambridge must be seen as much in terms of the development of the NST as the Mathematical Tripos. Although Part II was the main path to later eminence in the scientific world, failure in Part I was not an insurmountable barrier: among the failures there is one distinguished FRS—Sir Alfred Moritz Mond (Bart), "A Captain of Industry, an Apostle of Rationalism, a Leader of Zion, and an Imperialist."

[61] VENN, J. A. *op. cit.*, note [11], p. 65.

[62] SANDERSON, M. *op. cit.*, note [43].

[63] *University Papers* (Cambridge University Library), UP 57. "A Brief Statement of the History and Work of the Cambridge University Appointments Board" (1902).

[64] Graduates of the Mechanical Sciences Tripos found industrial employment easier to come by. A survey of graduates between 1894 and 1906 found that 81 were employed in some commercial or industrial capacity; 40 were in some government service at home or abroad; 23 taught in various educational institutions; and 31 had 'miscellaneous' occupations. *The Times Engineering Supplement*, 23 February 1910.

[65] Cf. JENKINS & JONES, *op. cit.*, note [18], pp. 100–102.

[66] Over 160 NST graduates lost their lives between 1914 and 1918.

[67] Cf. MACLEOD, ROY & KAY (1979) The contradictions of professionalism: scientists, trade unionists and the First World War, *Social Studies of Science*, 9, pp. 1–32.

[68] *Obituary Notices of the Royal Society*, 1, p. 331, 1934.

[69] STONE. L. *op. cit.*, note [1], p. vi.

[70] For criticisms of the 'aggressive' behavior of natural scientists, who were monopolising the university's public discussions, see *The Cambridge Chronicle*, 12 November 1886.

[71] ROACH, JOHN (1959) Victorian universities and the national intelligentsia, *Victorian Studies*, 3, p. 144.

[72] FOSTER, M. (1888) The science work of the late Rev. Coutts Trotter, in *Coutts Trotter: in Memoriam* (Cambridge University Press).

[73] CROWTHER, J. G. (1952) *British scientists of the twentieth century*, p. 21 (London).

[73] CROWTHER, J. G. (1952) *British Scientists of the Twentieth Century*, p. 21 (London).

[74] FLEXNER, A. (1968) *Universities: American, English, German*, p. 263 (Oxford University Press, 1930).

[75] ROTHBLATT, SHELDON (1968) *The Revolution of the dons* (London).

[76] BATESON, W. (1917) The place of science in education, in: BENSON, A. C. (Ed.) *Cambridge Essays in Education*, pp. 121–146 (Cambridge University Press).

[77] WILDE, OSCAR (1891) *The Picture of Dorian Gray*, p. 246 (London).

[78] Cf. CAMPBELL, L. (1901) *The Nationalisation of the Old English Universities* (London).

SYDNEY. P HALL

VIII

Breaking the Circle of the Sciences: The Natural Sciences Tripos and the "Examination Revolution".

Roy MacLeod *Russell Moseley*

> Amid the revolutions which have shaken thrones and overturned dynasties, we have not entirely escaped. A revolutionary movement which neither the experience of past ages nor the caution of the present age authorised us to expect, has startled the tranquil waters of the Cam and Isis. . . . The University of Cambridge, in Senate House assembled, has resolved at length that the time was come when its circle of knowledge must be enlarged.[1]

Introduction

Questions concerning the adaptation of the universities to new studies, the assimilation of these studies, and the manner in which transmitted knowledge has been institutionally determined, have in recent years assumed increasing importance for historians of education and of science.[2] As one of the principal routes leading to a degree in science in the late 19th century, the Natural Sciences Tripos (NST) at Cambridge occupies an important position.[3] Educationally, the discipline and rigour of examinations embodied in the Honours Schools of Oxford and the Triposes of Cambridge, were lifted from academic soil and transplanted by Macaulay, Jowett, Trevelyan, Gladstone and others, to a wide variety of different institutional settings. It has been argued, notably by Sheldon Rothblatt, that examinations were favoured in

Regency Oxford and Cambridge as a means of diminishing controversy on sensitive religious and political subjects; in this context, the introduction of the honours/pass distinction and the creation of new Triposes, can be understood as an appeal to 'higher motives' among students.[4] By the 1850's, as John Roach has put it, 'one of the major objectives of university reformers was to replace the ancient constitutions of Oxford and Cambridge by a new order based on merit and freedom of opportunity'.[5] In a comparable sense, the NST established standards of merit and achievement which became, with the examinations of London and the Scottish universities, a distinctive hallmark of British science.

Between 1851 and 1916, 4,223 men and 403 women[6] successfully took the NST; and by the decade before the outbreak of war in 1914, the 'naturals' as W. Aldis Wright called them, accounted for almost one-third of all Cambridge undergraduates awarded honours degrees. (See Figures 1 and 2). In creating this new 'elite', the NST mirrored the 'research' revolution which gradually transformed Cambridge life. The NST, as a university examination, ultimately gave the natural sciences a lever of institutional power. At the same time, it forced new accommodations between the University, as the examining authority, and the Colleges, as teaching bodies. Indeed, so close were these accommodations, welded symbolically by the arrangement of university laboratories and museums, that it proves impossible to make a clear distinction between the Tripos as a course of study and the Tripos as a set of examinations. The examinations both reflected and generated the syllabus. Moreover, new scientific disciplines could establish themselves in Cambridge student life and ultimately in University financial terms, chiefly by reference to the Tripos. The position ultimately enjoyed by the most powerful disciplines (notably physics and physiology) drew its strength principally through its operation. And just as the history of the Tripos mirrors the development of the different disciplines it examined, so the Tripos worked to shape the manner in which these disciplines were socially constructed.[7]

Finally, the NST commands attention in this period as an extraordinarily effective instrument in educating a large number of men and women who later held leading positions in education and academic research, in government service, the Empire, industry and medicine. This development of the Tripos accompanied and assisted the transition of late-Victorian Cambridge from the traditional ideology of *universitas* and the conception of a liberal education, in which the natural sciences would form a minor part, to an ideology of specialised knowledge, based on a new arrangement of disciplines, within which the natural sciences would form a culture of their own. This transition, the Universities assisted, often unconsciously, by breaking the 'circle of the sciences'[8], a process which would prompt the redefinition of 'liberal education' in professional terms.

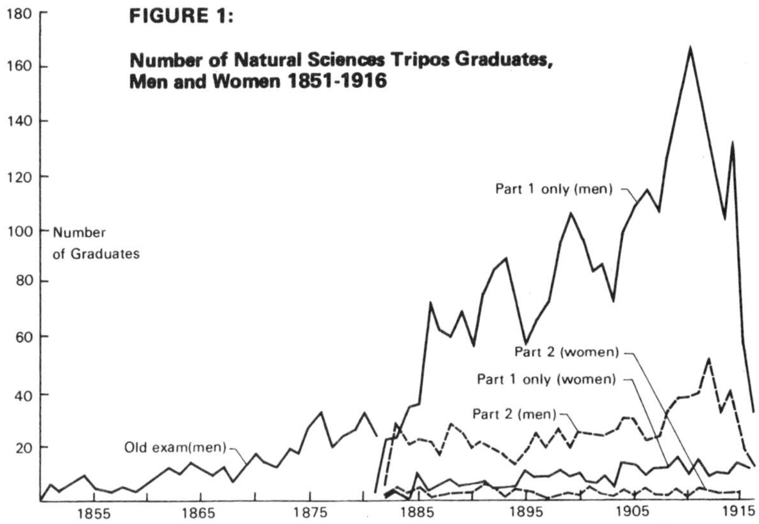

FIGURE 1:
Number of Natural Sciences Tripos Graduates, Men and Women 1851-1916
Number of Graduates
180
160
140
120
100
80
60
40
20
Part 1 only (men)
Part 2 (women)
Part 1 only (women)
Part 2 (men)
Old exam(men)
1855
1865
1875
1885
1895
1905
1915

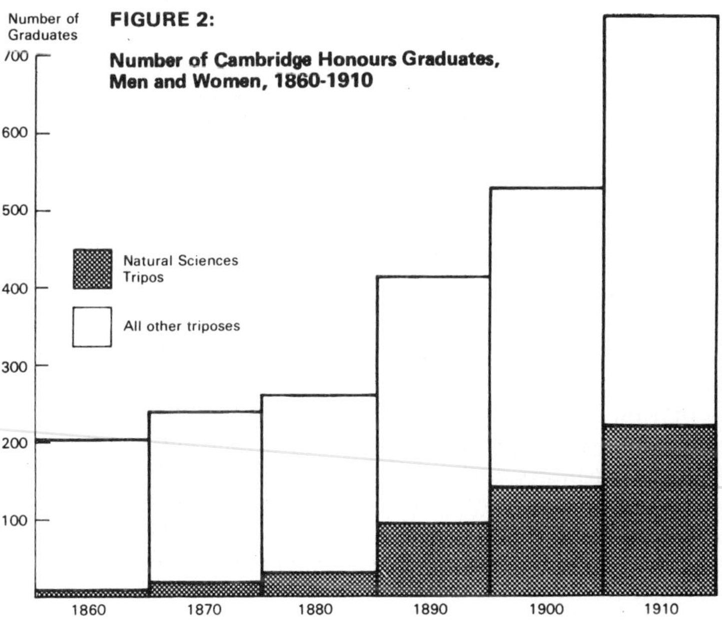

FIGURE 2:
Number of Cambridge Honours Graduates, Men and Women, 1860-1910
Number of Graduates
700
600
500
400
300
200
100
Natural Sciences Tripos
All other triposes
1860
1870
1880
1890
1900
1910

The Struggle for Acceptance

Between 1830 and the late 1840's, in parallel with agitation for reform in the learned societies and in the institutions of government, the universities of Oxford and Cambridge attracted widespread citicism.[9] Critics focussed upon antiquated modes of government, the inequities of the Test requirement, the absence of sound college teaching, the distribution of emoluments, and the limited range of university studies. Such criticisms were not new, but the economic and political circumstances of the day gave them added force. For how long could the universities remain unrepresentative of new middle class manufacturing, often dissenting, interests; how could they resist behind centuries' old statutes any attempt to refashion them as fully "national" institutions? Following an intellectual programme set out by William Whewell and Adam Sedgwick, and prompted by a combination of pressures operating within Cambridge to improve the University's image, a Syndicate was appointed in February 1848 to consider the introduction of studies other than mathematics and classics. Reporting later that year, the Syndicate recommended the establishment of triposes in the natural sciences and moral sciences. As conceived by Whewell, honours in the new NST would demonstrate a synthesis of natural knowledge, completing the liberal education of gentlemen and scholars. As such, they were available only to men grounded in mathematics and classics, who had previously taken the Mathematical Tripos.

Despite a stiff debate in the Senate, the proposal was accepted.[10] The model for these new studies was inevitably the Mathematical Tripos, which dated from the middle of the 18th century, and which had won Cambridge international reknown.[11] The NST was to be 'natural' in the sense of complementing the Newtonian (i.e., mathematical) sciences (notably physics and optics) represented in the Mathematical Tripos. And, like the Mathematical Tripos, the NST was based upon the assumption that all knowledge could be examined for similar purposes – that is, to discipline the mind and to develop the faculties of memory and reasoning. To contemporary reformers, the incorporation of the natural sciences into written university examinations would not only encourage the examination of new subjects themselves, they would also encourage undergraduate effort and achievement.

In 1851 the first class list for the new Tripos was published. Although only six names appeared, they included those of George Downing Liveing, later Professor of Chemistry, and F. J. A. Hort, later Lady Margaret Professor of Divinity. In the Botanical Lecture Room, candidates crushed next to each other on opposite sides of a narrow table, and in poor March light, found the experience, at best, uncomfortable. The examination was held over four days: on the first, questions were set in comparative anatomy and geology; on the second, in physiology and botany; on the third, in chemistry and mineralogy;

and on the fourth, 'mixed' questions including the history and philosophy of science. Candidates generally attempted questions in all subjects, and were expected to show competence in all of them; it was common for a mark of distinction to be awarded in one or two subjects. Following the model of the Mathematical Tripos, they also took a *viva voce* examination, and were classed, like the Wranglers, in order of final standing.

The rigours of the experience certainly impressed Hort, who wrote to his mother:

> Geology seems to have been tolerably done by all, brilliantly by none. If the paper had been a quarter of the length, it would have been more satisfactory to all parties. Fuller says Sedgwick boasted of having made it a 'very complete' paper, and got *all* geology into it, to be written out in four hours! We all (that is all the first class) did very well in the general paper. I was glad to find that Fuller thinks two subjects as much as anyone can manage.[12]

For the first decades of its existence, the NST, requiring an honours BA as a prerequisite, attracted few candidates. When the first college fellowship in natural science was won by T. W. Danby of Downing in 1867, a writer in the *Lancet* morosely observed:

> No wonder that the numbers who go in for the examination are few; no wonder that they do not show equal merit with the candidates for the other triposes if it be so. The wonder is that the Tripos keeps on its legs at all, and that any good men are found to devote themselves to natural science. Those who seek reward as well as honour, and they constitute the mass of hard-working students, see that the path of Natural Science will lead to little or nothing, and turn in the better-paid ways of Classics and Mathematics.[13]

Underlying the issue of recognition for aspiring undergraduates was the prejudice of parents and leading public schools against science. There was evidence that schools and colleges reinforced each others' conservatism, the former spending little effort on studies which promised few future rewards, the latter unable to find sufficient good candidates for the awards available.[14]

Despite these difficulties, the 'scientific lobby' at Cambridge was optimistic. In 1860 the regulations were changed to permit undergraduates to read for honours in natural sciences alone. A Board of Natural Science Studies (BNSS) was created to prepare regulations, appoint examiners, and give marks. After 1860 all candidates had still to pass in the Previous Examination (or 'Little-go'),[15] including the additional mathematics subjects expected of candidates for mathematical honours. Since candidates had to have kept at least seven terms (and have entered upon their eighth), the Michaelmas Term of their fourth year was the earliest opportunity for students matriculating in October to sit for the NST. There were few financial incentives, in the form of college awards. As late as 1872, scholarships in the natural sciences were offered by only Clare and Emmanuel, and in that year no elections were made.[16] But, while the President of the new Cambridge University Natural Science Society admitted in 1863 that the NST as yet 'did little or nothing towards encouraging the study of those branches of science included in the examination...,'[17] he hoped it would not be long before his

> members . . . will go forth into the world, not only to pursue Science as a recreation whilst engaged in other duties of the Church, the Bar or the Medical professions; but to rank amongst our Professed Naturalists and such as devote themselves exclusively to the advancement of the various departments of science, the encouragement of which the Society sets before it as its object.[18]

By the late 19th century the subsequent careers of early NST graduates had fully realised his hopes. The inauguration of an independent tripos was decisive. The first five years of the 1870's saw 79 candidates; the next five, 125. Gradually, schools began to send up more men with backgrounds in science and colleges began to increase the number of scholarships and exhibitions available to them.[19]

For its reputation in the schools and in the wider country, the first obligation of the new Tripos was to establish academic standards. The high demands set by the Mathematical Tripos were not forgotten. Of a maximum of 6,300 marks, few candidates were expected to gain more than 1,500. W. H. Miller, Professor of Mineralogy, examining in 1863, agreed that 'it would have been a breach of the trust reposed in us by the University if we did not use our best endeavours to make the standards of the First, Second and Third Classes of the Natural Science (sic) Tripos, correspond fully with those of Wranglers, Senior and Junior Optimes, and not in any degree to make them depend on the number of candidates'.[20] In 1861, a First Class was given for 2,217 marks, and the Third Class ranged from 628 to 331. The fact that there were no first class degrees awarded in 1862 supported Miller's firm policy. George Liveing, by then Professor of Chemistry, pointed out that 'there have been some strong opinions expressed at the Board (of Natural Science Studies) that the examiners have taken too high a standard ... [but] no one has thought that they have taken too low'.[21] The year 1865 was a 'better' year, with three Firsts (range 2670–1834), six Seconds, and two Thirds (739–604). The lower limit of the First Class was restricted to one third of the total marks. In consequence, it was possible to assert that a First Class in the NST was 'equivalent to that of a Wrangler, that it is as difficult to obtain, and that it ought to be so estimated'.[22]

Between 1851 and 1916 the structure of the NST examinations changed considerably. (A summary of these developments is given in Table 1). Although some changes were simply in the nature of technical adjustments, others reflected more fundamental concern over the proper function of the Tripos: was it intended to provide a general, liberal education in the natural sciences, or to cater for the new specialist interests of particular disciplines? If the former, how could the requirements of superficial and profound knowledge by reconciled?[23] If the latter, to what extent should 'breadth' give way to 'depth'? In attempting to answer such questions, the examiners sought repeated compromise, and in some measure met both objectives.

Breadth, Depth and Specialisation

Between 1861 and 1870 the examination consisted of twelve papers in six subjects and lasted six days. From 1871, the examination, was spread over eight days, with two days practical work in the middle. The division was carried further between 1876 and 1881 when two sets of examinations were held. Only those who obtained honours in the three days of the June examinations were admitted to the remaining five days of examination in December. Names were placed in alphabetical order in three classes, and the award of a "First" depended upon showing proficiency in one subject.

Undergraduates reading for honours in the natural sciences[24] do not appear to have been discouraged by the growing range and complexity of the Tripos. But the examinations repeatedly attracted criticism. As early as 1852, an anonymous author complained to the readers of *Frazer's Magazine* that:

> Young men of one or two-and-twenty are to be examined in half a dozen sciences at once, each of which may, or rather, if any progress is to be made therein, *must* be the study of a whole lifetime. You will get a delusive semblance of knowledge instead of knowledge itself; you will encourage a shabby superficiality which, so far from developing the collective powers, conceals their defects from oneself and from others, and so prevents their cultivation and improvement.[25]

Even sympathetic observers found the Whewellian prospect of a general education in all the sciences more than ordinary mortals could bear. As Alex Hill, later a University Lecturer in medicine, recalled, 'A candidate's place depended upon his aggregate of marks . . . and to secure a first class he must show that his knowledge, like that of the Master of Trinity, ranged from

Table I

Structure of the NST, 1851–1881

	1851–60	1861–70	1871–75	1876–81
1st Year		Candidates for the NST had to pass the Previous Examination including the additional mathematics subjects.	As previously	As previously
2nd Year				
3rd Year				June: first part of the revised exam held. Spread over three days with six papers, commencing on the Monday next but one before the general admission to BA degrees. Each paper consisted of questions in all subjects.
4th Year	Only BAs admitted. Examination held in March of each year. Papers grouped by subject. Examination held over four days.	The Exam began on the first Monday in December and lasted six days, with papers in both the morning and the afternoon. Of these twelve papers, eleven contained one or more questions in all the various subjects (chemistry, mineralogy, geology, botany, and zoology with comparative anatomy), while one paper consisted of questions in the history and philosophy of science.	As previously, but the Exam was now spread over eight days, with two days practical work in the middle. The first three days' questions were elementary, the last three more advanced. Twelve papers were set, plus practicals. Physics and Physiology were introduced as independent subjects in 1873.	December: those acquitting themselves sufficiently well in June were admitted to the second part of the exam. The exams commenced on the first Monday in December and were spread over five days or more (i.e. 10+ papers). Again, each paper consisted of questions in all subjects. In 1876 Human Anatomy was introduced as an independent subject, while in that year alone the first class was divided into two with names arranged alphabetically.

Table I *(cont.)*

Structure of NST Parts I and II, 1881–1916

	Part I Only		Part II	
	1881–98	1899–1904	1881–98	1899–1904
1st Year	Previous exam taken unless exemption obtained, usually through the Oxford and Cambridge Local Examinations.	As previously.	Previous exam taken (preferably in October or December) unless exemption granted.	As previously.
2nd Year	Part I could be taken in May. Consisted of three days written work (six papers) and two days practical.	As previously, but Part I was extended over seven days (or more), with two papers of three hours duration in each of eight subjects.	Part I	Part I
3rd Year	Alternatively, Part I could be taken in May of the third year.	As previously.	Part II taken in May if Part I taken the year before. Consisted of three days written work and a variable number of days practical. Three questions were set in each subject on each paper.	As previously, but Part II was extended to four days of written work with four separate papers of three' hours in each of the eight subjects.
4th Year			Alternatively, if Part I taken in 3rd year, then in 4th year Part II taken in May as above.	

zoology to mineralogy'.[26] Such a range of subjects inevitably led to charges of superficial learning, and with some justification: examiners' reports reveal that candidates answered questions in fields of which they had little knowledge, in an effort to pick up extra marks. This was particularly true in mineralogy and geology.[27]

In 1867, the examiners reported that:

> The knowledge shown by several of the candidates in their respective subjects was very creditable, evincing much labour, especially in bookwork. There was, however, a deficiency of the knowledge of principles, and a want of philosphical application of the subjects.[28]

Moreover, candidates were weak in practical work, and in mathematical applications:

> Numerical questions set with the view of testing the precision with which chemical and physical laws had been grasped, were not much attempted, and rarely with success.[29]

The most pressing problem, the examiners saw, was in 'the number and arrangement of the questions', which does appear to favour somewhat unduly those who acquire a superficial knowledge of several subjects'.[30] From the outset there was an almost annual battle on the issue of 'generality'. Invariably, the remedy was seen in the direction of greater specialisation.

In the meantime, the 'domain' of science at Cambridge was steadily expanding. Writing in *Nature* in March 1870, T. G. Bonney remarked on the NST's development and summarised the resources of Cambridge sciences. In addition to six university professorships, four colleges (Trinity, St. John's, Sidney Sussex and Downing) had appointed six lecturers in the natural sciences and were now providing ten to fourteen scholarships and exhibitions annually. Bonney reported that the 'Coldness and even dislike with which the study of natural science was once regarded here is rapidly passing away . . . and that the University is fully alive to the wants of the age. . . .'[31] This development, however, was not occurring without much painful reflection on the subject of examinations.

By the early 1870's, criticism of examinations was widespread throughout the country. Cambridge offered no exception. The 'business' of the university had become the business of testing. According to William Napier Shaw,

> The Cambridge man has the examiner's instinct in his bones. He is so accustomed to examining everything that his first impulse is to assign marks – but not too many. I should have given myself 75

> per cent for the sort of answer I would have written myself, a colleague examiner once said.[32]

The problems posed by examinations were not unique to the natural sciences. As Coutts Trotter of Trinity put it, their 'increasingly elaborate and professional character' was absorbing the whole energy of undergraduates to the detriment of education. Viewing this as a 'morbid development', he added:

> A man who reads for a high place in one of the Triposes will generally have studied much more of the details of his subject than is necessary in order to get the greatest educational results from the time spent upon it, and even the smaller class of men who aspire to be professional mathematicians or scholars or men of science will probably not have gone through exactly that course which a judicious adviser would recognise as the best possible preparation for their future career.[33]

By the late 1870's, the examination debate had acquired standard formulae, and ritual postures of attack and defence. Perhaps three general principles were established, which contributed in different ways to a fundamental revision of the Tripos in 1880/81.

The first principle held that, above all else, Cambridge was not to become wholly preoccupied as an examining body. In this respect, the experience of London unified Cambridge opinion.[34] The second principle was that the twin gods of 'depth' and 'breadth' had both to be appeased, but that each had to be served separately. The examiners attempted to do this by setting elementary questions for the first three days, and more advanced questions for the last three. But this innovation did not take into account individual strengths and weaknesses, and differences in the nature of college teaching. Indeed, the examiners were compelled to admit that, on balance, the established system of marking by aggregate points still gave 'undue encouragement to wide as opposed to deep study'.[35] The third principle was that, in order to reduce the unhappy prospect of intellectual fraticide, the 'order of merit' in examinations should be defined in such a way that there would be 'categories' of achievement, rather than individual placements along the linear scale. In retrospect, the competitive model represented in the Mathematical Tripos itself was much criticised and viewed as counterproductive.[36]

Armed with these general guides, the Board of Natural Science Studies and its examiners, reappointed every two years, proceeded on a basis of experiment and expedience.[37] In May 1873, *Nature* published two articles entitled 'A Voice from Cambridge'. In the midst of two Royal Commissions enquiring into the state of the universities, the editor of *Nature* ventured the view that science at Cambridge was dead. Of course, rumours of its death were greatly exaggerated. But there was growing dissatisfaction with the 'enormous examining engine, on the most approved Chinese model',[38] as a means of advancing the study of science. In that year, of the 23 candidates, ten answered questions in all seven subjects and seven in six subjects. Prompted by the Wrangler-inspired 'order of merit,' candidates continued to gather marks by attempting questions quite outside their learning. This gave a deplorable showing. For example, the average mark of the last nine men out of the sixteen taking mineralogy, in 1873, was only 14½ out of 100.[39] The examiner in that subject commented ruefully that 'these men appear to have committed to heart, for the most part unintelligently, elementary facts in the science likely to be asked in the examination'.[40]

In 1877, H. McKenny Hughes argued that even though 'The Natural Science examiners were at present, and perhaps for some time would necessarily be chosen from men not of the University', they should not have 'a deciding influence on the Tripos regulations'. There was, nonetheless, one suggestion that the 'externals' were enjoying power without responsibility:

> They were seldom present to hear the statement and discussions of practical difficulties, and were obliged to form their opinions from the letters of those with whom they happened to be in correspondence; and by their own letters they influenced the Board and the University as to the desirability of some change, while they did not assist in deciding how that change might best be carried out at all without more than counter-balancing injury.[41]

Their intellectual influence was notably in the direction of greater specialisation. In examining their individual subjects, however, their experience was cautionary.

Frederick Barff, examining as an 'external' from London in chemistry remarked that from what he had seen 'the system bids for 'cram'. This would not be the case if placing were done away with'. Others protested at the method of aggregating marks. James Clerk Maxwell, whose prolonged attempt to bring experimental physics into the Tripos requires a study in itself, believed that 'The questions in the last three days should . . . be so arranged as to give full occupation to the real student of physics during the hours of the examination while they afford no chance of marks to the man of hearsay information'. Michael Foster, another arch-critic of 'cram' complained that:

> Another year's experience has simply strengthened my conviction

> that the present system of the examination, the order of marks, is doing great mischief to the study of physiology at Cambridge, and I continue to believe that were a simple class system, with alphabetical order in each class, introduced, most beneficial changes would result in the work of the student. Such a change would necessitate the making of the first three days' examination a sort of pass examination.[42]

The prospect was particularly worrying for those undergraduates destined for medicine. Donald MacAlister, tutor of St. John's, pressed upon his students 'that the true and regular way of proceeding to the medical degree is through the Natural Sciences Tripos. Only thus, as I take it, are the full benefits of our Cambridge training to be gained'.[43] Until the 1870's, medical students who had read natural sciences were exempted from the second MB subjects included in the Tripos (comparative anatomy, anatomy and physiology, botany and chemistry), as well as the mechanics and hydrostatics examination in the first MB. Yet these men, numerically the largest group reading natural sciences, seemed to one observer 'more bent on picking up points and details which may best serve their purposes as future examinees, than in seeking that wider and sounder knowledge which will enable them successfully to cope with injuries and disease in future practice'.[44] Perhaps it is not fortuitous that the majority of early NST failures were also medical students.[45]

The *University Reporter* of 9 June 1874 carried a statement from the BNSS, describing the inducements which resulted in candidates spreading their reading 'over a wide area' rather than studying 'deeply a limited section of Natural Science'.[46] The BNSS felt that the object of the examination – 'to offer sufficient stimulus to exertion and at the same time to give encouragement to sound study' – could best be served by dividing the First (i.e., Wrangler) Class into two. Dividing the First Class was, in the words of Coutts Trotter of Trinity, a way of indicating that 'so far as the examiner could judge from one examination, the men above it were qualified for a Fellowship, while those below it were *prima facie* not qualified'.[47]

George Humphrey, Professor of Anatomy, claimed that this would 'prevent men from obtaining a high place in the Tripos by merely covering a great deal of ground, and . . . require really superior knowledge of some one subject'. The substitution of an alphabetical order within a First Class would 'remove a false stimulus, that of mere competition, and would substitute for it a real stimulus, the desire to obtain knowledge and power'.[48]

The regulations specified that, from 1876 onwards, First Class men should show general proficiency in the first three days of the examination and 'superior proficiency' in one, at least, of the various branches of science represented in the later part.[49] This more reasonable approach, according to

George Humphrey, would induce men to take the Tripos rather than the Ordinary examination.[50] Moreover, such an arrangement would ensure that 'preliminary' work was finished some months earlier, thus leaving the undergraduate free to deepen his knowledge thereafter. It was apparent that the traditional commitment to a general knowledge of a wide range of subjects was producing unforeseen and undesirable consequences. Seeking to avoid superficial 'breadth', the NST moved sharply towards 'depth'.

In the meantime, however, there was still room for debate. Of 44 candidates examined in Easter, 1876, only 31 were deemed to have acquitted themselves sufficiently well to proceed to the December examinations. Possibly the standard had been set too high. As G. T. Bettany remarked, 'In admitting them to such a qualification, tests should rather be applied which every student of the subject ought to be able to respond to; but it is questionable whether we can yet expect any student of zoology and comparative anatomy to 'state precisely the doctrine of evolution as employed in biology'.'[51] Indeed, it was feared that the new rule might be 'unduly favourable to men who have tried with some success to combine width with depth of reading, as the old rule was unduly favourable to those who sacrificed depth to width'.[52]

The Divided Tripos

Despite these difficulties, there was by the late 1870's widespread agreement that the claims for general knowledge and specialist expertise could only be reconciled by dividing the Tripos. Since the first important step in this direction had already been taken in 1874, it was natural that further proposals concerning the NST should be consolidated.[53] Following a Grace of the Senate on 26 February 1880, the NST was formally divided into a Part I and Part II, both parts leading to an honours degree, with admission to the latter being dependent upon success in the former. From 1881, Part I would test general knowledge of up to four subjects, while Part II, to be taken the following year, would require candidates to demonstrate 'special proficiency in one or more of the eight subjects', and (for a First) 'competent knowledge of some cognate subject'. Each part was to have a separate class list; and each was to be held over five days.[54] *The Athenaeum* remarked hopefully that this 'contest between the aims of general and special technical education [would] ... encourage real scientific study among medical students'.[55]

The examination standard of Parts I and II was also raised, partly by defining a lower 'limit of competency' and a minimum level of marks for an Honours degree. In 1884, a standard of 65–70% was set for a First. The proportion of Firsts increased (typically, during the 1880's and 1890's, about 25% took Firsts, and 30% took Seconds in Part I, while about 30% took Firsts in Part II), but the examiners were frequently uncomfortable about the level they required. In 1887, Henry Roscoe of Manchester, acting as an external reported 'considerable difficulty . . . in defining the meaning of 'competent knowledge'' in Part II.[56] Candidates with marks below the limits set by the Examiners were put separately to the Vote. Disagreements among the examiners remained common (their meeting in 1888 lasted nine hours) and there were cases of examiners refusing to sign the pass list if their best men were denied Firsts! Both teacher and student were locked inseparably in the gears of the examination machine.

Paradoxically, this machine, built in part to preserve general education, had the effect of spurring specialisation. In 1882, the BNSS reproduced by mitosis, producing Special Boards for Physics and Chemistry, and Biology and Geology. These bodies, composed of University staff, coordinated the selection of external examiners, and indirectly set expectations for college tutors. Once Part II was regarded as the domain not of liberal education, but of discipline-based knowledge, the establishment of disciplinary 'codes' and hierarchies became the province of the specialist.[57] The division of the Tripos into two parts ensured that the growth in the number of candidates could easily be accommodated: the majority of candidates sat only for the more elementary Part I while a much smaller number went on to read for Part II.

It is not perhaps surprising that, given the influence exerted by the Tripos, individual disciplines sought for a solution to examination difficulties in the introduction of yet more examinations. George Liveing of St. John's and Henry Latham, of Trinity Hall criticised the tendency which reduced undergraduate teaching to 'preparation for examinations'.[58] For some, the experience was clearly traumatic. Roger Fry, for example, went to Cambridge to read science; but later he wrote of 'the awful Tripos'. Despite his parents wishes, he abandoned science for art.[59] William Bateson, the biologist, later recalled:

> As a boy in Cambridge I learnt that if a man got a first class he might be happy; if he got a second class he would be unhappy; and if he got a third class, nothing but misery and a colonial life awaited him.[60]

Was there, in a University whose life was increasingly organised around the principle of competition, any alternative? As even Liveing admitted, 'a little pressure is needed to make men attend lectures which they admit to be beneficial'.[61]

For the next twenty years the new regulations remained virtually unchanged. By the turn of the century, the NST, particularly Part II, had developed from a device to test speed and memory, to an instrument for measuring an understanding of recent research. With the growth of university science, and a growing commitment to science among the colleges, the divided Tripos proved increasingly attractive. More undergraduates were choosing to read for honours degrees than for pass degrees, implying to the *Reporter* that 'an increasing number of those students who declare for science at Cambridge aim at thoroughness in their work and are not content with that superficial smattering of book knowledge which is considered sufficient in the examination for the Pass degree'.[62] A writer in the *Cambridge Review* speculated in early 1882 that so great had been the progress of the study of natural science in the recent past that 'before long it will be the leading study in the University, and that the list of candidates for the Natural Sciences Tripos will outnumber that of any other'.[63] Seven years later, candidates for the NST outnumbered for the first time candidates for the Mathematical Tripos.

It did not go unnoticed that 'the great popularity of the Natural Sciences Tripos had unquestionably drawn away men who a quarter of a century ago would have unhesitatingly chosen mathematics as their pathway to a degree'.[64] In order to meet competition from the NST, the structure of the Mathematical Tripos was frequently changed until it would have gone unrecognised by those who had studied for twenty hours a day, who had been coached to perfection and who prayed for a warm Tripos day to enable them to write faster.[65] In 1882, the Mathematical Tripos was also divided, but its structure was altered again in subsequent years. It was admitted that '(t)he general effect of these changes was to destroy the homogeneity of the Tripos'; Part I was too technical, while Part II 'was so constructed as to appeal only to professional mathematicians'.[66]

Indeed, a new Part II was devised in 1906, since 'the difficulty of getting candidates capable of sending in any answers was only equal to that of finding examiners capable of setting questions'.[67] As a result, the number of undergraduates reading mathematics fell rapidly, and the reputation of the Tripos suffered.[68] Subsequent efforts to reform the Tripos reflected not only the advances made by new knowledge in the 'Newtonian' sciences but also the inroads made by the NST.

By the late 1880's, the evident popularity of the NST, together with the mounting costs of science teaching were attracting critics. Forsaking its Whewellian compass, the NST, it was said, was being manipulated to advance 'professional' studies. In 1885, a Syndicate recommended to the Senate a substantial increase in the number of university lecturers and demonstrators in science. These recommendations were accepted, but similar proposals to build new science laboratories and lecture theatres generated hostile opposition. One Arts critic wanted to know why existing lecture rooms could not simply be ventilated if they were used as much as was claimed, and suggested that since science undergraduates had 'exceedingly small and inexpensive' libraries, they should buy their own microscopes.[69]

The cost of science provided ammunition for debate, but it was the demand for increased specialisation embodied in the natural sciences (although by no means confined to them) that posed the greatest threat to traditional Cambridge values. Much of this 'anti-science' sentiment was reflected in the *Cambridge Review*.[70] Some of the arguments were predictable: scientific study, as one critic claimed 'in some of its modern aspects . . . clashes grievously with the tone and the surroundings of this ancient university', while another argued that science 'should not be carried on under the shadow of buildings such as King's College Chapel, or within earshot of its organ'.[71] Objections that the study of science dulled the senses to beauty,[72] failed to engender a true appreciation of culture, and made men indifferent to criticism,[73] accompanied the growing recognition that the sciences were now disposing of a large proportion of college and university resources. By 1884, of the 39 University Professorships, 13 were in science (excluding 4 in mathematics); and of the 60 Readerships, Lectureships and Demonstratorships, 35 were in the natural sciences. In particular, the growth of science and medicine was seen as a dangerous development towards professionalism within the undergraduate degree. As a contributor to the *Cambridge Review* observed, the question was

> vital, not merely to the interests of this or any other generation, but to the very existence of the University as a focus of philosophic and literary culture in England; and it is one which cannot be settled without definite views of the function proper to a University, and the reason why such an institution should be maintained at all.[74]

The author of this passage believed that medical studies represented a transition to a 'technical education' and that science was being taught 'as a profession, rather than for its own sake'.[75] Science, properly pursued, could lead to a sound moral and intellectual training, but the manner in which both teaching and examining had developed was resulting in confused and ineffective thinking, narrowness and provincialism, listlessness and mean motives.[76]

These arguments were unwittingly reinforced by a report from the Special Board of Medicine in 1885, which drew attention to the fact that an unusually large number of candidates in physics and chemistry were rejected by their examiners.[77] A letter from Michael Foster appended to the report admitted that many candidates were rejected outright, and that of those who were accepted 'a considerable number acquit themselves in such a way that they do little more than escape rejection'.[78] For critics of the natural sciences the implications were clear: students were expected either to learn an impossible amount, or else became narrow specialists. In consequence, the Cambridge undergraduate reading natural science might as well live in 'the tawdriest lodgings in the newest and most vulgar commercial capital in Europe'.[79] To prevent the loss of liberal culture, it was essential that Cambridge should resist the demands of the new age, reject education 'as a means to money-making', and be jealous of its privileges.[80] At all costs the principle of utility must be rejected. Thus if the compulsory Prelim examination in Greek was abandoned, according to J. B. Bury, the first and decisive step 'would have been taken towards the abolition of University education altogether; for the principle of utility would be thereby accepted and that principle is by its very nature subversive of all such education'.[81]

Perhaps the science dons were, as had been claimed, indifferent to criticism. At most they would reply that 'while the University keeps a high ideal before its students, it also tries to prepare them for the active business of life'.[82] And if the business of life required adaptation in the curriculum, then so be it. *Over*specialisation was still deplored, but the steady growth of science was another matter, and in any case had assumed an air of inevitability. The Special Boards created in 1884 for Medicine, for Physics and Chemistry and for Biology and Geology, reported that the customary eight examiners could no longer suffice to examine the eight distinct subjects of the NST. Thenceforth, two examiners were appointed in each of the eight subjects.[83]

By the turn of the century the lines of debate had subtly altered. In the 1850's, the question had turned upon the difference between science and traditional studies; in the 1870's it had become a question of general versus deep knowledge – of 'science alone'; by the 1890's it had become an internal debate among specialists on the merits of 'a sound general scientific education, against a possibly sounder, certainly far narrower one'.[84] This debate revolved around the function intended for Part II of the Tripos. In 1900, a report from the two science Special Boards proposed awarding First Class honours in Part II to candidates who had only narrowly missed that standard in each of the two subjects they offered.[85] This proposal was intended to encourage candidates to read broadly in both their chosen subjects, rather than to concentrate on one and merely show 'competent knowledge' in the other. Adam Sedgwick, FRS, the great-nephew and namesake of Whewell's contemporary, then Reader in Animal Morphology and a graduate of the NST, hoped this would be 'one step in the direction of getting rid of the hide-bound condition into which things were getting through the immense increase of specialisation in the University.'[86] Others argued that a broader education in Part II was, in itself, desirable, and that Cambridge should have no truck with a specialised education which would be a 'preparation for professional and industrial careers'.[87]

Reaction to the proposal was swift. W. C. D. Whetham of Trinity strongly objected, claiming that the study of 'one subject in all its bearings and relations to others, by many methods and from different points of view',[88] was vital, and that a broader reading in two subjects would lead students to more work and to the loss of 'that keenness and originality so necessary for their research work or whatever career they intended to pursue'.[89] Similarly, A. C. (later Sir Albert) Seward, FRS, a graduate of the NST in 1885, later Master of Downing, advocated that Part II 'should be primarily designed to qualify a student for the prosecution of research'.[90] When the proposal was offered to the Senate, it was soundly rejected.

Thereafter 'depth' and specialisation were to be the hallmark of Part II. Indeed, in 1902, Senate discussed a recommendation that First Class degrees be awarded in Part II for one subject alone. This extreme proposal, which divided Cambridge scientists, precipitated what contemporaries saw as 'a crisis in the history of scientific education at Cambridge.'[91] Those who, clearly in the ascendant, supported specialisation, were themselves often recent graduates of the NST, and most aware of the increasing pressure of new knowledge. Those who, a few years earlier, had fought for an increase in 'breadth' now found themselves in the rearguard of reaction. Their argument was familiar:

> The width of attainment which is a peculiar feature of the Cambridge course is of the greatest benefit to our students of Natural Science. We regard therefore such a proposal as the present one, by which a final step in the direction of increased specialisation would be made, as wrong in principle, and we are convinced that this change would do grave injury.[92]

But their opponents dismissed their claims as anachronistic. In debate, Francis Darwin, FRS, (University Lecturer in Botany) described the distance that now separated him and his father's generation, over the space of three decades:

> Specialisation, we are told, is an evil to be strenuously opposed; a specious argument, [and] entirely at variance with the lessons of modern progress.... If our education is to be framed in the interests of scientific investigation, specialisation in a sense must be our aim.[93]

His arguments won the day. From 1904, it became permissible to offer only one subject in Part II. Cambridge science reaped rewards. Some 38 future Fellows of the Royal Society had taken the NST in the thirty years between 1851 and 1881; the number increased to 95 between 1882 and 1904.

The demands of specialisation finally triumphed, partly because of the research lobby, but also because of the inexorable logic of discipline development. The young physiologist Joseph (later Sir Joseph) Barcroft

(NST, 1897), writing to the *Cambridge Review* in 1902, pointed out that 'the quantity of information that the student must acquire has increased some 30–40% in the last half dozen years', with the result that even a coach was now obliged to take 'enormous classes to which he dictates 'peptonised' natural science'.[94] Clearly, it was becoming unrealistic to expect students to submit to examination in so great an area. As Barcroft recognised, it was 'the very breadth of the subject and the rapidity with which it grows (which) are to some extent its undoing'.

Sporadic attempts were made to reconcile the newer ideal of a liberal education in science with the requirements of research. For example, Sidney Harmer, the biologist, pleaded in 1902 'for the utmost width of scientific education that can satisfactorily be given in the time available', in the belief that 'a man will be better equipped for research work in proportion to the extent to which he can bring the results of other sciences to bear upon his own particular investigation'.[95] It could be argued that the teaching of science at Cambridge, buffered by the college system and by the powerful influence of sports and the other elements of the Cambridge student culture, produced a 'liberal' science graduate until well into the present century; and that, when 'specialised' graduates did appear, they were far more 'liberally' educated than their contemporaries in London or the provincial universities. But the days when an NST candidate would be expected to emulate 'the stupendous Master of Trinity' had passed forever. For some of the earlier 'naturals' this was a matter for regret. It was clear to Sedgwick and A. E. (later Sir Arthur) Shipley, tutor in natural sciences and later Master of Christ's, who took a First in the NST in 1884, that those wishing for change wanted 'to make their students Chemists, or Physicists, or Anatomists, or whatever it might be and nothing more. Surely', they appealed, 'this is not a high ideal of what a three or four year stay at Cambridge ought to do for a man?'[96]

Conclusion

Within the space of fifty years, certainly by the end of the Edwardian period, the NST had come to reflect features which were apparently antithetical to the mid-Victorian idea of a liberal education. Yet it represented at the time the crystallisation of a new ideal of service, based upon a new division of intellectual labour. The NST did not so much destroy as redefine the University's traditional duty to provide a 'due supply' of men serviceable 'in church and state'.[97] In judging its contextual significance, the revolution in Cambridge science cannot be separated from the many contemporary 'revolutions' Cambridge experienced in university finance and college customs, in the end of celibacy and the requirement of holy orders among dons, the abolition of Tests, the democratisation of university debate and the rise of new interest groups among 'demonstrators' and students.[98] However, against this background the NST stands out remarkably as focussing and mediating many changing traditional values of the university. In this sense the history of the NST not only symbolised a formative stage in the conflict of studies and purpose within the English educational tradition; it also reflected the gradual adaption of Cambridge to the new politics of knowledge which would characterise British universities in the twentieth century.

Acknowledgements

Research on which this paper is based was conducted with the assistance of a grant from the Social Science Research Council. The authors wish to thank Dr. Kay MacLeod for her comments and suggestions on earlier drafts.

Notes

1 C. J. Bayley, 'University Reform', *Cambridge Review*, **89,** (1849), 499, 501.

2 See Roy MacLeod and Russell Moseley, 'Breadth, Depth and Excellence: Sources and Problems in the History of University Science Education in England, 1850–1914', *Studies in Science Education*, **5,** (1978), 85–106.

3 See J. Adamson, *English Education, 1789–1902*, (London, 1930), pp. 174 et seq; D. A. Winstanley, *Later Victorian Cambridge*, (Cambridge, 1947), pp. 185–8; 190–206; V. H. H. Green, *The Universities*, (London, 1969), pp. 35, 42, 235–40; Michael Sanderson, *The Universities and British Industry, 1850–1970*, (London, 1972).

4 Sheldon Rothblatt, 'The Student Subculture and the Examination System in early 19th Century Cambridge', in Lawrence Stone (ed.), *The University in Society*, 2 vols., (Princeton, 1975), 1, pp. 247–303. By 1900 there were in Cambridge twelve Triposes outside Mathematics and Classics, including the Law and History, Historical, Semitic Languages, Indian Languages, Oriental Languages, Medieval and Modern Languages, Mechanical Sciences and Economic Triposes. (We have used, arbitarily, the Latin rather than the Greek plural of Tripos). Cf. J. O. McLachlan, 'The Origin and Early Development of the Cambridge Historical Tripos', *Cambridge Historical Journal*, **9,** (1947–49), 78–105; D. R. Stoddart, 'The RGS and the Foundation of Geography at Cambridge', *The Geographical Journal*, **141,** (July 1975), 1–24.

5 John Roach, *Public Examinations in England, 1850–1900*, (Cambridge, 1971), p. 31.

6 Particular issues concerning women at Cambridge are considered in Roy MacLeod and Russell Moseley, 'Fathers and Daughters: Reflections on Women, Science and Victorian Cambridge', *History of Education*, **8,** (1979), 321–33.

7 These influences are described in Roy MacLeod and Russell Moseley, *Science with Honours: The Natural Sciences Tripos and Victorian Cambridge* (in preparation). For an administrative account of the NST, see R. F. Holmes, 'An Historical Note on the Natural Sciences Tripos', *Cambridge Review*, 23 January 1965, 199–241.

8 To borrow the traditional formula, popular in contempory usage. Cf. William Somerville Orr's immensely successful series, published in 9 volumes, 1854–56.

9 For Cambridge, see Winstanley, *op. cit.*, note 3; and *idem.*, *Early Victorian Cambridge*, (Cambridge, 1940); for the reform period of British sciences, see Roy MacLeod, 'Whigs and Savants: The Royal Society of London and Reform, 1830–1848', in I. Inkster and J. B. Morrell, (ed.), *Metropolis and Province, (London, 1980)*.

10 Surprisingly, voting on the syndicate's proposals showed that older MA's (of more than seven years standing) were more in favour of the new studies than their younger counterparts. (*The Athenaeum*, 4 November 1848, 1102).

11 The classical account of the Mathematical Tripos is that of W. W. R. Ball, 'The Mathematical Tripos', reprinted in his *Cambridge Papers*, (London, 1918), 252–316. See also

C. Wordsworth, *Scholae Academicae: Some Account of the Studies at the English Universities in the Eighteenth Century*, (Cambridge, 1877).

12 A. F. Hort, *Life and Letters of F. J. A. Hort*, (London, 1896), p. 191.

13 Quoted in the *Cambridge Chronicle*, 9 November 1867.

14 On the relationship between the schools and the universities, see W. H. Brock, *supra*.

15 The Previous Examination was established in 1822. Its original purpose was to test a candidate's suitability for a higher education and to this end consisted of questions in classics, mathematics and on the tenets of the Established Church.

16 *Nature*, **5,** (1872), 429.

17 *Cambridge Papers*, Cambridge University Library (hereafter, CUL), J. 5990. Cambridge University Natural Science Society, President's Report for Easter Term 1863, p. 4.

18 *Ibid.*, p. 7.

19 By 1870, between ten and fourteen scholarships and exhibitions were offered annually by eight colleges: Clare, Christ's, Peterhouse, St. John's, Trinity, Sidney Sussex, Caius and Downing.

20 *Cambridge Papers*, *op. cit.* note 17, pp. 18–19.

21 *Ibid.*

22 *Ibid.*

23 Richard Yeo, *Natural Theology and the Philosophy of Knowledge in Britain, 1819–1869*. (Unpublished Ph.D. thesis, University of Sydney, 1977), pp. 103–61. *Idem* 'Superficial versus Profound Knowledge: Social and Cultural Dimensions of the Philosophy of Knowledge in Mid-Nineteenth Century Britain', *Proceedings of the Annual Conference of the Australasian Association for the History and Philosophy of Science*, (1979).

24 Namely, chemistry, mineralogy, geology, zoology with comparative anatomy, and botany.

25 'Modern History and other Matters at Cambridge', *Fraser's Magazine*, **45.** (1852), 173.

26 'Aspects of Science and the Universities', *Nature*, **104,** (1919), 256.

27 In 1871, of the eleven men answering questions in mineralogy, only one obtained sufficient marks (334 out of 1000) to count towards his final total, the other ten scores being so low as to be discounted by the examiners. These scores were 5, 10, 29, 15, 50, 38, 3, 52, 18, 30 and 95 again out of a possible 1000!

28 *NST Examiners' Reports, 1867–1909, Report for 1867*, p. 8.

29 *Ibid.*

30 *Ibid.*

31 T. G. Bonney, 'Natural Science at the University of Cambridge', *Nature*, **I,** (1870), 451–2.

32 W. N. Shaw, review of A. E. Shipley, '*J*': *A Memoir of John Willis Clark*, *Nature*, **91,** (1913), 527.

33 Coutts Trotter, *On Cambridge University Examinations*, (Cambridge, 1877), p. 3.

34 Norman Lockyer, recalling his own experiences at South Kensington, wrote in 1879, 'Go into a company of scientific men, and observe the most dogmatic, the most unfruitful and the least modest among them, you will find that this man is, as we may say, an examiner by profession', Lockyer, 'Technical Education', in *Education and National Progress: Essays and Addresses 1870–1905*, (London, 1906), p. 16.

35 *NST Examiners' Reports 1867–1909, Report for 1872.*

36 For criticisms of the Mathematical Tripos, see Ball, *op. cit.* note 11; see also, Isaac Todhunter, *The Conflict of Studies and Other Essays*, (London, 1873) and Sedley Taylor, 'Endowed Competitions at the Universities and their Results', *Macmillan's Magazine*, **29,** (1874), 380.

37 Reports of Examiners meetings are complete for 1861–73 and for 1882–96. Reports for the years 1874–1881 have either not survived or have been misplaced. Reports of Examiners Meetings beyond 1896 exist in the University Archives, but are not available to scholars under the provisions of an 'eighty-year rule'.

38 *Nature*, **8,** (1873), 21.

39 *NST Examiners' Report for 1873.*

40 *Ibid.*, Ironically, these men included Francis Maitland Balfour, later Fellow of Trinity and a

distinguished embryologist, who came second in the first class division, and the zoologist, N. M. Hartog, (third man in the same division). They obtained 17 and 24 marks respectively in mineralogy, out of their overall total of 3279 and 2448.

41 *NST Examiners' Report for 1877.*

42 *NST Examiners' Report for 1873.*

43 Donald MacAlister, 'Natural Science and Medical Practice', *Eagle*, **XII,** (1883), 10–33.

44 Sir William Priestley, 'The Sacrifice of Education', *Nineteenth Century*, **25,** (1889), 309.

45 See Roy MacLeod and Russell Moseley, 'The "Naturals" and Victorian Cambridge: Reflections on the Anatomy of an Elite, 1851–1914. *Oxford Review of Education*, **6,** (1980), 177–95.

46 See also, *NST Examiners' Report for 1873.*

47 *Cambridge University Reporter*, (27 March 1877), 342.

48 *Cambridge University Reporter*, (17 November 1874), 101.

49 The first part would be held at the end of the Easter term of the third year; the second part the following Michaelmas Term. Those who passed with credit in the first three days examination should be entitled to the BA degree'. *Cambridge University Reporter*, *ibid.* 100.

50 *Ibid.* In 1912 'poll' men still accounted for 45% of Cambridge undergraduates.

51 *Nature*, **14,** (1876), 216.

52 *Cambridge University Reporter*, (6 March 1877), 270.

53 Similar considerations affected the Mathematical Tripos; the subsequent Grace had bearing upon all eleven existing Triposes.

54 *Cambridge University Reporter,* (28 October 1879), 64–5. Between 1881 and 1898, Part I consisted of three days' written work (six papers) and two days' practical; from 1899 until 1916 it extended over seven days with two papers in each of eight subjects. Part II required success in either Part I of the NST or Part I of the Mathematical Tripos.

55 *The Athenaeum*, 1 November, 1877.

56 The question concerned the candidates in physiology – H. E. Durham, F. H. Edgeworth (later Professor of Medicine, Bristol), and W. S. Melsome – who were given less than first class marks in their knowledge of a second, cognate subject. Given the benefit of the doubt, all were admitted to firsts, but in and after the following year, the expectations were such that few men winning firsts in a given subject (65% for a First, 50% for a Second) could offer an equivalent level of knowledge in a second.

57 For a sustained analysis of the kind of 'collective curriculum' which developed, see Basil Bernstein, *Class, Codes and Control*, 3 vols., (London, 1971), vol. 1.

58 *Cambridge University Archives*, Cam. c. 877, 12. Professors' Lectures. Remarks addressed by George Liveing to the Vice-Chancellor as Chairman of the 'Professors' Lectures' Syndicate (1877). Henry Latham, *On the Action of Examinations Considered as a Means of Selection*, (London, 1877), p. 311.

59 Virginia Woolf, *Roger Fry, A Biography*, (London, 1940), p. 50.

60 William Bateson, 'Biological Fact and the Structure of Society', Herbert Spencer Lecture, Oxford, 28 February 1912, in B. Bateson, *William Bateson, FRS, Naturalist*, (Cambridge, 1928), p. 355.

61 Liveing, *op. cit.* note 58.

62 *Cambridge University Reporter,* (13 June 1887). At first, the number taking an ordinary degree in natural science was comparable with the number taking honours; in 1869, seven 'poll' men compared with thirteen NST men; in 1870, twenty-five compared with seventeen; and in 1878, thirty-two compared with twenty-three. After 1880, however, the number of poll men taking science fell sharply, while the number of honours men rose. In 1880, the figures were sixteen and thirty-two respectively.

63 Ernest Parkyn, 'The Natural Sciences Tripos', *Cambridge Review*, 8 March 1882, 211.

64 *Cambridge University Reporter*, (27 March 1887), 343.

65 See, for example, Noel Annan, *Leslie Stephen*, (London, 1951).

66 Ball, *op. cit.* note 11, pp. 305–6.

67 E. H. Pearce, Leader in *The Times,* 26 October 1906.
68 W. N. Shaw, 'Cambridge in the Nineteenth Century', *Nature,* **91,** (1913), 525.
69 Reported in *Nature,* **30,** (1884), 159.
70 For a fuller discussion of the 'anti-science' movement in England during the 1880's, see R. MacLeod, 'The Resources of Science in Victorian England', in P. Mathias, *Science and Society, 1600–1900,* (Cambridge, 1972), pp. 152–3.
71 Anon, 'Unnatural Science', *Cambridge Review,* (7 November 1883), 50.
72 'I am sure that the mind must be distorted which can contemplate the structure of an entozoon with greater satisfaction than the East Window of King's Chapel, and I cannot help fearing this class is on the increase'. *Ibid.*
73 This could give rise, claimed a critic, 'to an outward and visible expression of an inward superciliousness towards other people and their standards and manners and morals'. (Anon, 'A Criticism of the Discipline of Natural Science', *Cambridge Review,* (12 March 1884), 243).
74 'G', 'Natural Science at Cambridge', *Cambridge Review,* (3 February 1886), 178.
75 *Ibid.*
76 *Ibid.,* 179.
77 *Cambridge University Reporter,* (24 November 1885), 187–9.
78 *Ibid.,* 188.
79 *Op. cit.* note 74, 179.
80 *Ibid.*
81 J. B. Bury, 'Compulsory Greek', *Fortnightly Review,* **56,** (1891), 811–21.
82 M. M. Pattison Muir, 'Natural Science at Cambridge', *Cambridge Review,* (8 March 1886), 245.
83 Between the late 1870's and the First War, the number of external examiners steadily declined. The ratio of internal to external, of 4:4 in 1876, rose to 7:1 in 1879. After the doubling of examiners to 16 in 1885, it was rare to find more than four 'externals', and between 1905 and 1916 there were usually fewer than this.
84 'Argon', 'Natural Science at Cambridge', *The Cambridge Review,* (3 December 1896), 130.
85 *Cambridge University Reporter* (6 November 1900), 182–3. The proposal came about as a result of a circular sent out by Mr. M. M. Pattison Muir, Fellow of Caius and Superintendent of the College Chemistry Laboratories until 1908.
86 *Cambridge University Reporter,* (20 November 1900), 238.
87 *Cambridge University Archives,* UP. 60, Alfred Harker, 'Regulations for the Natural Sciencex Tripos, Part II', 3 December 1900.
88 *Cambridge University Archives,* UP. 60. 'The Natural Sciences Tripos', 30 November 1900, 240.
89 *Cambridge University Reporter,* (20 November 1900), 240.
90 *Cambridge University Archives,* UP. 60. 'Alterations in Regulations for Part II of the Natural Sciences Tripos,' 30 November 1900.
91 *Cambridge University Archives,* UP. 60. Various signatories, 'The Natural Sciences Tripos', 19 May 1902.
92 *Ibid.*
93 *Cambridge University Archives,* UP. 60. F. Darwin and A. C. Seward, 'The Natural Sciences Tripos', 19 May 1902.
94 J(oseph) B(arcroft), 'The Natural Sciences Tripos', *Cambridge Review* (22 May 1902), 330,
95 *Cambridge University Archives,* UP. 60. S. F. Harmer, 'The Natural Sciences Tripos', 21 May 1902.
96 *Cambridge University Archives,* UP. 60. A. Sedgwick and A. E. Shipley, 'The Natural Sciences Tripos', 21 May 1902.
97 Thomas Fowler, 'On Examinations', *Fortnightly Review,* **25,** (1876), 418–29.
98 See MacLeod and Moseley, *op. cit.* note 2; Sheldon Rothblatt, *The Revolution of the Dons,* (London, 1968); Gerald Geison, *Michael Foster and the Cambridge School of Physiology: The Scientific Enterprise in Late-Victorian Society,* (Princeton 1978).

IX

Scientific Careers of 1851 Exhibition Scholars

The patterns of education, career and migration of the Exhibition of 1851 Scholars provide a valuable case study for the endowment of research

by

ROY M. MACLEOD

and

E. KAY ANDREWS

DURING the past year, a group in the Science Policy Research Unit at the University of Sussex has been studying the criteria employed by public and private institutions concerned with the endowment of scientific activity in Britain, the United States and Western Europe since about 1850. We hope that this work will throw light on the development of the so-called "ethos" of science, the growth rate of different fields of science at different universities over time, the "sophistication factor" in scientific teaching and research, and the developing relations between the universities and government.

Part of our study has been devoted to the evolution of educational and professional career patterns in different national scientific communities. We have wished first to test the hypothesis that recent trends in the migration of high quality academic scientists have not been as exceptional as some might suggest; and second, to show that the "circulation of elites" between various changing centres of excellence, and within the rubric set by professional standards, may be represented qualitatively and quantitatively over time.

Historical information is at best difficult to obtain. One method consists in the use of pre-selected samples of scientists who, whether as students or as advanced scholars, held certain grants or endowed scholarships.

This article sets out to show the possibilities of this approach through the study of the Royal Commission for the Exhibition of 1851 awards—the first system of science research scholarships to embrace all the major universities of the British Empire. This particular study is made practicable by the excellent follow-up records kept by the Royal Commission for the Exhibition of 1851[1]. Under the guidance of men such as Kelvin, Rayleigh and Haldane, these awards came to provide a model which the British and American foundations and educational trusts later imitated. At a time of national self-searching, when new data are becoming available about the "brain drain"[2] and the "swing from science"[3], greater knowledge about the historical development of professional and institutional attitudes and assumptions is an important pre-requisite for scientific and educational planning. Historical sociology provides important tools with which to explore some of the "basic impulses in science and the critical conditions for its development"[4].

The Origins of the Scholarship Scheme

"Established at a time", wrote Sir Richard Glazebrook, "when the field was still untouched by any system that carried training beyond the limits of ordinary degree curricula, these scholarships have undoubtedly given a great and much needed impetus to postgraduate study"[5]. The 918 Scholars appointed by the Royal Commission for the Exhibition of 1851 between 1891 and 1960 have included no less than ninety-nine Fellows of the Royal Society (including Sir Robert Robinson, Sir Harrie Massey and Sir Harry Melville); two Presidents of the Royal Society; seven Nobel prize winners (including Sir Ernest Rutherford and Lord Todd); and two holders of the Order of Merit. The remarkable history of the Commission's scheme provides an unusual insight into the educational and career choices made by some of the most productive men in British science during the past three-quarters of a century.

By special charter in December 1851, the Royal Commission, which had supervised the conduct of the immensely successful Great Exhibition of 1851, was incorporated as a permanent body to administer the surplus of £186,000 arising from the Exhibition. Some of the Commissioners, however, expressed fears that the superiority of British manufacturers was in danger. A later report by the Commission showed the inadequacy of private funds committed to furthering science and art in Britain, in dramatic contrast to the resources which appeared available to science in Germany and France. To help redress the balance, the Commissioners proposed that national institutions should be established in London to promote and co-ordinate British efforts in scientific and industrial education.

The subsequent history of the scientific establishments on the grounds of South Kensington is well documented[6–8]. A number of national institutions were set up with the Commission's aid, including the South Kensington Museum and the Royal Colleges of Art and Science. Later, Commission funds assisted the development of the Research Department of the Imperial Institute, the Royal Albert Hall, and the National Physical Laboratory.

From the outset there was pressure on the Commission to distribute some of the income on the Exhibition's profits in such a way as to promote scientific education and industry in the provinces. The importance of this was urged by Lyon (afterwards Baron) Playfair, sometime adviser to Sir Robert Peel and one of the original Special Commissioners of 1851, who joined the Commission in 1869 as deputy chairman of its board of management. In 1851 he told his students at the Royal School of Mines that, despite Britain's triumph at the Exhibition, "if she continues to be in advance it will not be from the abundance of her coal or iron, but because, uniting science with practice, she enables her discoveries in philosophy to keep pace with her aptitude in applying them"[9]. Playfair's judgment seemed confirmed by the unspectacular British showing at the Paris Exhibition of 1867 and by the brilliant industrial successes of Germany in the Franco-Prussian war. That the Commissioners were influenced by these events is clear in their sixth report, published in 1878[10]. At the same time, the Commissioners announced their intention of establishing "a number of scholarships for the purpose of advising the development of scientific and technical training in the manufacturing districts of the country"[10].

For the next eleven years the Commission was obliged to devote its annual income to repaying the original mortgage on the South Kensington estate, and not until 1889, when the mortgage was paid and a surplus of £5,000 annually became available, was further action taken. Appropriately, Playfair was appointed chairman of a Science Scholarship Committee set up in 1890 to determine the form the new scheme should take. Following recommendations by Sir Norman Lockyer, the Committee adopted a programme along the lines developed by Dumas at the École Pratique des Hautes Études in Paris*[11]. Essentially the scheme was as follows: awards of £150 for two or three years of research were made available to British subjects younger than thirty years of age who had received university training, who showed promise of

* "The Committee had their attention drawn to the fact that there is a large number of scholarships in the country and if the Commissioners act on the same lines as those already occupied it is possible that education will gain little by their action, as the Commissioners may interfere with the establishments of new scholarships by private liberality. Hence it is desirable that these new scholarships . . . should be of a higher order than most of those now existing . . . This system has been adopted with excellent effect by the French École Pratique des Hautes Études." (Report to the Royal Commission in 1890, as quoted in ref. 11.)

capacity for advancing science or its applications by original research and who had been nominated by their home universities. An "1851 Scholar" could go to any university in England or abroad, except to that one institution by which he had been nominated. Scholars were required to submit annual reports to the Commission which, in turn, submitted them to authoritative assessors.

Playfair's warning, underscored by the reports of the Royal Commission on Technical Education in 1883–4, was repeated in Article II of Playfair's report to the board of management. This specified that the principal research of each Scholar must lie in some branch of science "the extension of which is important to the national industries". The intention of the founders was clearly embodied in the concept of an "officer corps" of scientific workers who, like their colleagues in Germany, would lead a growing army of skilled research workers[12].

Nomination Procedure

Because the Commissioners set out to aid the development of science in the manufacturing districts of Britain and the Empire*, Oxford and Cambridge were specifically excluded from the original list of nominating universities. All other English and Scottish universities, numbering thirteen in 1891 and seventeen in 1919, were invited to nominate students annually. The constituent colleges of the federal universities of Wales and Ireland were invited to nominate students in rotation. Eight scholarships were set aside each year for students younger than twenty-six from the universities of the Empire[13]. Two of each of the four major universities in Canada, McGill, Toronto, Queens University (Kingston), Ontario and Dalhousie, and two of the four principal universities in Australia and New Zealand, were invited to nominate students every year. Between 1891 and 1915, 415 scholarships were awarded, ninety-seven (or 23 per cent) of which went to Scholars outside the United Kingdom. Between the First and Second World Wars 295 awards were granted, 159 (54 per cent) of them to universities of the Empire. Between 1945 and 1960, 208 awards were granted, 146 (72 per cent) of them to Commonwealth Scholars.

The scheme was suspended during the First World War, when Scholars not in the Forces were required to work under the direction of the Ministry of Munitions. Scholarship awards were resumed in 1919, but a complete revision was necessary. On the one hand, British science faculties had expanded and multiplied and were urgently seeking more assistance than the Commission could give. On the other hand, the development of new governmental and

* For convenience, the terms "Empire" and "Commonwealth" are used interchangeably to designate the countries represented in the tables. The terms are not used rigorously except where the emphasis is obvious.

Table 1. DISTRIBUTION OF SCHOLARS BY NOMINATING UNIVERSITIES

	Total	1891–1914	1919–1939	1946–1960
Total for all universities	918	415	295	208
UK universities				
Total	531	318	136	77
Scotland	75	63	11	1
England	351	197	97	57
Wales	36	22	10	4
Ireland*	69	36	18	15
Overseas universities				
Total	387	97	159	131
Canada	151	46	74	31
New Zealand	54	13	20	21
Australia	117	32	44	41
South Africa	33	6	17	10
India and Pakistan	32	—	4	28

* In 1923, the National University of Ireland was transferred into the ' Overseas" (Empire and Commonwealth) category.

institutional schemes for encouraging science enabled the Commission to be more selective in its use of resources.

In 1918, the new Department of Scientific and Industrial Research (DSIR) began its research programme for postgraduate work, in many ways using the 1851 scheme as a model. In 1922, the Commissioners decided to devote their own awards to the encouragement of advanced research by experienced scientists under the age of thirty. The United Kingdom grants were therefore re-designated "Senior Studentships", and designed to offer "selected students of exceptional promise and proved capacity for original work the opportunity of devoting their whole time, for a period of not less than two years, to the prosecution of scientific research[14]". The number of annual awards was reduced from seventeen to thirteen (five for the United Kingdom and eight for the Commonwealth), but the stipend was increased to £450 and £250 for the United Kingdom and Commonwealth students respectively. The total budget remained unchanged at £5,000.

The distribution of Scholars by country of nomination is given in Table 1. Between 1919 and 1939 an average of six candidates applied for every Senior Studentship awarded. Among overseas Scholars, an average of twenty-five to thirty candidates applied for the eight awards available each year; three of these scholarships went to institutions in Canada, two to Australia and one each to New Zealand, South Africa and Eire. After the Second World War, two additional awards were made available for Pakistan and India. The highly competitive nature of the awards undoubtedly helps to explain the fact that so many of the Scholars later reached such heights of scientific achievement.

The Commissioners' post-war change of policy brought Oxford and Cambridge into the list of nominating universities in 1921. Almost immediately the science schools of Oxford and Cambridge began to furnish more scholars than any British university except London. Cambridge led the way with 28 per cent of the nominations between

1919 and 1939; between 1946 and 1960, Oxford and Cambridge nominated 42 per cent of the total number of United Kingdom Scholars.

Educational Patterns

Educational patterns prevailing among the Scholars in the pre-war interval and post-war periods are shown in Table 2. During the period 1891–1914, chemistry and physics were by far the most popular subjects, with chemistry absorbing between 45–50 per cent of the Scholars and physics absorbing 25–45 per cent of the Scholars every year. The life sciences were generally second in popularity, with engineering and geology, meteorology and medicine following in descending order. The predominance of chemistry and physics reflects contemporary activity in those sciences; it remains to be seen, however, whether the relative proportions reflected the actual proportions of all scientists who were involved in research during each period. The most significant change in the educational pattern of the period seems to be the relative increase of

Table 2. DISTRIBUTION OF SCHOLARS BY FIELD (in per cent)

	Total	Chemistry, bacteriology	Physics	Zoology/ botany/ biology, anthrop.	Engineering, mathematics	Meteorology, astronomy	Medicine, pharmacy	Geology/ mining
All scholars								
Total	100·0	43·1	30·8	13·2	4·8	1·4	0·4	5·0
1891–94	100·0	44·4	44·4	3·2	7·9	—	—	—
1895–98	100·0	39·1	27·5	11·6	15·9	1·4	—	4·3
1899–1902	100·0	47·6	38·1	3·2	3·2	4·8	—	3·2
1903–06	100·0	50·0	28·8	9·1	4·5	3·0	—	4·5
1907–10	100·0	49·3	25·3	9·3	6·7	1·3	2·7	5·3
1911–14	100·0	49·4	36·7	7·6	—	1·3	—	5·1
1919–21	100·0	50·0	23·2	14·3	1·8	3·6	—	7·1
1922–26	100·0	35·0	31·7	21·7	5·0	3·3	—	3·3
1927–31	100·0	29·8	40·3	22·4	1·5	—	—	6·0
1932–36	100·0	39·7	30·9	19·1	2·9	1·5	—	5·9
1937–39	100·0	43·2	27·3	20·5	2·3	—	—	6·8
1946–50	100·0	39·4	33·8	9·9	7·0	—	1·4	8·5
1951–55	100·0	50·0	26·5	16·2	2·9	—	—	4·4
1956–60	100·0	44·9	23·2	20·3	4·3	—	1·4	5·8
United Kingdom								
Total	100·0	41·4	29·9	15·6	5·3	1·6	0·8	5·3
1911–12	100·0	50·0	41·4	3·4	—	1·7	—	3·4
1919–21	100·0	52·9	23·5	11·8	2·9	5·9	—	2·9
1922–26	100·0	24·0	36·0	24·0	12·0	4·0	—	—
1927–31	100·0	44·0	28·0	16·0	4·0	—	—	8·0
1932–36	100·0	38·5	19·2	26·9	7·7	—	—	7·7
1937–39	100·0	42·9	35·7	14·3	—	—	—	7·1
1946–50	100·0	35·0	25·0	5·0	20·0	—	5·0	10·0
1951–55	100·0	20·0	40·0	30·0	5·0	—	—	5·0
1956–60	100·0	45·5	9·1	27·3	4·5	—	4·5	9·1
Empire and Commonwealth								
Total	100·0	43·2	31·4	17·2	1·5	0·6	—	6·2
1911–14	100·0	47·6	23·8	19·0	—	—	—	9·5
1919–21	100·0	45·5	22·7	18·2	—	—	—	13·6
1922–26	100·0	42·9	28·6	20·0	—	2·9	—	5·7
1927–31	100·0	21·4	47·6	26·2	—	—	—	4·8
1932–36	100·0	40·5	38·1	14·3	—	2·4	—	4·8
1937–39	100·0	43·3	23·3	23·3	3·3	—	—	6·7
1946–50	100·0	41·2	37·3	11·8	2·0	—	—	7·8
1951–55	100·0	62·5	20·8	10·4	2·1	—	—	4·2
1956–60	100·0	44·7	29·8	17·0	4·3	—	—	4·3

Scholars studying the life sciences. In 1914, these absorbed only 7·6 per cent of the Scholars. By 1921 the percentage doubled, by 1939 it had almost trebled among United Kingdom Scholars; after 1950 the life sciences accounted for 30 per cent of all United Kingdom Scholars.

Table 3 shows the distribution of Scholars by subject, with reference to the country in which they chose to do their research. This table is supported by the evidence of Table 4, which demonstrates the obvious concentration of Scholars in certain universities. Tables 3*a*, *b*, and *c* illustrate the distribution of Scholars during each of the three periods. Table 3*a* shows that 167 (53 per cent) of students chose at some stage to study in Germany during the years 1891–1914. Of this number, 74 per cent were chemists and 18 per cent were physicists. Of the total number of chemists, 73 per cent studied at some stage in Germany during this period. As chemists sought out the rigorous training of German teachers and laboratories, 58 per cent of the total number of physicists studied exclusively in Britain, and physicists composed 46 per cent of the total number of Scholars studying in the United Kingdom. This distribution of United Kingdom Scholars among British and European universities according to subject of research is reflected, too, in the distribution of overseas Scholars; 50 per cent of the overseas Scholars who came to study in the United Kingdom were physicists, while 38 per cent of the chemists went to Germany.

The years between the wars introduced a very significant change in the migration patterns. By 1919 the voluntary

Table 3*a*. DISTRIBUTION OF SCHOLARS BY SUBJECT AND PLACE OF STUDY, 1891–1914

Scholars from the UK

Place of research	Subjects of research							
	Chem.	Phys.	Life sciences	Engineering	Geol.	Astron.	Med.	Total
United Kingdom	34	55	11	12	4	2	1	119
Germany	76	6	—	2	1	—	—	85
UK/Germany	29	20	4	1	1	—	—	55
Empire/UK and Empire	—	—	—	1	—	—	—	1
USA/UK and USA	—	—	—	1	2	—	—	3
Europe	6	3	—	—	—	—	—	8
UK/Europe	7	7	4	—	—	2	—	20
Germany/Europe*	20	3	3	—	—	1	—	27
Total	171	94	22	17	8	5	1	318

Scholars from the Empire/Commonwealth

	Chem.	Phys.	Life sciences	Engineering	Geol.	Astron.	Med.	Total
United Kingdom	7	23	4	5	3	3	—	45
Germany	5	6	—	1	1	—	—	13
UK/Germany	4	2	3	—	1	1	1	12
Empire/UK and Empire	1	6	—	1	—	—	—	8
USA/UK and USA	5	7	2	2	3	—	—	19
Europe	—	—	—	—	—	—	—	—
UK/Europe	—	—	—	—	—	—	—	—
Germany/Europe*	—	—	—	—	—	—	—	—
Total	23	44	9	9	8	4	1	97

* This category represents students who studied in both German universities and one or more other European (that is, non-British) universities.

Table 3*b*: 1919–1939

Scholars from the UK

	Chem.	Phys.	Life sciences	Engineering	Geol.	Astron.	Med.	Total
United Kingdom	47	28	21	5	6	3	—	110
Germany	—	3	—	—	—	—	—	3
UK/Germany	—	1	1	—	—	—	—	2
Empire/UK and Empire	1	—	1	1	—	—	—	3
USA/UK and USA	3	2	—	1	—	—	—	6
Europe	—	—	—	—	—	—	—	—
UK/Europe	—	—	—	—	—	—	—	—
Germany/Europe	—	—	—	—	—	—	—	—
Total	51	34	23	7	6	3	—	124

Scholars from the Empire

	Chem.	Phys.	Life sciences	Engineering	Geol.	Astron.	Med.	Total
United Kingdom	59	52	2	31	11	2	—	157
Germany	1	2	—	—	—	—	—	3
UK/Germany	3	—	—	—	—	—	—	3
Empire/UK and Empire	2	—	—	1	—	—	—	3
USA/UK and USA	—	2	—	3	—	—	—	5
Europe	—	—	—	—	—	—	—	—
UK/Europe	—	—	—	—	—	—	—	—
Germany/Europe	—	—	—	—	—	—	—	—
Total	65	56	2	35	11	2	—	171

Table 3*c*: 1946–1960

Scholars from the UK

	Chem.	Phys.	Life sciences	Engineering	Geol.	Astron.	Med.	Total
United Kingdom	17	14	11	5	4	—	2	53
Germany	—	—	—	—	—	—	—	—
UK/Germany	1	1	—	—	—	—	—	2
Empire/UK and Empire	—	—	2	—	—	—	—	2
USA/UK and USA	2	—	—	—	—	—	—	2
Europe	1	—	—	—	—	—	—	1
UK/Europe	—	—	—	—	—	—	—	—
Germany/Europe	—	—	—	1	1	—	—	2
Total	21	15	13	6	5	—	2	62

Scholars from the Commonwealth

	Chem.	Phys.	Life sciences	Engineering	Geol.	Astron.	Med.	Total
United Kingdom	69	42	17	4	8	—	—	140
Germany	1	1	—	—	—	—	—	2
UK/Germany	—	—	—	—	—	—	—	—
USA/UK and USA	—	—	1	—	—	—	—	1
Europe	2	—	1	—	—	—	—	3
Total	72	43	19	4	8	—	—	146

flow of students to Germany and to other European universities had almost completely ended because of the war and the rapid deterioration of research conditions in German laboratories. A very important factor, however, was the development of new research facilities and doctoral research programmes in Britain. As shown in Table 3*b*, after the First World War, 88 per cent of the United Kingdom Scholars and 89 per cent of the overseas Scholars chose to stay in the United Kingdom. From Table 3*c* we see that after the Second World War, 85 per cent of the United Kingdom Scholars and 93 per cent of the overseas Scholars studied exclusively in the United Kingdom. The flow of overseas research Scholars to the

Table 4. DISTRIBUTION OF UNIVERSITY ATTENDANCES AT THE TEN LEADING UNIVERSITIES

	Total: Number of attendances	Total: % of total attendances	From United Kingdom	From Empire and Common-wealth
1891–1960				
Total	1,026	100·0	549	477
Ten leading universities	795	77·5	449	346
1891–1914				
Total	448	100·0	317	131
Ten leading universities	315	70·3	254	61
Cambridge	82		54	28
London	50		41	9
Leipzig	45		36	9
Göttingen	35		30	5
Berlin	30		31	2
Manchester	23		16	4
Zurich	17		17	—
Paris	14		14	—
Bristol	11		9	2
Liverpool	8		6	2
1919–1939				
Total	341	100·0	158	183
Ten leading universities	285	83·6	140	145
Cambridge	148		72	76
London	44		18	26
Oxford	32		13	19
I.C.S.T.	33		21	12
Manchester	15		8	7
Edinburgh	4		2	2
Bristol	3		2	1
Leeds	3		2	1
Nottingham	2		1	1
Glasgow	1		1	—
1946–1960				
Total	237	100·0	74	163
Ten leading universities	195	82·3	55	140
Cambridge	91		26	65
Oxford	34		12	22
I.C.S.T.	23		10	13
London	17		2	15
Birmingham	7		—	7
Bristol	6		2	4
Edinburgh	6		1	5
Manchester	5		—	5
Liverpool	3		1	2
Leeds	3		1	2

The total number of attendances over the period 1891–1960 is 1,026, while the total number of Scholars is 918.

universities of the USA also dried up. Whereas before the war 18 per cent of the overseas Scholars, including many Canadians, had studied in the United States, this trend had almost entirely disappeared between 1919 and 1939 This seems to suggest that the policy of the Commissioners to encourage overseas students to come to the United Kingdom had been successful.

Turning to Table 4, we see the distinctive popularity scores of the top ten "research universities". We find that the incidence of research attendance corresponds to the presence of internationally distinguished researchers. There was a tendency of chemists to visit Germany before the First World War; Leipzig, Göttingen and Berlin absorbed 75 per cent of the 162 Scholars who studied wholly or partly in Germany, and the same universities accounted for 30 per cent of the total number of attendances recorded at all universities during the period. Between 1891 and 1895, Leipzig alone received four Scholars every year. From the fact that 41 of the 45 Scholars who went to Leipzig before the war were chemists, we may reasonably infer that they went to study with the famous chemist, Wilhelm Ostwald, Professor of Chemistry between 1887 and 1906. Certainly the number of British Scholars at Leipzig declined sharply after Ostwald's death; between 1907–1914 there were only five Scholars there.

The number of physicists in England were largely concentrated at the Cavendish at Cambridge which, by the end of the century, had already established a brilliant reputation. Of the ninety physicists educated in England during this period, fifty-four studied at the Cavendish. Furthermore, of the eighty-two Scholars researching at Cambridge during this period, fifty-four were physicists, twelve were life scientists, eight were chemists, and eight were engineers.

The increased concentration of Scholars in the United Kingdom after 1919 reflects the emergence of "centres of excellence" at Oxford, Cambridge and London (particularly Imperial College, London). Between the wars Cambridge absorbed 43 per cent of the total number of Scholars studying in the United Kingdom, and the four "top" universities—Cambridge, Oxford, the University of London, and Imperial College—accounted for 70 per cent of the total. Between 1946–1960, 38 per cent of the total number of Scholars studied at Cambridge, and the four leading universities accounted for 74 per cent of all the Scholars. If this trend is indicative, it may suggest something about the direction and rate at which British universities were coping with this flood of scientific manpower and were attempting to "balance" their scientific faculties. In 1914, only twenty of the Scholars at Cambridge were chemists; by 1925 almost half of the Scholars there were chemists.

The "1851 Scholars" were not specifically required to publish their research, nor to work for a higher degree. Nonetheless, of the 389 Scholars receiving awards during the period 1891–1914, 335 (86 per cent) took "higher degrees". Seventy-nine (22 per cent) of the Scholars took German PhDs, of which seventy-two were in chemistry. Twenty-six of these degrees were awarded from Leipzig alone. Before the war the PhD was a German phenomenon. Only a few British universities awarded a research doctorate and this was usually called the DSc. During the period 1891–1919, 64 per cent of the total Scholars studying in the United Kingdom were awarded British degrees of which 75 per cent were DScs, the remainder being MScs. About 12 per cent of the Scholars took both

British and German degrees. After the war, in correspondence with the change in the educational pattern, the quality and type of degree altered. Of the Scholars between 1919–1939, 92 per cent were awarded British degrees.

Career Patterns

The careers chosen by the Scholars also show some interesting trends which invite further research. The data given in Table 5 show that about half of all the Scholars ever appointed chose to pursue academic careers. In this connexion two points are worthy of notice. First, despite the avowed intention of the original Commissioners to use the Scholarship scheme to stimulate the improvement of industrial science, it is manifest that this goal was never realized. The percentage of Scholars who entered industry fell sharply from 25 per cent between 1891–1914 to 16·3 per cent between 1919–1939 and to 5·8 per cent between 1945–1960. The inter-war shift was partly influenced by the development of DSIR studentship schemes specifically designed to meet the needs of industry, but it also seems to have reflected a tendency on the part of the Scholars themselves to move away from industrial connexions. The distribution of scientists by discipline shows that, between 1891 and 1914, academic life attracted 50 per cent of the physicists, 47 per cent of the chemists, 56 per cent of the life scientists, 51 per cent of the engineers and all the mathematicians. In the same period, industry absorbed 14 per cent of the life scientists, 15 per cent of the physicists, 38 per cent of the chemists and a third of the engineers. Government scientific institutions and laboratories absorbed 16 per cent of the physicists, 5 per cent of the chemists, 16 per cent of the life scientists, and 8 per cent of the engineers.

Again, in the period 1919–1929, into university life went 58 per cent of the physicists, 42 per cent of the chemists, 43 per cent of the life scientists, and all the mathematicians. Industry took only 10 per cent of the physicists, 26 per cent of the chemists, and 10 per cent of the life scientists. Since the Second World War, 61 per cent of the chemists have entered academic life, together with 62 per cent of the physicists, 58 per cent of the life scientists and 90 per cent of the mathematicians. By 1960, industry had accounted for only 3 per cent of the post-war Scholars in physics, 10 per cent of the chemists and 10 per cent of the mathematicians, although these figures are slightly suspect, because many of the group were still working on research scholarships at that date.

If these statistics are representative of the best of British science as a whole, they suggest that the trend away from industrial science and toward academic careers gathered momentum just after the First World War. This was the time when American scientists began entering specialized industrial research institutions in large numbers[14]. Naturally, any interpretation of the Scholars' experience must allow for the consequences of the Commissioners' altered award policy after 1922, and further research is clearly needed before the repercussions of this policy can be fully established.

Migration Patterns

Perhaps the most intriguing set of hypotheses to be drawn from the Scholars' experience have reference to the vexed question of international migration among "pure" scientists. From Table 6 we can form a number of interesting speculations about this aspect of the "brain drain".

Following the 1914–1918 war, the proportion of United Kingdom Scholars emigrating to the USA increased from 5·0 per cent to 9·7 per cent. Although recent data are incomplete this proportion seems to have remained at about 9 per cent between 1946 and 1960. During the period 1891–1914, the proportion of United Kingdom Scholars who chose to emigrate to various parts of the Empire (13 per cent) was higher than that which emigrated to the United States (5 per cent); likewise, between 1919–1939, 16 per cent of the United Kingdom Scholars emigrated to the Empire as opposed to 9 per cent who went to the United States. After the Second World War, however, only 6 per cent of the Scholars chose to emigrate

Table 5. DISTRIBUTION OF SCHOLARS BY SUBSEQUENT CAREERS (in per cent)

	Total	1891–1914	1919–1939	1946–1960
All scholars				
Total	100·0	100·0	100·0	100·0
Academic (university posts)	51·3	49·2	48·5	59·6
Industrial posts	18·2	25·8	16·3	5·8
Research in Govt. establishments	14·9	12·3	20·0	13·0
Government administration	3·4	4·3	3·7	1·0
Professional posts (law, medicine, etc.)	3·6	5·1	3·7	0·5
Army	0·2	—	0·7	—
Continuing research	5·8	—	4·7	18·8
Unknown	1·5	1·2	0·7	1·4
Dead	1·1	1·2	0·7	1·4
United Kingdom				
Total	100·0	100·0	100·0	100·0
Academic (university posts)	50·8	47·5	54·8	59·7
Industrial posts	22·8	28·9	14·5	8·1
Research in Govt. establishments	10·9	10·4	13·7	8·1
Government administration	4·2	5·0	4·0	—
Professional posts (law, medicine, etc.)	4·8	5·3	4·8	1·6
Army	0·2	—	0·8	—
Continuing research	3·2	—	2·4	21·0
Unknown	2·0	1·9	3·2	—
Dead	1·2	0·9	1·6	1·6
Empire and Commonwealth				
Total	100·0	100·0	100·0	100·0
Academic (university posts)	51·9	54·6	43·9	59·6
Industrial posts	12·6	15·5	17·5	4·8
Research in Govt. establishments	19·8	18·6	24·6	15·1
Government administration	2·4	2·1	3·5	1·4
Professional posts (law, medicine, etc.)	2·2	4·1	2·9	—
Army	0·2	—	0·6	—
Continuing research	8·9	—	6·4	17·8
Unknown	1·0	3·1	0·6	—
Dead	1·0	2·1	—	1·4

Table 6. DISTRIBUTION OF SCHOLARS BY PLACE OF ULTIMATE RESIDENCE

		Students from UK			Students from Empire and CMW		
	Total	1891–1914	1919–1939	1946–1960	1891–1914	1919–1939	1946–1960
Total	918	318	124	62	97	171	146
Analysis by country of residence							
United Kingdom	424	231	89	43	15	25	21
Germany	—	—	—	—	—	—	—
Empire/Commonwealth	310	43	20	4	50	113	80
USA	93	16	12	6	22	20	17
Other countries	28	12	—	1	5	6	4
Dead / Continuing research / Unknown	63	16	3	8	5	7	24
Per cent total	100·0	100·0	100·0	100·0	100·0	100·0	100·0
Analysis by country of residence							
United Kingdom	46·1	72·7	71·5	69·4	15·5	14·6	14·4
Germany	—	—	—	—	—	—	—
Empire/Commonwealth	33·8	13·5	16·3	6·5	51·5	66·1	54·8
USA	10·1	5·0	9·8	9·7	22·7	11·7	11·6
Other countries	3·1	3·8	—	1·6	5·2	3·5	2·7
Dead / Continuing research / Unknown	6·9	5·0	2·4	12·9	5·2	4·1	16·4

to the Commonwealth, while the proportion emigrating to the United States remained at about 9·5 per cent.

The number of overseas students who failed to return to their home countries was proportionately higher than those students emigrating from the United Kingdom. The number of overseas Scholars emigrating to the United States declined from 22·7 per cent of the Scholars during the period 1891–1914 to 11·7 per cent during the two consecutive periods. This is partly a reflexion of the increasing numbers of overseas Scholars who were coming to the United Kingdom to continue their research work rather than going to American universities. Throughout the period the proportion of overseas students who chose to remain in the United Kingdom remained about level at 14–15 per cent.

These figures suggest that the relative proportions of both United Kingdom and overseas Scholars who migrated to and remained in another country appear to have been fairly constant. The numbers are too small to make comparisons by discipline meaningful, but there is perhaps some significance in the fact that in each period between 16–24 per cent of the United Kingdom Scholars and between 25–37 per cent of the overseas Scholars chose to make their permanent homes outside the United Kingdom or their country of birth.

It is interesting to analyse briefly the after-careers of those Scholars who emigrated from the United Kingdom during the three periods. During the 1891–1914 period, of the sixteen emigrants to the United States, four went into the scientific service of the government, eight took up academic posts, and four went into industry. During the second period 1919–1939, of the twelve emigrants to the United States, six were academics, three went into industry and three worked in government scientific establishments. Between 1946–1960, of the six emigrants, four became academics and two went into industry.

Of the forty-three United Kingdom emigrants to the Empire during 1891–1914, thirty were academics, nine went into industry and four went into the scientific service of the government. Of the twenty Scholars who emigrated to the Commonwealth during the period 1919–1939, fourteen were academics, five took up positions in the government scientific service and only one went into industry. During the final period, 1946–1960, two of the four emigrants to the Commonwealth were academics, one went into government scientific establishment and one went into industry. The general trend away from industry and towards university teaching and research or government scientific service is also reflected in these figures.

Although it is dangerous to generalize from small figures, this study does raise a number of questions which should be pursued further. This evidence suggests, for example, that there have been definite trends in the popularity of different fields of science chosen by the best British scientists, and that the variations in these can be quantified. This evidence also shows that certain "centres of excellence" developed, and shifted in popularity, with the passage of time. The creation of such centres throughout Europe may be correlated with specific policy decisions taken by different European governments and universities; their study may suggest conclusions about the manner and speed with which institutional forms and scientific techniques were taken up and applied.

This study shows, moreover, that although the 1851 awards were originally intended to assist the development of industrially orientated science in the provinces, this criterion was eventually abandoned. The causes underlying the gradual shift in award policy from "oriented" to "fundamental" research are undoubtedly complex. It remains to be seen whether the modification of award policies may have been partly due to the increasing participation of "fundamental" academic scientists on the nominating and award committees. It is clear, however, that after 1919 the largest proportion of United Kingdom and Commonwealth Scholars chose and, by implication were encouraged, to go into academic life.

Finally, the evidence suggests that there may be certain patterns of continuity and change in the relative proportions of scientific "emigrants" over the survey period. After 1900, however, there is evidence that the natural propensity to search out world leaders was deliberately reinforced by the various scholarship committees and by the creation of first class research facilities by the universities, governments and industries abroad. At the same time this evidence must be used with caution; it is too early to apply any similar generalizations to the engineers and applied scientists whose movements have occasioned so much recent concern.

It seems very likely that both migration and career patterns among academic scientists have been determined by a mixture of personal, professional and financial factors which themselves have been influenced by a growing consensus among men of science about the kinds of science worth exploring. Much more remains to be learned about the factors which have shaped this consensus. Certainly, if public institutions such as the 1851 Commission could influence, by either positive or restrictive actions, the development of different individual sciences, their actions assumed consequences of vital importance to the conduct of fundamental research. The study of such institutions remains extremely important to our understanding of the professional development of modern science.

This investigation was supported in part by a grant for research in science policy from the Department of Education and Science. We thank W. D. Sturch, present Secretary to the Royal Commission for the Exhibition of 1851, for his assistance.

[1] *Record of the Science Research Scholars of the Royal Commission for the Exhibition of 1851* (London, 1961).

[2] *The Brain Drain: Report of the Working Group on Migration*, Cmnd. 341 (London, HMSO, 1967).

[3] *Enquiry into the Flow of Candidates in Science and Technology into Higher Education*, Cmnd. 3541 (London, HMSO, 1968).

[4] *Second Report on Science Policy*, Cmnd. 3420, paragraph 18 (London, HMSO, 1967).

[5] *Record of the Science Research Scholars of the Exhibition of 1851*, 3 (London, 1930).

[6] *The Science Museum: the First Hundred Years* (London, HMSO, 1967).

[7] Reeks, Margaret, *The Royal School of Mines, Register of Old Students, 1851–1920 and History of the School* (London, 1920).

[8] *Centenary of the Imperial College of Science and Technology: a Short History of the College, 1845–1945* (London, 1945).

[9] Playfair, Lyon, in *Records of the Royal School of Mines and of Science Applied to the Arts*, 1, Part 1, 26.

[10] *Sixth Report of the Commissioners for the Exhibition of 1851*, 49 (London, 1878).

[11] Reid, W., *Memoirs and Correspondence of Baron Playfair*, 442 (London, 1899).

[12] *Ninth Report of the Commissioners for the Exhibition of 1851*, 10 (London, 1935).

[13] *Record of the Science Research Scholars of the Exhibition of 1851*, 105 (London, 1930).

[14] McKeen Cattell, J., *Science*, **66**, (1717), 513–516 (1927).

X

The Genesis of *Nature*

Science in Grub Street

By the mid-nineteenth century, biologists found there was no journal suited to their needs, but attempts to start one ended in failure.

For the best part of the past hundred years, *Nature* has held the position, in Europe and elsewhere, of 'the leading international weekly scientific periodical' and the 'main vehicle for prompt publication of recent advances in science'.[1] For years, it was unquestionably the single most comprehensive scientific journal in the world. Since the beginning it has had an international audience. But to the development of British science it has had a special and uniquely instructive relationship. It has not only mirrored faithfully the progress of science and the growth of scientific institutions - it has also, to a subtle degree, shaped the standards by which science is known in Britain and throughout the English-speaking world. Its fundamental anti-philistinism, literary integrity and editorial candour have presented to a world watchful of British methods and values an image of science in its broadest cultural usage. Arising against a vivid background of intellectual controversy, political reform and social change, *Nature* crystallised the spirit of empiricism, of idealism, of imagination and observation and breadth of vision which Victorian thought has bequeathed to modern science.

For scientific publishing, *Nature* has other lessons as well. Its greatness lies not so much in its originality as in its endurance. The secret of its success has been its constancy of purpose. That purpose came to life in the work of a small missionary band of men in the decade preceding *Nature*'s birth. In their vision one sees the reasoning which in 1869 brought a young scientist to the door of a friendly prince in the publishing courts of Covent Garden.

Since the invention of printing and the scientific revolution of the seventeenth century, the publication of scientific news and memoirs has been one of the chief objects of science and one of the chief functions of scientific societies.

---

1 John L. Thornton and R.I.J. Tully, *Scientific Books, Libraries and Collectors* (London: Library Association, 1971), 244, 301.

In Britain scientific literature followed the patterns set by the *Philosophical Transactions of the Royal Society* or the *Journal des Scavans*. Learned societies found the first suitable for the objective publication of experimental results; the second became the fashion for 'scientific periodicals'. Both forms of publication had value, but it was to the 'periodical' rather than the 'transaction' that Victorian scientists and interested amateurs increasingly looked. There were many reasons for this. By the middle of the nineteenth century the publication habits of the learned societies were often like the societies themselves, slow-moving and exclusive. Many 'publications committees', by their pronouncements or apparent bias, were set against iconoclast men of science with unorthodox ideas. Finally, learned 'transactions' were becoming increasingly specialist and incomprehensible to all but a few. By the 1850s, the world of science was being partitioned from within, and communications between geologists, naturalists, chemists, botanists and astronomers - each group of whom now had its own society - had become unsystematic and unpredictable. The only unifying social force in British science was the Royal Society, but it attempted neither to cultivate this unity nor to supply the news service that might reduce this sense of professional disintegration.

This void was partly filled by the 'academic' newspapers, like the *Athenaeum*, the *Saturday Review*, the *Spectator* and, to a lesser extent, *The Times*. The informal network provided by the quarterlies helped to mask isolation, but these columns were largely restricted to a handful of well known (if anonymous) contributors. But these did not pretend to answer the need of men of science to publish at some length fairly quickly. Despite obvious risks, some individuals, with the aid of a few friendly publishers, had already experimented with private journals appearing quarterly or monthly. The monthly *Philosophical Magazine*, begun in 1798 by Alexander Tullock (later joined by Richard Taylor), was a good example of a reasonably frequent journal that survived. Scottish scientists initiated several others. In 1819 David Brewster and Robert Jameson, Professor of Natural History at Edinburgh, began the *Edinburgh Philosophical Journal* which was left in Jameson's hands in 1826. Jameson was succeeded first by Edward Forbes and then by a board of chiefly Edinburgh editors who began the *Edinburgh New Philosophical Journal in 1855*. This series ended in 1864. *Nicholson's Journal*, which lasted from 1802 to 1813, was absorbed by the *Philosophical Magazine* in 1814. Thomas Thomson's *Annals of Philosophy* (1813-24 and 1875-1926), a London monthly described by Thomas Graham as 'the leading scientific journal of the country', was also absorbed by the *Philosophical Magazine*.[2] Meanwhile,

---

2 R.A. Smith, *The Life and Work of Thomas Graham* (London, 1884), 11.

Brewster began his own journal, the *Edinburgh Journal*, which lasted from 1824 to 1832, and then joined forces with the *Philosophical Magazine* to produce the *London and Edinburgh Philosophical Magazine and Journal of Science* in 1832. With alterations, this journal has survived to the present.

The *Philosophical Magazine* served a need for natural philosophers and chemists but naturalists had no similar periodical for the swift readable treatment of new discoveries. In 1833, Richard (later Sir Richard) Owen began the *Zoological Magazine*, but this proved too demanding on his time and resources, and ended after a few months.[3] In 1838, Richard Taylor of the *Philosophical Magazine* began the *Annals and Magazine of Natural History* with William Francis, Fellow of the Linnean Society and formerly editor of the *Chemical Gazette*. In 1840, the *Annals* merged with the *Magazine of Natural History*, begun in 1828, and survived. But most naturalist periodicals did not. The *Entomological Magazine*, which lasted from 1833 to 1838, and the *Entomologist* from 1840 to 1842, are cases in point. Despite the risks, attempts did not end, and general enthusiasm for the idea remained alive. This was particularly true among young biologists seeking some sort of corporate identity.

By 1850 many young naturalists like Huxley and Darwin had left medicine to follow science, but the few scientific appointments which existed in English universities or in government service were confined to astronomy, chemistry, geology, natural philosophy and medicine. There were, by and large, no places for biologists. The Linnean Society, the natural haven for biologists in London, was among the poorest of the major learned societies and neither it nor the Zoological Society wielded great influence. With the exception of marine biology, even the British Association had paid little attention to the claims of naturalists. Apart from a few government posts, naturalists as a group lacked status and position in society. The classic social remedy in Victorian Britain lay in association.

Isolation at Kew weighed on the young Joseph Hooker, who complained to Huxley that naturalists never met 'except by pure accident and seldom then as naturalists and if we want to introduce a mutual friend it is only by cut and thrust into one another's business hours'.

> ... Without some recognised place or resort that will fulfil the conditions of being a rendezvous for ourselves, an inducement to our friends ... and at the same time, a profitable intellectual resort - we shall always be ignorant of one another's whereabouts and writings.

---

3 G.F. Barwick, 'The Magazines of the 19th Century', *Trans. Bibliog. Soc.*, XI (1911), 241.

'My impression is', he continued,

> that we shall make no great advance in teaching Natural Science in this country except by some joint effort of Botanists and Zoologists who should pave the way ...[4]

There was a pressing need 'to bring about more unity in our efforts to advance science'. Despite the *Annals*, there was no unifying medium of scientific communication. 'Our publications', Hooker wrote, 'are sour bread cast upon the barren [wastes] of journals and other periodicals, which none of us can afford to buy and then to weed'. Part of Hooker's demands were met by the Natural History Society of London, formed in 1858, but the second task was much more difficult. The high infant mortality of young scientific periodicals was common knowledge. Whether Huxley shares Hooker's sense of isolation is not clear. Another motive had already stimulated his interest in scientific publishing.

Where private journals could pay for contributions, there was always a willing pool of scientific men to write. This was partly the result of the changing economic and class structure in British science and partly the effect of zealous British amateurs willing to write for a popular readership. Hooker's wealth removed him from the need of popular writing. But for any young man of science without private means, scientific writing - more or less popular in style - was often a chief means of support. 'By the powers!', wrote Thomas Graham to his sister in 1827. 'A very good method of adding two or three guineas a month to one's name'.[5] Sometimes the journals paid very well. *Chambers Journal*, for example, paid six guineas a sheet; the quarterlies (for those in the inner circle), even more.

For many years Huxley and Tyndall were greatly dependent on journalism and reviewing for a living. Both of them, friends since the British Association meeting at Ipswich in 1851, began their collaboration with a review article on 'Contemporary Science' in the *Westminster* for 1854. Chapman the publisher chose the books for review and paid six guineas a half sheet. When Tyndall pleaded overwork, Huxley urged him to 'be a brick and split the difference and say you will be ready for the April number ...'. But on reflection he added, 'What idiots we are to toil and slave at this rate. I almost repent me of tempting you after all'.[6] From time to time, there were attempts to interest particular weeklies in science. After trying to launch a *Scientific Review*, a

4 *Huxley Papers*, vol. 3.23, Hooker to Huxley, 4 April, 1856.

5 Smith, *The Life and Work of Thomas Graham*, 16.

6 *Tyndall Papers*, vol. 11, f. 2868, Huxley to Tyndall, October 22, 1854.

'delicate flower which budded only to fade', Huxley got wind of a new scheme, in early 1858, 'to work up the public in science'.

This scheme was the conception of Nevil Story-Maskelyne, professor of mineralogy and chemistry at the University of Oxford, and John Douglas Cook, editor of the *Saturday Review*. Since the *Review* began in 1855, Cook had regularly solicited scientific articles but had not been wholly successful. He had tried, like the *Athenaeum*, to report the contests of scientific papers given before the Royal Society, but this had proved of limited interest. Popular reviews of technical books were still well received, but men willing to do them were difficult to find. Books that could be dealt with briefly gave little excitement, brought small fees and interrupted individual research. The most successful part of Cook's plan was his scientific 'middles'. But few men of science could write lucidly regularly for a popular readership. In 1858, therefore, Cook asked Maskelyne, a frequent contributor, to form a 'panel' of seven or eight scientific authors. In return for a guaranteed article once a fortnight, the panel could select its own subjects and set its own limits.

Maskelyne got Huxley, then lecturer at the Royal School of Mines, for biology; J.J. Sylvester, then at Woolwich, for mathematics; Edward Frankland, professor at the Royal Institution, for chemistry; and Andrew Ramsay and Warrington Smyth from the Royal School of Mines and the Geological Survey, for geology. All were to form a 'corps (d'elite) each member of which should pledge himself to supply at least one article in three months, the subject to be at the pleasure of the individual in his own branch of science'.

Huxley inevitably asked for Tyndall's help:

> The plan seems to me so feasible that I promised at once to write to you and Hooker. It seems to me to offer the advantages without the trouble of a *Scientific Review* and it is not too much of a pull on one's time.[7]

Tyndall accepted and the news went around. 'The thing seems very feasible in my eyes', he confided to Darwin, 'I think I could promise to perform my share of the business. I think it may become a very important agency and exercise a salutary influence in this quack ridden country'.[8]

Hooker, as Huxley discovered, was less easily convinced. The proposals, he feared, would 'seek uncertain ill, uncertain good'.[9] He would not do reviews because these would have to be done 'at the expense of my scientific work' (Hooker was then engaged as a regular reviewer for Lindley's

7 *Tyndall Papers*, vol. 11, f. 2868, Huxley to Tyndall, April 20, 1858.

8 *Tyndall Papers*, vol. 11, f. 2889, Tyndall to Darwin (n.d.).

9 *Huxley Papers*, vol. 2.33, Hooker to Huxley, April 20, 1858.

Gardeners' Chronicle); he knew little of Maskelyne and had a dim view of the *Saturday Review,*[10] which unhappily acquired the sobriquet 'Pale Mall Gazette'.

In July and August 1858, their first fortnightly articles appeared. But the scheme was not a complete success. Only nine articles appeared during 1859 and only nine middles and eleven reviews in 1860. Except for a slight surge of effort in 1864, science slowly passed from the *Saturday*.[11] Only reviews continued to come, but these were not, by and large, written by Huxley, Hooker or Tyndall. Indeed, their energies had turned in a different direction. The *Saturday* had not the measure of independence, consistency and balance of presentation that the scientists wished.

In 1860 Huxley was attracted by the possibility of transferring the Dublin *Natural History Review* to London. The *Natural History Review* had acquired a sound reputation but had suffered from a small circulation in Ireland. Huxley's motives were partly economic, partly missionary. To Hooker, he wrote, 'if I choose to join as one of the editors, the effective control would be pretty much in my hands'. Once again Hooker was pessimistic. From his own sad experience with the short-lived *Journal of Botany* at Glasgow and Kew, Hooker warned Huxley against adding heavy editorial tasks to his existing academic commitments. Private publishing was expensive and could not be kept on schedule. It was too much work for one man, or even a panel of part-time editors, and reliance on unpaid contributions was illusory. 'I do not believe', Hooker added dryly, 'in the prolonged vitality of any unpaid review'.

Moreover, a successful journal needed a sense of accepted authority. 'You must ... be in a position to command assistance, not to ask it', Hooker continued, 'and without this you cannot do much'.[12] Huxley nonetheless persisted. With the agreement of Williams and Norgate, the scientific publishers, and in the midst of Darwinian controversy, the first number of the new *Natural History Review*, 'A Quarterly Journal of Biological Science', appeared in January 1861. Hooker, typically the last to agree, was the first to congratulate. 'The only fault with the *Review* is its brevity', he wrote, and promised to contribute himself.[13] An editorial board was formed of eleven

---

10 'The Review', he wrote, was 'dreadfully sententious and priggish and has lately become so pompous and vapid in style that several of my friends have discontinued it. The ignorance of the India articles is astounding, and its general politics are amateur and twaddling'.

11 Merle M. Bevington, *The Saturday Review, 1855-1868* (New York: Columbia University Press, 1910), 277.

12 *Huxley Papers*, vol. 2.67, Huxley to Hooker, July 17, 1860; *Huxley Papers*, vol. 3.119, Hooker to Huxley, July 18, 1860.

13 *Huxley Papers*, vol. 3.81, Hooker to Huxley, January 4, 1861.

distinguished naturalists, including George Busk, FRS, secretary of the Linnean Society; Francis Curry, FRS and FLS; R. MacDonnell, FLS; P.L. Sclater, FZS and FLS; W.B. Carpenter, FRS and FLS; J.R. Greene; John Lubbock, FRS and FLS; Daniel Oliver, FLS; Wyville Thomson, FRSE; E.P. Wright, FLS; and Huxley himself. Of these eleven, nine would later contribute to *Nature*.

The need for unity which Hooker and Huxley had discussed in 1856 was set out in the preface of the new journal. Without interfering with other journals, or with other than 'scientific' considerations, it offered 'to all whom it may concern a means of discussing the general problems suggested by the progress of biological investigation in a scientific spirit'.[14] It would be as authoritative as the 'ordinary quarterly'. Its theme Huxley took from Goethe - as he was to do again eight years later - 'Everything in science is become too much divided into compartments'. It would be specialised but not so as to 'demand a place in the Transactions of a scientific society'.[15]

But the *Review* did not last. It could not pay authors, so it did not get contributors. It was authoritative without an audience, and agnostic when most naturalists were still persuaded by natural theology. Despite Hooker's warning, Huxley assumed editorial responsibility and wrote most of the unsigned articles. Two voluntary assistants came to help, but they were undependable and counter-productive. 'I find, unless I review every page of it, it goes wrong', Huxley finally admitted.[16] When at last he decided to appoint a paid editor, the journal's financial position could not bear the expense. The last blow came when American journals 'pirated' articles without permission and thus limited American circulation to about twenty copies.[17]

The *Natural History Review* proved, as John Murray once remarked, that 'quarterlies did not pay'. For Huxley and his colleagues, and for natural science, the need for an effective means of expression remained abundantly clear. It was also clear, however, that most men of science had not the time, the expertise or the financial resources to get a scheme under way.

There were a few exceptions. One was the *Quarterly Journal of Science*, begun in 1864 by James Samuelson, a barrister, and William Crookes, FRS, the chemist. Crookes was a student of Hoffman's at the. Royal College of Chemistry in Oxford Street. After serving in various minor scientific posts,

---

14 *Natural History Review*, 1, notices (1861).

15 *Natural History Review*, 1, editorial (January, 1861).

16 Leonard Huxley, *Life and Letters of Thomas Henry Huxley* (London: Macmillan, 1900), vol. 2, 211.

17 See *Lubbock Papers*, Add Mss. 49639, f. 61, copy of letter to the editors of the *Journal of Science*, from Williams and Norgate, November 7, 1862.

Crookes found himself out of work but wanting to do science. It is true that he had an excellent chemical training, but, as his biographer recalls, 'in those days a scientific training was no passport to a public appointment...'.[18] With the help of Wheatstone, Crookes got £20 from the Royal Society to aid his researches and found odd jobs to support himself. Journalism beckoned. In 1856, he edited the *Liverpool Photographic Journal* and in the following year he became editor for the London Photographic Society. In 1859, buying the copyright of the old *Chemical Gazetteer* and making arrangements for printing with William Spottiswoode, the scientific house, he began *Chemical News*. The weekly was addressed to chemical manufacturers but insisted on including 'abstract science', provision for advertising and a 'Weekly Summary of all matters of a scientific or practical nature' from eminent correspondents abroad.[19] Its success was duly noted; 'with the exception of the *Phil. Mag*. and proceedings of learned societies no other English scientific journal enjoys this universal diffusion'. Crookes's earnings went to keep up his laboratory and family, but *Chemical News* paid him only three and a half guineas a week. In 1861, he therefore began to contribute to the new *Quarterly Review of Popular Science* edited by James Samuelson. As he reminded Angus Smith, 'a stationary income will not do with an increasing family and domestic necessities are apt to make scientific men very mercenary'. In January 1864, Crookes, recently elected to the Royal Society, joined with Samuelson in the *Quarterly Journal of Science*. It was to be 'a review of the progress of science ... and a medium of communication between students in various branches of Natural and Physical Science, as well as between Scientific Observers and the Reading Public'.[20] 'We have been told by men in every walk of life', Crookes announced, that 'the time has come when Science may claim for herself a special organ'. The journal boasted the support of fifty-eight listed men of science 'ready to avail themselves of these pages'.[21] Of this list, eighteen were later to write for *Nature*. Among them occurs the name of J.N. Lockyer, FRAS.

---

18 E.E. Fournier, *The Life of Sir William Crookes* (London: T. Fisher Unwin, 1923), 34.

19 *Ibid*., 48-49.

20 *Ibid*., 89-90.

21 Introduction, *Quarterly Journal of Science*, 1 (1864), 4.

Macmillan and the Scientists

> **... Just as we have got to learn what the capacities of Literature are as a profession, we still have to realise what publishing may be and what an immense power a Publisher might wield for good without the least neglect of business conditions.**[22]

While Huxley, Hooker, Tyndall and Crookes pondered the plight of scientific communication, two important events were taking place in the London publishing world. During the early 1860s, British publishing awakened to the potential market for scientific material. Traditionally, John Murray, with its proximity to the Royal Institution and, after 1857, the Royal Society, had enjoyed an unrivalled reputation in the field, and had published Lyell, Geikie, Murchison, Du Chaillu, Mary Somerville and much of Darwin.[23] Cambridge and Longmans also enjoyed a long history of scientific work. But young firms like Williams and Norgate began to usurp some of the more likely texts and the climate was favourable for expansion.

In parallel, the 1860s also saw a great expansion in periodical publishing, called forth by the character of the time - the emphasis on reading for moral importance, the spread of literacy, rising prosperity, the custom of family reading and the technological opportunities of mass production. Periodicals became one of the nineteenth century's most characteristic products.[24] Between 1821 and 1830, 100 new periodicals began in the United Kingdom and the figure increased to a peak in 1861-1870, when 170 periodicals were issued, or one new magazine every three weeks.[25] The abolition of compulsory newspaper duty in July 1855 and the quickening pace of Victorian life brought new drive to the daily press and success to weeklies like the *Saturday Review*. Only the great quarterlies did not share in the increase.[26]

The idea of science and the attraction of frequent publication appealed to the young house of Macmillan. Alexander and Daniel Macmillan had moved to Cambridge about 1845, where they soon flourished. In eighteen years they

---

22 Charles Graves, *The Life and Letters of Sir George Grove* (London: Macmillan, 1903), 303.

23 J.L. Thornton and R.J.J. Tully, *Scientific Books, Libraries and Collectors* (London: Library Association, 1962), 300.

24 R.D. Altick, 'Nineteenth Century English Periodicals', *Newberry Library Bulletin* (May 1952), 255.

25 G.F. Barwick, 'The Magazines of the Nineteenth Century', *Trans. Bibliog. Soc.*, XI (1911), 238.

26 A. Ellegard, 'The Readership of the Periodical Press in Mid-Victorian Britain', *Goteborgs Universitet Arsskrift*, LXIII (1951), 3.

gathered round their shop in Trinity Street a collection of influential and productive scholars who were to remain firm friends for life. In 1863, with the success of Kingsley's *Westward Ho!* (1855), and Thomas Hughes' *Tom Brown's Schooldays* (1857), the brothers moved from Cambridge to London, leaving behind in Cambridge Alexander's nephew, Robert Bowes. (The firm became Bowes and Bowes of Trinity Street.)

Macmillan in his novels had caught the crest of Victorian middle class enthusiasm for adventure and exploration. Memories of the hungry forties and the struggle of Chartism had passed; movements of 'social reform' through association and due process had become the nation's salvation from the philosophical death of 'dismal science' and the horrors of revolution. Competition to abolish corruption, industrial progress to erase poverty, idealism to invigorate utilitarianism in social thought - there seemed no end to the moral purpose of 'higher journalism'.

In Cambridge, Alexander Macmillan had considered beginning a journal, more a 'magazine' than a newspaper, selling at a popular price of 1*s* or 1*s 6d* which could be built up 'into an enormous sale'. Isaac Todhunter, the Trinity mathematician, had mentioned the idea as early as 1855 and Alexander had in principle agreed. '(Todhunter says) he is determined it shall be done, so if that is the case we might as well have a share in it', Alexander told his brother.[27] By 1860, the abolition of paper duties, together with postal reforms and changes in the law of libel, had made the prospect 'in some respects more important and interesting in the history of journalism than any portion of the preceding two hundred and fifty years'.[28]

In London, Alexander soon found himself in contact with the men who could best create a journal with him. Macmillan's weekly 'evenings' at his home in Balham, affectionately recalled as 'Talk, Tobacco and Tipple on Thursdays', became regular events between 1858 and 1864.[29] Holding evenings where 'science, art and letters are fairly represented', Macmillan forged relationships that helped him build his reputation. Kingsley, Huxley, Spencer, Holman Hunt and Tennyson all became habitués. Macmillan's integrity and willingness to risk short-run commercial losses for good causes (or at least for personal satisfaction) won him the fidelity of his 'list'. As Tyndall once confessed to Huxley, while accepting a particularly onerous writing task,

27 George A. Macmillan, *Letters of Alexander Macmillan* (London: privately printed, 1908), Introduction, xxvi.

28 H.R. Fox-Browne, *English Newspapers* (London: Chatto and Windus, 1887), vol. 2, 261.

29 Macmillan, *Letters of Alexander Macmillan,* 173, Macmillan to P.G. Tait, June 18, 1864.

'I would do anything in reason to oblige Macmillan'.[30] Macmillan would also do much for his friends.

The idea of the new periodical had attractions. Macmillan consulted his old friends John Malcolm Ludlow and David Masson, who, with Arthur Hugh Clough, had been in the 'goodly band' that began the Christian Socialist movement between 1848 and the early 1850s. It was David Masson, in fact, who had recruited F.D. Maurice to the group. Alexander wrote in the first volume of the *Christian Socialist* while Daniel had contributed to Ludlow's *Politics for the People* (1848). After the Macmillans moved to London, in 1855, the strongest ties were broken, but Ludlow and Masson remained firm friends. John Ludlow had long wished to edit a journal that would carry his own message of social reform. For years he had also been urging Macmillan to produce a monthly magazine modelled on the *Revue des Deux Mondes*, and calling upon the remarkable group of reform-minded contributors Ludlow had to hand. Macmillan at first preferred a quarterly,[31] but the competition of Bentley's new political quarterly helped to change his mind. In November 1859, just two months before the rival *Cornhill Review*, *Macmillan's Magazine* was issued - the first shilling monthly in England. Science, literature and the arts would be unified under one banner and Huxley and Tyndall would be well employed. The vision of authoritative science shared with the public was one step nearer realisation.

Much to Ludlow's disappointment, Alexander appointed David Masson editor of the new magazine. However sympathetic Macmillan might be, Ludlow's politics were against him. Alexander's strong moral sense was combined with a shrewd sense of commercial propriety. His method of publishing reflected this. 'I believe', he once wrote in approving some articles, 'it would do us good besides enabling us to do good'.[32]

Macmillan's Magazine was quickly enveloped in waves of controversy. The publication of Darwin's *Origin of the Species* alone threw Macmillan into the raging question of man in nature. From the first, *Macmillan's Magazine* opened its pages to the Darwinians. In December 1859, Macmillan's second number published Huxley's *Time and Life* and asked others to explicate Darwinism for his readers. Macmillan could scarcely avoid becoming a neutral 'middle ground' between his scientific, literary and theological friends. Macmillan after all published Gladstone. Macmillan described a Thursday in

30 *Tyndall Papers*, XI, f. 3087, Tyndall to Huxley, December 16, 1878.

31 Macmillan, *Letters of Alexander Macmillan,* 4-7, letter to James MacLehose, October 27, 1858.

32 *Macmillan Archives*, Masson Correspondence, f. 33, Macmillan to Masson, March 3, 1860.

July 1860 when Huxley came in and 'Darwin and conundrums and general jollity [were] pleasantly intertwined'. Alexander was not blind to its implications:

> That Darwin controversy has awakened interest in races and origins to a large extent. Every original investigation as to the relics of bygone races is looked for with the keenest interest. Unfortunately a good deal of theological asperity has got mixed up with the controversy on both sides. But I think a true scientific spirit is on the whole gaining ground.[33]

As Kingsley's Dean told Alton Locke, Macmillan believed 'there is no lie in Nature, no discord in the revelation of science, in the laws of the universe. Infinite, pure, unfallen, earth-supporting Titans - fresh as the morning of creation, those great laws endure'.[34] The notion of evolution, at least as it involved geological change and the uniformity of nature, was intellectually acceptable to him; only the nagging possibility that evolutionary theory might destroy the uniqueness of man and the existence of free will caused him real concern.

> The scientific enquiry is too immature for us in any way to identify ourselves with that theory. I feel, too, most strongly that the plain, distinct recognition and maintenance of the Bible doctrine of the Divine origin of the man, moral or spiritual ... is right for us to hold - because it is the truth and the only view that would deliver us from the ghastly consequences of Darwin's book, whether true or false.
> I quite contentedly and calmly abide the issue of physical enquiry even if it should turn out that my first physical ancestor was a black slug or a polyp, while I have this as a background. But I have no hope with any theory of creation or generation if this is ever lost sight of, for the progress of my own species. With it, I have hope even for the poorest polyp that crawls or sticks to a stone all its poor little existence here.[35]

To his death, Macmillan refused to admit what he called the 'ghastly' logical consequences of Darwinism for the nature of man. But, this apart,

---

33 Macmillan, *Letters of Alexander Macmillan*, 57, Macmillan to Wilson, July 25, 1860.

34 Charles Kingsley, *Alton Locke* (London, 1854), 214.

35 *Macmillan Archives*, Masson Correspondence, f. 18, Macmillan to Masson, January 29, 1860.

Alexander remained responsive to argument on either side. 'True', he wrote to Tennyson after reading *The Origin*, 'I wish someone would bring out the other side. But surely the scientific men ought on no account to be hindered from saying what they find are facts'.[36] Reading the *Athenaeum*'s account of the famous duel between Wilberforce, Owen and Huxley at Oxford in 1860, Macmillan, more in sorrow than anger, wrote: 'Why will you scientific men quarrel? Your little hands were never made to tear each other's eyes - but to dig fossils and dissect tadpoles' tails and suchlike wise and harmless objects'.[37]

But scientific men would quarrel and a means of regular authoritative and frequent communication was needed. It required, however, a publisher with a world view and social purpose that could accommodate debate within the intellectual ferment of his day. From the early 1860s, Macmillan was moving in this direction, as much a pilgrim as a commercial traveller, excited, mystified and enchanted as any enquiring mind by the intellectual adventure of unfolding Nature.

Seeds of Competition

The failure of the *Reader* in 1867 after a lifetime of four years nevertheless taught a number of important lessons.

Macmillan's appointment of Masson in 1859 had one particularly unexpected result. Ludlow's self esteem had been damaged and Macmillan wanted to keep in close contact - 'I hope you will be one of our contributors', he wrote to Ludlow, '... like a good fellow as you are - with a dash of bogie which requires discrimination and even exorcising now and then'.[38]

But Ludlow was not content to let matters end with Macmillan's kindly dismissal. In late 1862, joining forces with Thomas Hughes, his close friend and neighbour in Wimbledon, Ludlow began a new weekly review of literature, science and art. The enterprise was to be independent of any publishing house. Hughes wished a free hand to 'supply the long felt want of a first class literary paper equal in literary merit and general ability to the

36 Macmillan, *Letters of Alexander Macmillan*, 29, Macmillan to Alfred, Lord Tennyson, November 26, 1859.

37 Macmillan, *Letters of Alexander Macmillan*, 123, Macmillan to Frederick McCoy, November 25, 1862.

38 *Macmillan Papers*, Macmillan to Ludlow, November 7, 1859.

political press of London'. To be their science editor, Ludlow and Hughes asked another Wimbledon neighbour, a young amateur astronomer and War Office clerk who shared their compartment on the morning train to London. The young clerk's name was Norman Lockyer.

The *Reader*, as the new review was called, promised to be 'candid and impartial and liberal in social affairs'.[39] Its first publisher was James Bohm and its first number appeared on January 3, 1863. The *Reader*[40] brought to a focus the weekly *belles lettres* tradition of the *Athenaeum* and the *Saturday Macmillan's Magazine*, the missionary purpose of the *Quarterly Journal of Science* and the moral energy of Christian Socialism. 'Its goal was to serve men of culture'.[41] It was but one step removed from *Nature*.

In its first few months, the *Reader* generated excellent reviews. The *Annual Register*[42] classed it with the *Athenaeum* and it is still remembered for its excellence.[43] All the 'Young Guard of Science' (as Hooker called them[44]) were represented, including Huxley, Lubbock, Lockyer, Galton, Tyndall and Hooker, as well as George Airy, H.C. Bastian, W.R. Birt, T.G. Bonney, John Browning, George Busk, W.G. Carpenter, Charles Darwin, W.R. Davies, W.H. Flower, Edward Frankland, J.P. Gassiot, J.R. Hind, William Huggins, James Hunt, William Pole, Andrew Ramsay, Henry Roscoe, Warren De la Rue, H.C. Sorby, William Spottiswoode, George Stokes, Edward B. Tylor and James Glaisher. All of the thirty-eight men of science who contributed to its reviews later wrote in *Nature*.

Macmillan observed Ludlow's handiwork with obvious admiration. 'The Reader is doing very well now, I understand ...', he wrote to F.J. Hort:

---

39 Advertisement, *The Reader: A Review of Literature, Science and Art*, Saturday, January 3, 1863.

40 For a recent literary analysis of the review, see John F. Byrne, *The Reader: A Review of Literature, Science and the Arts, 1863-1867* (Unpublished PhD thesis, Northwestern University, 1964). Dr Byrne's thesis arrived in this country just in time to be included. Its research will repay closer study.

41 D. Roll-Hansen, 'The Academy, 1869-1879', *Anglistica*, VIII (Copenhagen: Rosenkilde and Bagger, 1957), 30.

42 *Annual Register for 1863*, 359.

43 Smith, *The Life and Work of Thomas Graham*, 331.

44 Leonard Huxley, *Life and Letters of Sir Joseph Dalton Hooker* (London: Murray, 1918), vol.1, 541.

> It undoubtedly is the best literary paper we ever had in England, and will undoubtedly improve. You should tell everybody to take it instead of the *Athenaeum*.[45]

Nevertheless, Ludlow could not organise the paper to perform commercially. Dr Kingsley complimented him warmly on getting the review 'into first rate fettle in a very short time'[46] but, according to his autobiographical notes, he found his manager, publisher and printer 'impossible'. Finally, in March 1863, 'after having gathered a list of contributors such as was never known for a newspaper', Ludlow resigned.[47] Ludlow henceforward confined himself to contributions for *Macmillan's*, the *Spectator*, the *Contemporary* and the *Fortnightly*, and to the development of his Working Men's Colleges.

On Ludlow's resignation, David Masson, professor at University College, assumed control. Since 1859, Masson had been a good friend of Huxley's circle[48] (it was indeed through the intervention of his scientific friends, Thomas Hirst and William Sharpey, that he was elected to the *Athenaeum*) and it was natural that, following his successful work with *Review*, the intelligence sections of the *Chemical News*, the controversy of *Macmillan's Magazine*, Huxley should turn to him. It is curious that twice in five years Masson should assume control over something Ludlow wished to start. But even Masson's presence did not save the journal, and its financial position failed to improve. In the meantime, its fate was shared by another review called *The Scientific Record* which had appeared briefly in April 1864. The *Record*, published by Frederick Mathieson on Telegraph Street, promised to be a 'Weekly Journal of Scientific Progress' and a record of the 'progress of man in conquest of nature'.[49] As usual, it included book reviews, notes of the proceedings of learned societies and miscellaneous news and intelligence. It also boasted a weekly editorial which could well have been written by the Young Guard. 'The editors hope', it proclaimed, 'to render the *Brotherhood of Science* a real service, and to make their journal ever welcome, not only amongst scientific men, but also in those happy English homes where the light

---

45 Macmillan, *Letters of Alexander Macmillan*, 156, Macmillan to Rev. F.J.A. Hort, October 23, 1863.

46 N.C. Masterman, *John Malcolm Ludlow: The Builder of Christian Socialism* (Cambridge: Cambridge University Press, 1963), 159.

47 J.M. Ludlow, *Biographical Notes to My Life*, 23.

48 *Masson Papers* (Edinburgh University), DC.3.99 (14), Masson to Sharpey, March 11, 1868.

49 *The Scientific Record* (April 9, 1864), 14.

of Nature prevails'.[50] But its proud hopes had no chance. Once again expenses sapped the journal's strength.

In its second issue, the editors admitted defeat and sadly observed that 'science is not sufficiently advanced in England, notwithstanding the labours of our scientific men, to ensure for such a paper as the *Record* at present popular appreciation and patronage'. The editors had faith 'that a weekly scientific journal is much wanted and when subscribers are prompt, they will be equally prompt in reissuing the *Record*'. Reviews like the *Reader* would not suffice. As one correspondent exclaimed: 'Science is too important and its objects are too multifarious to be adequately disposed of in what I may call the Postscripts of Literary Journals'. This letter hailed the *Record*'s articles in astronomy. It was signed 'J.R.', but it embodied the sentiments of Norman Lockyer.

> Politics and social questions very properly claim priority of place and importance in the daily newspapers. A weekly scientific journal, therefore, in which 'all important discourse in science will be faithfully recorded' containing also 'Scientific Notes and Queries' and 'admitting correspondence on general scientific topics of interest' was certainly a desideratum which your journal will opportunately supply.

In the summer of 1863, the future of the *Reader* was uncertain. Masson left the editorship. Among the sub-editors, Norman Lockyer was especially anxious that it should not stop, and he asked Herbert Spencer's advice. Spencer canvassed support for 'a new weekly paper of literature and science' which would 'eclipse the existing one'. 'If a few selected men were to combine', Spencer said, 'each to write one article weekly on a subject within his own province, a periodical might be produced that would have great weight and authority'. The paper could be published for £2,250, so forty shares of £100 were issued.[51] Tyndall said, however, that there were only thirty-five shares issued altogether.

All the scientists concerned canvassed their friends, 'We sadly need a good weekly scientific organ', Tyndall wrote to Herschel, 'and the *Reader* ... promises to supply this want. It is exceedingly well supported. Would you object to adding your name to the list of shareholders? ...There never was a better opportunity for establishing a healthy weekly scientific organ'.[52] But Herschel declined to pledge himself. With his best wishes he added: 'I wish it

---

50 Francis Galton, *Memories of My Life* (London: Methuen, 1908), 167.

51 *Tyndall Papers*, vol. II, f. 793, Tyndall to Bence Jones, November 18, 1864.

52 *Tyndall Papers*, vol. II, f. 534, Tyndall to Herschel, November 21, 1864.

could be accomplished to write the best matter which crops up, not being in the form of memoirs presented to the great scientific bodies of the country but into one really good monthly or bi-monthly journal of science, but that, I fear, is hopeless'.[53] Past experience augured ill for the fate of any such enterprise, but by November things looked brighter; in the end, about thirty-five shares were sold and the Young Guard was strong enough to bid £2,000 for the paper and to have working capital besides. Mr Huth agreed to invest £500 on condition that it be maintained 'an organ of free opinion', and Hughes held a meeting to settle questions of policy. Octavius Smith, a philosophical correspondent of Spencer and proprietor of one of the largest distilleries in England, bought several. Henry Huth took five; James Campbell, another friend, took two; Huxley, Francis Galton, John Cairns, Sir Frederick Pollock and Spencer each took one. Bence Jones of the Royal Institution bought another and William Spottiswoode was proposed as printer. 'The paper is not yet quite paying its expenses', Spencer wrote to Mill, 'but it can scarcely be doubted that with the concentration of faculty now about to be engaged upon it, it will soon do so, and may not improbably become a good investment'. By December, Mill had promised his support and Darwin and Lubbock soon followed.[54]

On November 3, 1864, the reconstituted journal was discussed at the first meeting of the X-Club. The nine members - Huxley, Tyndall, Spottiswoode, Hooker, Lubbock, Busk, Spencer, Hirst and Frankland - were among the most influential scientists in the country. There was still some question about a general editor. On November 22, Thomas Hughes expressed to Huxley his strongest belief that 'Lockyer can do the general editing and will be the best man for us. He knows the machinery, having been there from the first, has been in constant relations with such men as Ludlow, etc. ... has the science already in the right grooves and is not above taking advice, is a real good worker and above all has his heart in the business ... He will do the work too gladly at a lower figure than any other competent man, a consideration to be regarded at the present until we get more capital and know where we are'.[55] John Dennis, a noted literary scholar and critic, was apparently appointed. But in the end Sir Frederick Pollock (whose brother, Walter Herrie, edited the *Saturday Review*) was literary editor-in-chief. Pollock's father, George Frederick, was a Trinity mathematician and a close friend of Faraday. The Pollock family, in turn, was close to Tyndall and his neighbours in Hindhead. Tyndall and Huxley agreed to edit science. Huxley specialising in physiology and Tyndall in physics.

53 *Tyndall Papers*, vol. II, f. 535, Herschel to Tyndall, November 21, 1864.

54 *Lubbock Papers*., 49,639, f. 40, Spencer to Lubbock, November 12, 1864.

55 E.C. Mack and T.H. Armytage, *Thomas Hughes* (London: Ernest Benn, 1952), 125.

John Llewellyn-Davis did theology and Galton travel, ethnology and natural history. Lockyer assisted in astronomy, Spottiswoode in physics, and John Cairns in political economy. G.H. Lewes did fiction and poetry, and Spencer did philosophy and psychology.

A programme advertising the new *Reader* was drawn up, listing 75 men of science, and, on February 4, 1865, a new prospectus was issued. One passage in the new prospectus was especially significant:

> The very inadequate manner in which THE PROGRESS OF SCIENCE AND THE LABOURS AND OPINIONS OF OUR SCIENTIFIC MEN are recorded in the weekly press and the want of a weekly organ which would afford scientific men a means of communication between themselves and with the public, have long been felt. They have been the subject of special consideration by some of the leaders of Science in London.[56]

To remedy this, the *Reader* expanded its science space to eight pages out of twenty-five. The new prospectus was issued and new series launched. Spencer told Mill: 'If this opportunity of establishing on a safe footing an organ of scientific thought and of conscientious literary criticism is lost, it may be long before this very desirable object can be achieved'.[57]

In November 1864, Francis Penrose, archaeologist and astronomer, enthusiastically told Lockyer that he had seen Ludlow 'from whom I learnt that the *Reader* was casting off its old slimy skin and coming out renewed in better style. I hope it is for your sake - you were very despondent about it when I saw you'.[58] For about ten months the journal attracted scientific work; Lockyer published Croll's paper on the physical causes of climatological changes during geological time,[59] and a paper by William Huggins on nebulae.

Unhappily, these good omens were short lived. Unexpected difficulties of communication and disagreements about commitments arose among the team. Moreover, according to Spencer, John Bohm, the paid sub-editor and the only man with journalistic experience, found himself out of sympathy with his scientists. 'In the end their aims as well as the expectations of our subscribers were balked'. In April 1865, Spencer significantly tried to arouse wider interest by getting, 'so far as possible, occasional brief letters from the leading men of science announcing such interesting novelties as admit of being

56 *Reader*, V (February 4, 1865), 152.

57 Herbert Spencer, *Autobiography*, (London: Williams and Norgate, 1904) vol. II, 119, Spencer to Mill.

58 *Lockyer Papers*, Penrose to Lockyer, November 23, 1864.

59 *Lockyer Papers*, Croll to Lockyer, January 14, 1865.

understood by the general public and one of fit nature to be quoted for our columns'.[60] In August 1865, Thomas Bendyshe[61], a professional author, editor and anthropologist, purchased the paper and the management improved a little, but personal relations between the editor and the staff remained poor and the journal ran aground in stormy literary disputes.

There were other disappointments. Huxley, struggling with the *Natural History Review*, found the strain of one collapsing journal too much and he wrote nothing for the *Reader*. In January 1867, the last issue appeared and the *Reader* died soon afterwards. The paper had suffered, according to Galton, from continuing inefficiency in operation and dullness in style 'notwithstanding some really good articles'.

> The management was naturally too amateurish; promised articles were delayed and the time of the Committee was too much wasted in frequent discussions about first principles upon which Spencer loved to dilate.[62]

But there were other reasons for its failure. Karl Pearson noted first that, while its scientific list 'was a tremendous force to bring together', because 'there was no one man who would devote his whole life to the projected task, the *Reader* came to nought'. Second, the powerful scientific guns who had given tacit support remained silent when asked for reviews. The great men of the day - Lyell, Darwin and Herschel, for example - were solicited but did not reply. Lacking this appeal, the journal's circulation could only suffer. Third, the journal was dissipated by divided responsibility among the editors. 'The ship had too many first rank commanders aboard and no one whose livelihood depended on successful voyages. It is small wonder that it never reached port'.[63] As Pearson adds, *Pereat lector, Natura resurgat.* Indeed, at its death the *Reader* was the closest approximation yet to the wishes of the London scientific circle.

---

60 David Duncan, (ed.), *Life and Letters of Herbert Spencer* (London: Methuen, 1908), vol. I, 153-154.

61 Thomas Bendyshe was educated at Eton and King's College, Cambridge (M.A. 1852 and Fellow of King's). Called to the Bar, Inner Temple, 1857. Translated and edited treatises from Latin, German and French for the Anthropological Society. (Venn, *Alumni Cantabrigiensis*).

62 Francis Galton, *Memories of My Life* (London: Methuen, 1908), 168.

63 Karl Pearson, *The Life and Letters of Sir Francis Galton* (Cambridge: Cambridge University Press, 1914), vol. II.

Macmillan and the Young Guard

The search continues for a journal which would be the 'advocate of the cause of Science and the interests of scientific men in England'.

In a sense, the *Reader* had failed. But during the hard years between 1863 and 1867, Huxley, Tyndall and Lockyer had forged a lasting set of working friendships. Equally important, they had kept alive Macmillan's interest. The *Reader*'s effort to combine scientific and cultural questions within a weekly news journal was exemplary. Only the shortage of capital, the difficulties of attracting 'great men' and the absence of a single responsible editor with commercial experience had brought it down. Among the most active in its defence and the loudest mourners at its death was Lockyer.

In December 1866, after ten years as a junior clerk with the War Office, Norman Lockyer discovered that impending administrative reorganisation made future promotion impossible. A long and bitter struggle ensued which moulded the attitude towards officialdom which was to dominate the next fifty years of his life. A change of job seemed inevitable and desirable; it was only a matter of time. The same year, he delivered his first paper to the Royal Society and achieved the first important results of his studies on solar prominences and sunspots. Scientists turned to Lockyer, and Lockyer looked around him.

Some time during 1866, Lockyer, having essentially quit the *Reader*, met Alexander Macmillan. It is unclear where or in what circumstances they met, but the close contact Lockyer had gained over the past four years with Ludlow, Hughes, Masson, Huxley and the whole Young Guard, together with his reputation in the *Saturday Review* and the *Athenaeum* in his spare moments from the War Office, made the prospect likely. Mutual admiration for Tennyson might have provided the catalyst. (In 1866 Lockyer moved from Wimbledon to Finchley. Lockyer was then doing research which would result in the discovery of solar helium and had a telescope in his garden. Tennyson, who was at that time interested in astronomy, may have visited him). In any event, some time during 1866, Macmillan asked Lockyer to write a popular book on the Sun. This book was not completed at the time although Lockyer's book on *Solar Physics* may have been the result. Their association ripened; as Huxley and Tyndall were busily involved in teaching and research, Macmillan, not unnaturally, came to regard Lockyer as his 'consulting physician' in scientific matters. His place in the firm became analogous to that held by George (later Sir George) Grove who met Macmillan in 1863, succeeded Masson as editor of *Macmillan's Magazine* in 1868 and became literary adviser in 1873.[64] Over the next two years, Macmillan and Lockyer became good friends. In April 1867, as Lockyer was recovering from what appeared to be a nervous breakdown, they travelled to France together.[65]

Lockyer, however, was still looking for a job, and the young scientists, for a mouthpiece. In June 1868, Lockyer tried to persuade the *Daily News* to establish a 'scientific column'. About the same time, Lockyer's arrangement with Macmillan became formal and he was retained as the firm's chief scientific expert 'to encourage sound work' and to discourage what Macmillan called the 'frippery of science'. In the meantime, the search for a satisfactory scientific newspaper went on. In November 1868, Charles Wyman, publisher of Queen Street, began a weekly called *Scientific Opinion*, including the now familiar intention to provide a 'Weekly Record of Scientific Progress at Home and Abroad'. The journal, strongly reminiscent of the *Record* of 1864 and the *Reader* of 1865, promised to fill the 'want of a weekly published periodical devoted to discoveries of current topics of interest to the scientific world'. Its prospectus summarised the discussions and the experience of the last four years:

> At a time when Science enters more and more into the commonest operations of daily life, and when an enormous impulse has been given to the cause of scientific education and literature, it seems but consistent that the numerous workers in the wide and important domain of Science should be afforded the advantage enjoyed by other professions of possessing a journal of more frequent issue than the quarterly or monthly.

More than this, the journal was going to take the policy of the young scientists and the X-Club (which were largely intertwined) one step further by being the 'advocate of the cause of Science and the interests of scientific men in England, to enforce ... the claims of science upon the general public, to secure her followers their proper need of recompense and social distinction and to help them in their daily pursuits ...'.[66]

Scientific Opinion lasted until February 1, 1870, and carried leading articles that could only have come from the London reformers. In the seventh number, in December 1868, there was a leading article entitled 'How

---

64 Charles Graves, *The Life and Letters of Sir George Grove* (London: Macmillan, 1903), 96, 156, 228.

65 Charles Graves, *The Life and Letters of Alexander Macmillan* (London: Macmillan, 1910), 266, 296.

66 *Scientific Opinion*, November 4, 1868, 1.

Government deals with Science'; in the twenty-first, in April 1869, on 'Mr Lowe's Estimates of Science'. The journal raised demands for improved representation, increased State assistance and the better organisation of science through a collective and centralised Academy of Science. The mast-head and composition of the journal were to provide a model for *Nature*. It is highly probable that Lockyer found his way into its columns.

Several events in the year 1867 proved decisive to scientific publishing, and Macmillan's in general. First, the British Association Committee on the Public Schools Commission reported and gave evidence from Huxley, Tyndall, William Sharpey and W.A. Miller to the House of Lords on the extent to which physical science might be introduced into the Public Schools. The proposed Bill did not pass, but made a great stir. Second, Macmillan published two critical books. The first was *Essays on a Liberal Education* by F.W. Farrar and contributors which demonstrated the difference between the 'humanities' and the 'classics' and made it clear that there was no longer an intellectual, aesthetic or practical justification for the dominance of the classics. One of its contributors was James Wilson, the first science master at Rugby, who was in the vanguard of scientific teaching. These arguments set forth the main lines of educational development for the latter half of the century.

The third event was a collection of essays by twenty-seven leading scientists and scholars, edited by E.L. Youmans, and published by Macmillan, called *Modern Culture: Its True Aims and Requirements*. On balance, its authors demanded more space for scientific studies in education and in national life. Other significant publications appeared at the same time. Matthew Arnold's essay in the *Cornhill*, later to be collected in *Culture and Anarchy* (1869), presaged a new phase in the struggle to extend education, especially scientific education, to the level of the working man. These events, taken together with the appearance of Mark Pattison's *Essays on Academical Organisation* at Oxford in 1868 and the sensational British Association paper by Colonel Alexander Strange on the necessity for state support for science, brought the issue to a major turning point. Macmillan had the good sense to appreciate what was happening. On the one hand, standards of education and standards of science were going to change and textbooks and lectures were soon to be needed. But, more important, the place of the scientist was commanding more public attention than ever before. The publisher who, for whatever reason, recognised a vocation in science, assumed (however unconsciously) a role of unprecedented importance for his own future and for the conduct of scientific activity. It was time to begin the journal which Lockyer and others had struggled to produce for more than eight years.

The New Journal

Macmillan agrees to publish a new scientific journal with Lockyer as editor, to be called *Nature*. The first publication date is set.

In early 1869, Lockyer approached Macmillan with a proposal to establish a general scientific journal. In the three years of their friendship, Macmillan had greatly broadened the base of his firm. In 1866, he had begun the *Journal of Anatomy and Physiology*. In 1868, to meet growing demands for a general medical journal for G.P.s, he had begun the *Practitioner*. A steady stream of works in medicine and surgery was established and new scientific books were beginning to appear - partly in an attempt to fight the transatlantic practice of pirating English editions. Moreover, in 1869, Macmillan began his publishing house in America. 'I have good accounts from our agents in New York of the prospects of business there', he wrote Henry Roscoe.[67] Macmillan's policy became characterised by investment in 'long-term publishing' in books and journals which at first sold slowly but which had the capacity of endurance. Whether in medicine, science or literature, 'classics' were the keystone - classics in any field would command the attention for years. In the nature of things, this policy was not immediately profitable. It required the ability to take initial losses and place complete confidence in the assumption that change and development would only bring progress and profit. It was this policy which guided Macmillan in assenting to publish another new journal.

The experience of the *Reader*, the persistence of Huxley and Tyndall at the X-Club and the visible demand for a professional voice, a growing market for a frequent and authoritative review, were catalysed by the presence of an ambitious young scientific journalist of intellectual distinction who had already won the confidence of an aggressive publisher with sufficient capital, an appreciation of science, a willingness to take risks and a strong sense of social mission. Some time near the middle of 1869, Macmillan took the initiative, and agreed to produce the new periodical, with Lockyer as editor.[68] The new periodical was to combine the critical and literary standards of the *Reader* with the timeliness of *Chemical News*, and the format of the *Scientific Record* and *Scientific Opinion.*

The year 1869 was auspicious for Lockyer. In August 1868 he had observed the solar prominences and had discovered helium spectroscopically. In February 1869 he gave a lecture in a series at University College, London, led by Huxley and Ruskin; in April he was elected to the Royal Society. In

67 *Macmillan Archives*, Alexander Macmillan to Henry Roscoe, September 14, 1869.

68 Charles Morgan, *The House of Macmillan* (London: Macmillan, 1943), 84.

May he gave a Friday evening lecture in the Royal Institution and had become a young lion in scientific circles. Lockyer, aged 33, had reached a crest in his new career.

The year was also auspicious for a new journal. The only real competitor, *Scientific Opinion*, was losing money and would not last long independent of a publishing house. There were, however, rumours of plans by John Murray to launch a new journal under Charles Appleton, to be called *The Academy*. Having made the decision in principle, Macmillan decided to act swiftly. The deadline for the first issue of his new journal was set for October 1869.

Macmillan and Lockyer began to enlist promises of support and subscriptions. Lockyer had the advantage of more personal experience than anyone who had preceded him. His list of eighty-six intending contributors was virtually ready. It had only to be made up from the *Quarterly Journal of Science*, the *Natural History Review*, the *Record*, the *Opinion* and - the easiest of all - the *Reader*. The prospectus was virtually identical with that in the 'revised' *Reader* of 1865. The pitfalls awaiting any young scientific journal were not new. Editorially it had to avoid religion and politics and an appeal to any particular class of research or researchers. It had to be a weekly and appear on time. For this, it required a responsible paid editor and paid contributors. It should aim to educate as well as inform; it should be suitable for science masters in schools throughout the Empire as well as for scientific men in the universities. For this it required co-operation from the United States, the continent and the colonies. For this, in turn, it needed 'contacts' and an aggressive distribution system. Here, of course, Alexander Macmillan was in his element.

While Lockyer was offering free subscriptions to foreigners such as Spencer Baird, the director of the Smithsonian, if he would allow his name 'to appear among the supporters of the journal and inform us from time to time of the progress of science in [his] neighbourhood',[69] Macmillan was cultivating his friends in the schools, like James Wilson of Rugby, in the universities of England and Scotland, in India - in short, wherever Macmillan's books were read. Thus he wrote to William Thomson (later Lord Kelvin):

> ... Lockyer is going to start a weekly journal of Science which we are to publish. It is meant to be popular in part, but also sound, and part devoted specifically to scientific men and their intercourse with each other. Balfour, Steward, Wilkinson, Tyndall, Roscoe and almost everyone else who is about London have given him their names, and he

---

69 Smithsonian MSS. 46283, Lockyer for Macmillan to Director, Smithsonian Institution, August 2, 1869.

> very greatly wishes yours as among those who promise support. May I tell him you consent?[70]

Similar letters went to Sir George Airy, Lyon Playfair (who approved but declined to contribute), to Sir Benjamin Brodie, William Crookes, Thomas Graham, Nevil Story-Maskelyne, Charles Pritchard, P.G. Tait and Alexander Williamson, all of whom offered their support. Of his scientific friends, only one, Joseph Hooker, was outspokenly pessimistic. Ironically, Hooker was among the few who had actually tried to run a journal himself. 'By all means make public my goodwill to Mr Lockyer's periodical', he wrote to Macmillan,

> I fear, however that scientific support is a broken reed and that it will be difficult to supplant the *Athenaeum* ... as it is, the failure of scientific periodicals patronised by men of mark has been dismal - e.g. the *Reader*, *Natural History Review*, my father's *Botanical Journal* and so forth ...

But the essence of his despair was this:

> I do not see how a really scientific man can find time to conduct a periodical scientifically; or brains to go over the mass of trash that is communicated to it and requires expurgation.[71]

An editor would either sacrifice his science for his journalism or else ruin the journal by giving it second place to science in his life. Lockyer presumably knew this but was not to be put off. After all, Hooker had not considered the one possibility remaining - that a journal could be so representative of the whole community of science that it could actually be run by that community. The editor's job was not that of a clerk but a clearing house.

By June, the name of the new journal was decided. Where the decision took place or who made it is not clear. The word Nature appeared often enough in Lockyer's own writings. It was common usage and thus, in the *Reader*, he referred almost instinctively to the 'open book of Nature'. But the use of the word as a title was inspired.

Commercially the word was easily understood and translated into every major European language. Philosophically, it synthesised the Carlylean image of Nature which had profoundly influenced Huxley, Tyndall and G.H. Lewes,

---

70 Graves, *Life and Letters of Alexander Macmillan*, (London: Macmillan, 1910), 299, Macmillan to Thomson, July 20, 1869.

71 *Lockyer Papers*, Hooker to Macmillan, July 22, 1869.

the Baconian view of scientific method in unveiling natural uniformity and the symbolic separation of matter from theology. As it gently lifted science from threadbare controversies with religion, it offered the hope of fuller philosophical insight into the process of creation. Though it stressed the 'natural world' of the naturalist, the word also carried a message of unity in investigation, of shared understanding between all different branches of natural (and potentially social) science. Its effect was exhilarating. As J.J. Sylvester wrote:

> What a glorious title, *Nature*, a veritable stroke of genius to have hit upon. It is more than a Cosmos, more than universe. It includes the seen as well as the unseen, the possible as well as the actual, Nature and Nature's God, mind and matter. I am lost in admiration of the effulgent blaze of the ideas it calls forth.[72]

Its immediate application may have arisen from the first meeting of the little known but highly important Metaphysical Society in April 1869. The society, of twenty-six original members, was formed to confront men of science with men of faith, to discover what philosophical reconciliation was possible between religion and the principle of uniform, evolving Nature. The scientific part of the society included Huxley, Tyndall, Pritchard, Lubbock and Carpenter - all Macmillan's men and Lockyer's friends. The society, however, also included R.H. Hutton, editor of the *Spectator*, James Knowles, editor of the *Contemporary* (and later the *Nineteenth Century*), and Wilfred Ward, editor of the *Dublin Review*. It must have been all too clear to the scientific men that they had no authoritative medium of opinion in which they could present a united front. The chance of a new journal by Macmillan, edited by a young, successful and (so far) uncontroversial man, was too good to miss. At the first meeting of the society in April 1869, Tennyson's new poem 'The Higher Pantheism' was read. Tennyson had been on the continent recovering from illness and depression and had turned to Wordsworth. In some sense his strange abstract poem was a reconciliatory reply to the fear, despair and the materialism he saw in Tyndall and revealed in *Lucretius*, published in 1868. Whatever may be the mysteries of matter there was promise of reconciliation:

> God is law, say the wise; O Soul, and let us rejoice,
> For if He thunder by law the thunder is yet His voice.[73]

---

72 Quoted in Mary T. Lockyer and Winifred L. Lockyer, *Life and Work of Sir Norman Lockyer* (London: Macmillan, 1928), 48.

73 *The Works of Alfred, Lord Tennyson* (London: Macmillan, 1902), 239.

Where to look for the new natural theology - the 'higher pantheism'? Perhaps to Wordsworth:

> ...To the solid ground
> Of nature trusts the Mind that builds for aye;
> Convinced that there, there only, she can lay
> Secure foundations ...[74]

These lines from the revised version of an obscure sonnet became the new journal's motto.[75] Significantly, in the journal, the capitalisation of 'nature' and 'mind' are interchanged and remain thus until 1920. It is unlikely that this reflects a mere typographical error. A mast-head was chosen, perhaps under Huxley's or Tyndall's influence, with a cosmical representation of the Earth in the context of the universe - a striking and unique reflection of *naturphilosophie*, even of Goethe, in British scientific thought, and perhaps also a wishful plan for the journal's universal distribution.

On November 3, Macmillan innocently betrayed to James Maclehose the thrill of first night anticipations:

> *Nature* is to be published in London at 7.30 ... Lockyer was preemptory that our publication day should indicate the point to which our information is brought up. The fallacy of a Saturday publication with a Thursday actual information he doesn't think right ... We start with 10 pages of advertisement. ... I think we will look nice.[76]

The First Issue

> **'First, to place before the general public the results of Scientific Work and Scientific Discovery and to urge the claims of science to a more general recognition in Education and in Daily Life ...'**

The journal did, in fact look nice as it appeared on November 4, 1869. Beneath the conservative dress it shared with the *Athenaeum* and the *Academy*, it

74 W. Wordsworth, *Miscellaneous Sonnets*, 21 (London, 1827), 5-8.

75 The original text, published in 1823, does not contain these lines. The influences which brought Wordsworth to make the revision are surely most intriguing. I thank Professor John Hollander for his illumination of this point.

76 Graves, *Life and Letters of Alexander Macmillan*, 302, Macmillan to MacLehose, November 3, 1869.

conveyed a dynamic view of science in human affairs. The stated objects of *Nature* were

> First, to place before the general public the results of Scientific Work and Scientific Discovery and to urge the claims of science to a more general recognition in Education and in Daily Life; and secondly, to aid Scientific Men themselves by giving early information of all advances made in any branch of natural knowledge throughout the world and by affording them an opportunity of discussing the various scientific questions which arise from time to time.

These two objects really embodied four distinct activities: the announcement of fresh results, the public promotion of science, the diffusion of scientific information and the airing of controversy. Part professional and research; part educational and amateur.

The first issue beautifully blended the London scientists' fascination with German *naturphilosophie*, Carlylean idealism, British empiricism and English romanticism. The first article was Huxley's translation of Goethe's rhapsody on Nature, a fitting introduction for a journal which proposed 'to mirror the progress of that fashioning by nature of a picture of herself in the mind of man'. Some thought that Huxley, 'in an interval of lunacy', had written the rhapsody himself. But it was, he said, 'the best expression of the modern aspect of Nature known to me'.[77] Other articles attacked the view of science as dry calculating Benthamism - 'the Dullness of Science' - and appealed to flights of creative imagination. Soon followed such diverse contributions as accounts of winter-flowering plants, reports of the German scientists' meeting at Innsbruck and a note of the recent solar eclipse.

The appearance of a new weekly paper, as the *Saturday Review* remarked, was not generally a rare or remarkable event. But there was peculiar importance in this one. At last science, as such, had received 'representation' in the fourth estate by a journal promising depth, 'universality and promptitude'. To succeed, it would need to encourage co-operation and to avoid specialisation. But it was vitally important that it should succeed. 'To include within its territory every separate class of the great domain of science [and] to have intimate relations with the little band ... of nature's servants in every civilised region of our planet ...' *Nature* promised to be as universal as its title.

---

77 *Macmillan Archives*, Huxley Correspondence, f. 134, Huxley to Macmillan, October 30, 1892.

XI

The Social Framework of *Nature* in its First Fifty Years

Securing the Foundations

> ***Nature* soon became the recognised forum for scientific debate in Britain. Few of the competitors of *Nature* from the nineteenth century have survived, but for many years *Nature*'s finances left much to be desired.**

In *Nature*'s first fifty years, Lockyer and his assistants produced 2,600 weekly numbers in 103 volumes. Between 1869 and 1909 there were two volumes per year and six months per volume (November-April and May-October). For sheer endurance, it set a record. But it had a good deal of help, even from those who might have been set against it. The *Academy* was among the few who threatened at first to steal *Nature*'s thunder. This weekly, published at sixpence, was begun by the house of Murray, printed by Williams and Norgate and edited by Charles Edward Appleton. Appleton was a leading Oxford reformer in the 1850s and 1860s who probably coined the term 'endowment of research' as it became recognised in England. 'Research', in his sense however, was not the sole preserve of natural science but the well-spring of every 'science' - of '*wissenschaft*'. Thus the *Academy* carried work, not only by Tyndall and Huxley, but also by Arnold, Morris and Rossetti.[1]

In the *Academy* appeared the only negative review of *Nature*'s first issue. This exquisitely observed that the new journal was not the 'Zeitschrift' Appleton had expected.

> The eminence of the names attached to the preliminary notice of its appearance naturally led to the conclusion that the work would be a great contribution to science, as containing original essays on subjects of scientific interest, as a guide to the student as to the value of recent publications and as a record of scientific investigation; but ... it does not, in its present form, supply the want of those interested in 'Nature'.[2]

---

1 'The Academy: A Retrospect', *The Academy* (November 14, 1896), 375.

2 *The Academy* (November 13, 1869), 45.

A few days after the review, however, Appleton cordially reassured Lockyer that 'some of my friends had upbraided me for the notice of *Nature* in the *Academy*, thinking it severe. I hope you won't think so'. .

> It is, in fact, the opinion of one of the very best scientific workers we have up here into whose hands I put the first number. In pointing out deficiencies, I hope in no unfriendly way we have done very much what the French *Revue Critique* had done for us - in an article which I, for one, value much above all the foolish and indiscriminate laudation that we, like you, received from the English press. I wish you all possible success.[3]

Appleton's blessing reflected the kinship and sympathy which the new paper had almost by birthright. It was not only the culmination of years of experimentation and the embodiment of Macmillan's ideas of journalistic innovation; *Nature* was also the favourite godchild of the scientific community. While Lockyer guided it and Macmillan gave the help of an indulgent uncle, its development was anxiously watched and recorded by the country's most important men of science. 'I am so glad you like *Nature*', Macmillan proudly exclaimed to Roscoe in mid-November, 'Number three will be the best, I hope, and we mean to go on improving. Thanks for your article. You will give us more'.[4] 'I will be very glad to do what I can', William Thomson would later write from Glasgow, 'and I have no doubt that we shall supply papers almost continuously'.[5] Meanwhile, Macmillan spread the good news wherever British journals were read.[6]

Early sales were promising. 'I have been told 18,000', Ray Lankester wrote to Macmillan enthusiastically. 'If that keeps on, or near it - it must be a great success'.[7] He must have been wildly optimistic, but the spirit of hope was with *Nature*.

The price had been set purposely low at fourpence (most weeklies were sixpence) to attract subscribers. Advertising revenue was expected to help meet costs, and much of this came from scientific instrument makers and publishers. Special rates were offered for institutional advertisements, which made *Nature* a profitable instrument for the academic profession.

---

3 *Lockyer Papers*, Appleton to Lockyer, n.d., but early November 1869.

4 *Macmillan Archives*, Add. Mss. 35.3.90, vol. 17, Macmillan to Roscoe, November 16, 1869, 17.

5 *Macmillan Archives*, Thomson letters, Thomson to Macmillan (*c.* 1872).

6 George A. Macmillan, *Letters of Alexander Macmillan* (Glasgow: published privately, 1908), 259., Macmillan to C.B. Clarke, December 16, 1869.

7 *Macmillan Archives*, Lankester Letters, Lankester to Macmillan, November 16, 1869.

There was certainly, both for the sake of economy and the interests of general readability, a need to keep the journal 'popular'. The unsigned biographical sketches under the title 'Scientific Worthies', together with the space devoted occasionally to regular obituaries, were attempts of this kind. The engraved portraits accompanying them seem to have been an idea of Macmillan, who waited until a set was ready and then launched the series with a sketch of Faraday by Tyndall, in September 1873.[8]

In the absence of printing records, it is difficult to calculate production costs or distribution figures for *Nature*'s early years. Some clues can, however, be found in the experience of *Nature*'s contemporary - *Chemical News*. According to William Crookes, its average cost in the 1860s, excluding editorial and contributors' fees, was about £14 13*s* 0*d* per issue, or about £730 p.a. *Nature* was typographically similar to *Chemical News* but had more illustrations, which would probably have cost at least £200 more. Lockyer's salary as an editor is not certain, but it was perhaps £400. (His successor's salary was £500 p.a. and his assistant's £200 p.a.).[9] Assuming there were at least two (and probably three) assistants at a likely salary of £100 p.a., this suggests that *Nature*'s editorial costs alone reached at least £600 p.a. If *Chemical News* is a good guide, the annual cost of producing *Nature* would have come to £1,530 before fees to contributors. But these amounted to about ten shillings per column in 1921 and were probably near seven shillings in the nineteenth century. Calculating on the basis of 15 pages of text at fourteen shillings each, contributors' fees would approach £11 per issue, or £572 per year. Assuming that Macmillan met the overhead costs of stationery and clerical help, this gives a hypothetical budget along the following lines:

Estimated production costs	£930
Estimated editorial costs	£600
Estimated contributors' fees	£572
Total	£2,102 p.a.

Nature's income is obscure, but in 1869-70 each issue averaged about four pages of advertisements which, at the rate of five guineas per page, must have amounted to £1,092 per year (one guinea=21 shillings). By 1876, the number of pages of advertisements had doubled to about 400 per year, which must have been worth about £2,000 per year. This figure seems to have remained nearly constant until the turn of the century, which helps to explain *Nature*'s

8 *Macmillan Archives*, vol. 22, f. 294, Macmillan to Tyndall, August 21, 1873.

9 *Macmillan Firm Letters*, R.A. Gregory to Maurice Macmillan, October 8, 1921.

precarious balance and nagging deficit.

It is difficult to calculate *Nature*'s circulation for its first forty years. In 1870, Lockyer boasted that there were nearly 5,000 subscribers, and 15,000 readers, but this seems exaggerated.[10] Subscriptions must have been much lower. As late as 1913, according to house records,[11] there were only 970 subscribers, of whom 241 were Americans and 494 were colonial or foreign.[12] Only 24.3 per cent of the circulation was British. There is no way to estimate 'counter sales', but Macmillan staff suspect that they, too, were quite small. Circulation reached a peak in 1913, but during the First World War, subscriptions fell off and began to recover only in the 1920s. Total subscriptions rose from 1,072 in 1920 to 2,660 in 1930, and 2,780 in 1939. British subscriptions were 376 (35 per cent) in 1920, 847 (32.6 per cent) in 1930 and 807 (29 per cent) in 1939. At least until the beginning of the Second World War, British readers accounted for no more than 35 per cent of *Nature*'s regular circulation, although probably for most of its advertising revenue. Projecting backwards from 1913, and assuming an acquisition rate of at least twenty subscriptions a year, there were probably no more than 100-200 subscribers at the birth of *Nature* and no more than 300-500 before at least 1895.

One has the impression there were 'practical' journals which would have snuffed out *Nature* if they had the chance. Such a journal was the penny weekly *English Mechanic*, which advertised itself as a 'literary Aaron's rod',[13] and which by 1870 had swallowed the *Mirror of Science*, the *World of Science* and the *Penny Mechanic*, and soon acquired *Scientific Opinion* and *The British and Foreign Mechanic* as well. This journal addressed itself with laudable catholicity to the entire 'working and scientific community not only among those who speak the English tongue but foreigners'. It claimed a circulation of 10,000, 'wider than all the English scientific papers combined', but there is no evidence that it dampened Macmillan's enthusiasm.[14]

On the contrary, *Nature*'s scientific competitors (see Table 1), including the *Philosophical Magazine*, the *Chemical News*, the *Journal of Science* (both edited by William Crookes) and the *Quarterly Journal of Microscopical Science* (edited after 1868 by Ray Lankester) offered active co-operation and

---

10 'To Our Readers', *Nature*, 2 (1870), 1.

11 C. Morgan, *The House of Macmillan, 1843-1943* (London: Macmillan, 1943).

12 Mr A. Clark, of Macmillan (Journals) Ltd, kindly supplied the twentieth century figures. The estimates for the nineteenth century are highly speculative, and subject to correction.

13 *English Mechanic*, advertisement, 10 (1869), 2.

14 *Macmillan Archives*, Lankester Letters, f. 102, Lankester to Frederick Macmillan, July 22, 1919.

the means of entry for new *Nature* recruits.[15] William Crookes, a friend of Lockyer since the eclipse expedition of 1870, welcomed Lockyer's support in the struggle for the advancement of science. There was often a certain deference to *Nature*. When, for example, in 1895, Crookes sent a proof of his own account of the helium spectrum for insertion in *Nature* soon after Ramsay evolved helium gas, he added revealingly:

> ... it will appear in the *Chemical News* on Friday, but my circulation is not to the same class of researchers as that of *Nature* and having taken a great deal of trouble about it I want the results to get to the right people.[16]

Table 1 Nature's peers

		Cost per issue				Pages per issue			Pages per year		
Weeklies:		1869	1876	1890		1869	1876	1890	1869	1876	1890
The Academy	1869-		*6d*		Text						
	1922				Ads						
The Athenaeum	1828-	*4d*	*8d*	*4d*	Text	12	10	11	668	520	600
	1921				Ads	4	6	1	208	312	50
English Mechanic	1865-date				Text						
and World of Science					Ads						
became *Mechanics* in											
1942											
Nature	1869-date	*4d*	*4d*	*6d*	Text	21	18	2,7	1,128	936	1,312
					Ads	4	8	2	208	416	432
Monthlies:											
Annals and Magazine	1838-date	*2s6d*	*2s6d*	*2s6d*	Text	76	84	84	926	1,088	1,000
of Natural History											
Scientific Review	1865-	*6d*	*6d*	-	Text	22	14	-	256	163	-
	1883				Ads	2	2	-	24	24	-
Quarterlies:											
Popular Science	1861-	*2s6d*	*2s6d*	-	Text	112	112	-	457	450	-
Review	1881										
Quarterly Journal of	1864-	*5s0d*	*5s0d*	-	Text	258	144	-	638	576	-
Science	1885										
					Ads on covers			-	on covers		-

15 The scientific objectivity of Crookes's *Quarterly Journal of Science* suffered from his own work on spiritualism, which was published in the 1870s when Crookes was sole editor; the journal was sold, and died in 1885.

16 E.E. Fournier, *The Life of Sir William Crookes* (London: T. Fisher Unwin, 1923); Crookes to Lockyer, August 20, 1895, 348 - 9.

Other journals bore, if not as happy, at least a vital relationship to *Nature*. Although *Nature* was not, for example, closed to the mathematical biologists, and Macmillan certainly cultivated Galton, there was some feeling that it did not meet the needs of the new field.[17] In 1901, after an angry struggle with Bateson and other British biologists, Karl Pearson, Francis Galton and W.F.R. Weldon began *Biometrica.* The statistical chances against its survival were overwhelming, but *Biometrica* finally crossed the £200 income threshold required by its publishers.

Lankester's journal, which he owned himself, had a regular subscription of no more than 450-470, but to its staff Lankester brought W. Thiselton-Dyer (Hooker's son-in-law and successor at Kew), Edward Klein (pathologist and Lankester's colleague in Vienna) and J. Frank Payne (the biologist, fellow of Magdalene and Lankester's colleague in Oxford).[18] Almost all Lankester's authors contributed to *Nature* and joined Lockyer's intellectual aristocracy.

Looking to the continent, *Nature* soon had several (chiefly short-lived) competitors. Among them were, in France, *La Nature*, begun by Gaston Tissandier, in 1873 and, in another sense, *La Revue Scientifique*, begun in 1871; in Italy, *La Natura*, a 'giornale illustrato di cognizioni utili', edited in Florence between 1877 and 1880 and in Milan between 1881 and 1885. There were also the Belgian *Natura*, edited in Ghent between 1883 and 1885; the Dutch *Die Natur*, a popular 'geillustreerd maandschrift', in Utrecht 1881-1894; and the Norwegian *Naturen*, begun in Bergen in 1877. There was even a Mexican *La Naturaleza* begun (second series) in 1887. In India, *Current Science*, which began in 1931, carried a mast-head that clearly proclaimed its debt to *Nature*.

These competitors do not seem to have materially affected the journal's economic progress. Between 1870 and 1877, the journal gained in popularity (or at least in popular currency); it gained in contributors who, in turn, swelled the ranks of subscribers. In May 1878, after nine years of financial sacrifice, Macmillan felt confident enough about the future of this happy cycle to raise the price from fourpence to sixpence (annually from eighteen shillings and sixpence to twenty-eight shillings) and enlarged each issue from eighteen to twenty-eight pages. The firm's announcement cited the generous support it had received from the profusion of new scientific societies throughout the land and the increasing aid of readers 'some of them in the more remote and least known spots on the earth's surface'. Even in America, the firm boasted, *Nature* was circulated 'more widely than any other English journal'. It would

17 See, for example, *Galton Letters* (University College London), Alexander Macmillan to Galton, September 4, 1879, and George A. Macmillan to Galton, October 30, 1883.

18 See Gilbert Bourne, 'Fifty Years of the Q.J.M.S.', *Quarterly Journal of Microscopical Science*, 64 (253), (1919), 1.

be difficult to corroborate this claim, but it was already clear that *Nature* could, 'when we deem it advisable', be the medium of communication of the opinions of the most eminent men of science in Europe and America on any important subject attracting general attention.[19]

In a leading article entitled 'Retrospect and Prospect' Lockyer proudly proclaimed a new stage in the journal's career.[20] 'At last we are compelled to enlarge it', he said, 'to find space for the stream of communications that week after week come pouring in upon us from all parts'. As the stream continued, he added, 'the editor's function will, as heretofore, be a modest one'. The stream did continue; by 1881 an advertisement for *Nature* in *Knowledge* claimed that its correspondence columns formed a 'medium of scientific discussion ... among the most distinguished men of Science and had become the recognised organ for announcing new discoveries and new illustrations of scientific principles among observers of nature all the world over ...'.[21]

American journals, though not much read in England, had a good deal to do with the circulation of English ones. *Science* (begun in 1883) and, to a lesser extent, *Scientific American*, were *Nature*'s chief American competitors, but it is unclear how they affected *Nature*'s distribution. Lockyer could always expect respectful piracy from the *Popular Science Monthly* which, in return gave *Nature* a rare form of advertisement. The *Popular Science Monthly* was begun, rather accidentally in New York by Spencer's old friend and publisher E.L. Youmans, as a means of presenting Spencer's essays to America in periodical form.[22] Youmans at this time sent only one copy to England; one to Spencer and one to Tyndall. For years it thrived by reproducing articles which had first appeared in European or British journals. According to Youmans himself, it was a 'practical concern' of which he was 'duly ashamed' but it met an important need. 'I have always believed', Youmans wrote to Lockyer in 1872, 'that the American public would sustain a well-conducted periodical of this type, but that they should have accepted anything so crude as that which we have offered them ... is ... a matter for surprise'.

Youmans's deference towards *Nature* further enhanced its image abroad. *Nature*, he told his readers in 1889, was the 'foremost scientific journal in Europe'. Its columns were 'faithful mirrors' of scientific thought, its correspondence department was a 'free parliament for discussion'; it had

19 'Notes', *Nature*, 17, (1878), ccxii.

20 'Retrospect and Prospect', *Nature*, 18 (1878), 1.

21 *Knowledge*, 1, (1881), iv.

22 *Lockyer Papers* (Courtesy of Prof. A.J. Meadows), Youmans to Lockyer, July 18, 1872.

become 'the one journal indispensable to students in every branch in every land'.[23]

Indispensable or not, the journal's finances left much to be desired. In May 1889, with *Nature*'s budget still in doubt, there were rumours from J. Wemyss Reid, who reminded Lockyer that he was 'having *Nature*'s estimates prepared anew ...', of a takeover bid by Cassell and Co.[24] What became of this is not clear, but in November 1889, Lockyer, with the aid of A.W. Rücker, wrote a celebratory piece on *Nature*'s twentieth birthday that confirmed its place on Macmillan's list. 'As regards the journal itself', they wrote,

> we may claim that it has not disappointed the hopes of its founders, not failed in the task it undertook ...[25]

Once again Lockyer claimed,

> its achievement was due not to his own efforts but to the unfailing help we have always received from the leaders in all branches of natural science. This help has not been limited to their contributions to our columns, but has consisted also of advice and suggestions which have been freely asked and as freely given.

The policy worked - but very slowly. Not until 1899 did Macmillan's ledgers show a profit. As Craik wrote, 'I am glad that your diligence and tenacity has been at last rewarded by money besides all the honour that *Nature* has brought you. We have waited many years'.[26]

---

23 E.L. Youmans, 'The Twentieth Birthday of *Nature*,' *Popular Science Monthly*, 36 (October, 1889), 413.

24 *Lockyer Papers*, J. Wemyss Reid to Lockyer, March 13, 1889.

25 'Twenty Years', *Nature*, 41 (1889), 1.

26 Mary T. Lockyer and Winifred L. Lockyer, *Life and Work of Sir Norman Lockyer* (London: Macmillan, 1928), 173.

Private Army of Contributors

***Nature* benefited from an army of contributors co-ordinated by Lockyer, but disputes were not unknown.**

Nature enjoyed an uncommonly good start. The question of access to scientists was solved almost instantly. First, Lockyer knew, or had access to, all former contributors to the *Natural History Review*, *The Academy* and the *Journal of Science*. Moreover, as Macmillan's scientific editor for two and a half years, Lockyer had at his editorial fingertips Macmillan's entire stable of scientific authors. Of the 163 men listed as Macmillan's authors in 1891, forty-five made contributions to *Nature* in its first 50 years. For scientific news he became, like Henry Oldenburgh, 'a clearing house of Science'.[27] He was in a pivotal position to lever the weight of scientific opinion across the entire literary front.

Despite his access to outside help, Lockyer needed help within. 'I wish *Nature* had a strong enough hold upon you', he told Hooker, 'to make you find and name a botanist who would treat her lovingly and represent her faithfully. The difficulty of getting the right men to write is so great that I think of casting you all off to the pole where both days and nights are longer to give you time'.[28]

The immediate result of Lockyer's plea seems to have been Alfred Bennett (1833-1902), a botanist from University College, London, who was appointed biological assistant in 1870 and seems to have stayed until 1874. Bennett, who knew German well and did most of *Nature*'s biological texts, was among the first of Lockyer's private army of sub-editors. Others soon followed. Alfred H. Garrod (1846-79), son of Sir Alfred Garrod (physician to the Queen) and himself a physician and anatomist - first in the Natural Science Tripos at Cambridge in 1871, the first Fellow of St John's to be elected in natural science, and cousin of both Charles Keane, editor of *Punch*, and Meredith Townsend, editor of the *Spectator* - became Lockyer's zoological and medical sub-editor from 1869-79. Another recruit was John (later Sir John) Scott Keltie, a graduate of St Andrews and Edinburgh, who had come from the staff of William Chambers to Macmillans in 1871, and became Lockyer's educational geographical editor in 1873. Later to join was Richard Lydekker (1849-1915) from Trinity College, Cambridge (where he came second to Garrod in the Natural Science Tripos of 1871); he returned from the

27 A.R. Hall and Boas M. Hall (eds), *The Correspondence of Henry Oldenburgh* (Madison: University of Wisconsin, 1965), vol. 1, xix.

28 *Hooker Papers*, 10, f. 221, Lockyer to Hooker, September 1869.

staff of the Geological Survey in India (1874-82) to write books and reviews on geology and palaeontology. Between 1869 and 1872, John O. Earwaker (1847-95), later FSA, a graduate of Owens College (Manchester), Pembroke College (Cambridge), Merton College (Oxford) and sometime deputy keeper of the Ashmolean (1873-74), wrote on genealogy and anthropology. John Brett (1831-1902), a landscape painter, member of the pre-Raphaelite brotherhood, close friend of Ruskin, Associate of the Royal Academy and a keen Fellow of the Royal Astronomical Society, provided Lockyer with notices of astronomical meetings. John Russell Hind (1823-95), director of the Regents Park Observatory and veteran of the *Reader*, contributed astronomical columns between 1872 and 1882, and Prince Kropotkin, an émigré from Czarist Russia, brought Russian texts to *Nature* readers from 1876. When Kropotkin returned to Russia in May 1917, he warmly recalled the day he found in England and *Nature*'s offices 'the most cordial reception - and work'.[29]

A second source of informal (and often unpaid) editorial talent came from the Royal College of Chemistry and the Normal School of Mines at South Kensington (after 1890-91, the Royal College of Science). In 1874, Raphael Meldola, a chemical assistant to Edward Frankland at South Kensington, where he helped Lockyer and Frankland in their spectroscopic studies, became sub-editor. His hours were ten to five on weekdays and ten to two on Saturdays and his salary was £100 a year. Among the other young students Lockyer shared with Frankland were Richard J. Friswell (1849-1908),[30] a veteran of the India eclipse expedition of 1871, who joined Lockyer about 1872; Herbert MacLeod, later professor (at Cotton Hill); and Alexander (later Sir Alexander) Pedler (1849-1918) who succeeded Herbert MacLeod in assisting Lockyer in spectral research between 1871 and 1873.

Lockyer's friends recommended candidates to him.[31] James Wilson (then Canon Wilson), former master at Rugby, nominated George Seabroke, a young solicitor from Rugby. A few years later, a young factory boy from Bristol, whom Wilson had once nominated to a job at Clifton, took up a place at South Kensington and became Lockyer's assistant. In May 1893, this man, Richard Gregory, became the chief sub-editor of *Nature*.[32]

Lockyer's staff not only helped to assemble the journal. They also became his 'eyes and ears' in the councils of scientific societies everywhere. Some

---

29 Lockyer and Lockyer, *Life and Work of Sir Norman Lockyer*, 221.

30 'Notes', *Nature*, 77 (1908), 349.

31 'Notes', *Nature*, 101 (1918), 208.

32 Richard Gregory, 'Science in a Changing World: Recollections and Reflections', *Nature*, 137 (1936), 981; 'Autobiographical Notes', *Gregory Papers* (by courtesy of Lady Gregory).

even helped in his research - the dividing line was never very clear. Almost without exception, Lockyer's assistants, both scientific and editorial, became famous in their own right, a fact which only magnified *Nature*'s reflected glory. Friswell, for example, was elected to the Council of the Chemical Society in 1871 and was a charter member of the Institute of Chemistry in 1877. Meldola became professor at Finsbury Technical College in 1885 and President of the Entomological Society (1895), the Chemical Society (1905-07), the Society of Dyers and Colourists (1907-08) and the Society of Chemical Industry (1908). Many of Meldola's students, including William Pope, became *Nature* stalwarts. Pedler became professor of chemistry at the Presidency College, Calcutta, and later vice-chancellor of the University of Calcutta. Seabroke became director of the Wilson Observatory at Rugby, an FRAS and an active member of the British Astronomical Association.[33] Bennett became a Fellow of the Royal Microscopical Society (1879) and later editor of its journal (1897-1902).[34] Keltie later became librarian (1885), assistant secretary (1892), and secretary (1896) of the Royal Geographical Society and, in 1893, began the *Geographical Journal*. Through his sub-editors, Lockyer remained closely associated with the national press. Keltie and Hind became correspondents on *The Times* in 1874, while E.S. Dallas, a member of Delane's staff and correspondent at the Paris Exhibition in 1867, wrote occasional pieces for *Nature*.

Lockyer was not always easy to please. 'His bursts of enthusiasm', recalled Alfred Fowler, his assistant in the early 1890s, 'sometimes proved too much for some of his assistants, who especially objected to work on Saturday afternoons, on weekends and vacations'.[35] He added: 'Lockyer rarely allowed me to forget that I was his assistant and spoke of little outside the work in hand'.[36] The regular appearance of the journal, with a careful eye on advertisements and space, did not make life easier. Lockyer's ruthlessness cost him in friendship. In May 1870, for example, John Brett, Lockyer's astronomical reporter who had offered his services freely, threatened to resign if Lockyer did not keep a proper sense of priorities:

---

33 'Notes', *Nature*, 101 (1918), 91.

34 'Notes', *Nature*, 65 (1902), 34.

35 Alfred Fowler, 'Memories of Sir Norman Lockyer', in Lockyer and Lockyer, *Life and Work of Sir Norman Lockyer*, 461.

36 Idem., 457.

> I have taken a good deal of pains for some years now to give you accurate accounts of the monthly meetings of the Royal Astronomical Society in a very condensed form. They cannot be so condensed, and yet give a true representation, without extremely careful writing. I have taken pleasure in doing it both for the sake of *Nature* and the [Royal Astronomical] Society. It is to no advantage to me: I don't even get a copy of the paper.
> Now, if after I have taken all the trouble, you cut some of my most careful paragraphs, as you did last report, for the sake of a trumpery advertisement, you must dispense with your reporter and keep your advertiser; and I must conclude that his 3/6*d* is more important to *Nature* than correct scientific information.

'You are bound to hear a little truth sometimes', Brett concluded. 'I only hope [the cutting] was done by some of yr. understrappers'.[37] But Brett's gentle sympathies were the stuff of which scientific journals had been made and which had brought them to collapse. Lockyer was determined to survive. Not surprisingly, Lockyer's greatest personal contributions to the leading articles came in the earliest years. In the first decade, 1869-79, he wrote twenty-four, while in the next three decades he wrote between twelve and fourteen. Two of his leading articles were on Science and Art (May 1878 and May 1883). Sixteen of his leading articles between 1869 and 1886 were on astronomical subjects. From 1887, with a series of three articles on the Jubilee, there is a shift in his preferences. Virtually none of these articles was on astronomy and most are connected with educational or scientific institutions.

Lockyer's house staff often wrote on subjects outside their own professional experience. This gave many editorials a generalist style and also left room for issues which academic scientists would not have chosen. A.W. Bennett, for example, wrote leading articles in December 1870 on such varied topics as the Eclipse Expeditions; in November 1870 on 'Science and the Working Classes' and January 1870 on 'Dust and Disease'. In November 1871 and again in September 1874 he returned to 'Science for Women' and 'The Education of Women'. In between, he wrote pieces on the Royal Institution (March 1872), 'Newspaper Science' (April 1872), 'Botany in the Oxford Natural Science School' (May 1872), 'Economic Entomology' (July 1872) and 'Agassiz at San Francisco' (October 1872).

John Brett's contributions, for example, were more in *Nature*'s social notices. Only one review - 'Natural Science at the Royal Academy' (June 1870) - is attributed to him. John S. Keltie wrote on 'Technical Education'

---

37 *Lockyer Papers*, John Brett to Lockyer, May 20, 1870.

(March 1882) and 'Science and Industry' (January 1874). He was also *Nature*'s correspondent on the Arctic Expedition of 1874. Gradually, as Table 2 shows, the academic hand became dominant, but this took some time. Between 1869 and 1879, for example, Lockyer, Bennett, Garrod and Keltie wrote most of the leaders. In the following years the picture changed and the participation of Lockyer and his men declined, but by the third decade Lockyer and Richard Lydekker were at the top of the list. Only with the fourth decade a gradient towards the universities appeared. Following Lockyer (with 66 leaders between 1869 and 1919), the most prolific were T.E. Thorpe (with 55), Archibald Geikie (with 35), G.H. Bryan (with 25), Georges Romanes (with 23) and G.B. Matthews (with 22). Certain men became inevitably associated with certain topics. G.H. Lewes, for example, wrote a piece (his only leader) on 'Instinct' in April 1873. Some important names are scarcely represented at all. Among the leader writers, Charles Darwin appears only once, preceding Lewes in April 1873, on 'The Origin of Certain Instincts'. Given Darwin's well known dislike of public exposure this is not surprising. But Hooker, despite Lockyer's pleading, produced only one leading article and only eight reviews between 1870 and 1893. By far the greatest number came from William (later Sir William) Ramsay (professor of chemistry at University College, London), who wrote prolifically on science and the state. Regional specialists included H.B. Brady, who wrote anonymously about the College of Physical Science at Newcastle (Armstrong College, later the University College of Newcastle) in April, May and July 1871, and Henry Roscoe wrote from Manchester.

Table 2 Most productive contributors to *Nature* : leading articles in each decade 1869-1919

1869-79		1880-89		1890-99		1900-09		1910-19	
J.N. Lockyer	24	G Romanes	23	J.N. Lockyer	12	G.H. Bryan	14	T.E. Thorpe	26
A.W. Bennett	16	A. Geikie	19	R. Lydekker	11	R. Lydekker	14	J.A. Thomson	20
A. Geikie	11	J.N. Lockyer	14	W.E. Plummer	10	R. Lockyer	14	R.A. Gregory	14
A.R. Wallace	10	E.P. Wright	13	E.B. Poulton	10	R. Meldola	13	G.H. Bryan	10
A.H. Garrod	9	R. Lankester	9	R. Lankester	9	R. Glazebrook	12	E.J. Russell	10
Col. A. Strange	8	J.W. Judd	8	T.E. Thorpe	9	J.A. Thomson	11	G.B. Matthews	9
J.S. Keltie	7	J.S. Keltie	7	A.R. Wallace	9	W.E. Plummer	10	W. Ramsay	9
R. Proctor	7	E.B. Tylor	7	A.G. Greenhill	9	J.B. Falmer	10	E.A. Minchen	6
J. Tyndall	7	T.E. Thorpe	7	H.E. Armstrong	8	G.B. Matthews	9	A.E. Clawley	6
R. Lankester	7	A. Rücker	7	T.G. Bonney	7	T.E. Thorpe	8	J.H. Reynolds	5
		H.E. Armstrong	7	A.C. Haddon	7	W. Ramsay	8	T.C. Allbutt	5
		A. Newton	7			E.B. Poulton	8	J.L.E. Dreyer	5
						F.A. Dixey	8	W.A. Tilden	5
								F. Soddy	5
								A.B. Randel	5

As inscriptions on the file copy testify, Lockyer often worked with his experts in blending 'composite leaders' covering several different fields of science. Such work, recalled T.E. Thorpe, was usually done in long sessions late at night, when he helped compress Lockyer's 'too ready flow of ideas into a single editorial page ... and found his way home in the small hours of the morning'.[38]

At the beginning, Lockyer followed the conventional mid-Victorian practice of having signed scientific contributions but anonymous leading articles and reviews. Lockyer's own astronomical articles were almost always signed; his educational editorials never. But as most leading articles were assumed to be written with the known bias of the editor, reviews were more easily attacked.

In September 1869, Hooker had expressed the hope that *Nature* would

> give us better analyses of books than reviews in general afford us. We have no end of reviews, but they are generally the author's views on the subjects of the books supposed to be reviewed and convey no precise information as to the books themselves. This is a crying evil and arises from the fact that reviewers find it easier to descant on a given subject than to read a book on it.[39]

By the mid-1870s Lockyer was confronted by a need to decide whether *Nature* should be an editorial participant in scientific debates arising from reviews or merely a dispassionate vehicle for them. There was, of course, no real choice; had *Nature* lost its objectivity it could not satisfy everyone.

The two different viewpoints are well illustrated by letters from William Pole and W.H. Preece, both frequent contributors. William Pole preferred the more 'open' approach: 'You are very hard on your poor Reviewers', he wrote to Lockyer in 1876,

> in allowing criticised authors to pitch into them in your columns, particularly as I conceive it would be infra dig for a reviewer to justify his opinions in the publication which professes to adopt them. It seems impossible to criticise freely under this condition.
> I think in future I shall ask you to let me *sign* my articles and to use the pronoun *I* instead of the editorial *We*. It will then, of course, be quite fair for the author to reply if he thinks fit, and I shall be prepared for it ...[40]

W.H. Preece disagreed. As he sent an initialled review under protest, he remarked, 'They lose all their force'.

---

38 'Memorial Tributes to Sir Norman Lockyer', *Nature*, 106 (1920), 24.

39 *Hooker Papers* (Kew), vol. 10, f. 220, Hooker to Lockyer, September 23, 1869.

40 *Lockyer Papers*, William Pole to Lockyer, May 1, 1876.

> One does not care twopence for the individual opinion of W.H.P. or J.T. The most disparaging of one individual would not cause me a moment's discomfort, but the anonymous strictures of an unknown, mysterious editor carry weight and produce effect ... A reviewer, however, does not like to speak the truth when he criticises a friend's book ... whereas he can speak out when he is shielded by the aegis of an unresponsible [sic] editor.[41]

Fairly early on, Lockyer decided that the problem of 'free and just' criticism was more a matter of personality than principle and the choice of signature or anonymity was left to his authors. Some, like T.G. Bonney, signed every leading article they wrote. Others, like Victor Horsley and Lockyer himself, were almost consistently anonymous. Still others, like T.H. Huxley, Ray Lankester, William Ramsay, T.E. Thorpe and Henry Roscoe, signed some but not all. Fourteen of John Brett's fifteen leaders were anonymous.

There were 105 'chief' leader writers (those who wrote more than five leaders) in the period 1869-1919. The numbers of chief authors in the period 1869-1919 increased from 43 in 1869-79 to 56 in 1880-89 and to 62 in 1890-99 (Table 3). The physical sciences and the life sciences are almost equally represented in the first decade, while the latter tend to increase during 1890-99 and the former in 1900-09. By the second decade, the early influence of the astronomers had begun to wane.

Table 3 Representation of chief leader writers* giving leading articles 1869-1919

	CONTRIBUTORS			LEADING ARTICLES		
	Total known contributors	Chief contributors†	% of total	Total	By chief contributors	% by chief contributors
1869-79	145	43	29.7	528	211	40.0
1880-89	180	56	31.7	520	· 256	49.4
1890-99	212	62	29.2	520	265	51.0
1900-1909	223	60	22.0	520	274	52.7
1910-1919	219	47	21.0	520	220	42.3
1869-1919	685	105	15.0	2,608	1,226	47.0

* Leader writers=Persons who wrote five or more leaders.

† Including Norman Lockyer, Richard Gregory and the *Nature* staff.

41 *Lockyer Papers*, W.H. Preece to Lockyer, July 9, 1884.

With the growth of this élite, there was naturally a 'democratic' or participatory element. The 'chief authors' represented only about fifteen per cent of the 685 who contributed leading articles during the whole period, and about thirty per cent of those writing at any one time. This fifteen per cent accounted for about forty per cent of all leading articles.

The backgrounds of this productive élite show, as expected, large numbers of Scottish graduates at first, followed by the graduates of the University of London (particularly University College) and the Royal College of Science (see Table 4).

The number of Cambridge men suggests that the new Natural Science Tripos had begun to work its effects on the output of graduates who ventured into science despite the absence of jobs. In the first decade, however, there were still a good many authors who (like Lockyer) had been privately educated and were not university men. Until the turn of the century this helped *Nature* keep its generalist appeal and satisfied the amateur approach to science. Afterwards, the university men - particularly along the London-Cambridge axis that Macmillan himself had helped create - came into power. A quarter of the sixty chief contributors in the decade 1900-09 were Cambridge graduates and another twenty-two per cent were from the Science School in London.

An occupational analysis of the sample of 105, places even greater stress on the role of the academics - especially Lockyer's colleagues at the Royal College of Science. It seems that the University of London provided employment for at least a quarter of *Nature*'s most prolific men. Whether this implies that university salaries were inadequate; that staff from the University of London were more literary; or merely that they were more 'available', leaves room for conjecture. It is significant, however, that scientists in civil servant posts - for whose prospects *Nature* always had a ready plea - grew proportionately during the period. Most of these men, of course, were based in London. By contrast, the participation of Scottish academics never became significant, and provincial university men were absent, unsought or unavailable. Scientific amateurs with means still provided a strong wedge, but only one industrial scientist was among this 'élite'.

Table 4 Characteristics of chief leader writers 1869-1919*

	1869-79		1880-89		1890-99		1900-09		1910-19		Totals
	New Chief Authors	Stock of Chief Authors									Chief Contributors
Total No. of FRSs	**31**	**31**	**16**	**52**	**15**	**47**	**5**	**45**	**3**	**36**	
	A	**B**	**A**	**B**	**A**	**B**	**A**	**B**	**A**	**B**	
Field											
Physical sciences	**12**	**12**	**7**	**16**	**2**	**12**	**7**	**15**	**-**	**12**	
Life sciences	**13**	**13**	**7**	**17**	**8**	**24**	**1**	**20**	**2**	**15**	
Earth sciences	**6**	**6**	**3**	**8**	**2**	**9**	**1**	**8**	**1**	**8**	
Astronomy/maths	**5**	**5**	**2**	**5**	**3**	**7**	**1**	**7**	**2**	**6**	
Medicine	**1**	**1**	**-**	**-**	**-**	**-**	**-**	**-**	**-**	**-**	
Engineering	**1**	**1**	**1**	**2**	**2**	**4**	**1**	**4**	**-**	**2**	
Other	**1**	**1**	**3**	**4**	**1**	**4**	**1**	**3**	**1**	**4**	
Not known	**4**	**4**	**2**	**2**	**1**	**2**	**-**	**3**	**-**	**4**	
Total	**43**	**43**	**25**	**56**	**19**	**62**	**12**	**60**	**6**	**47**	**105**
Education											
Oxford	**5**	**5**	**3**	**6**	**2**	**8**	**3**	**10**	**1**	**9**	
Cambridge	**5**	**5**	**6**	**10**	**5**	**15**	**3**	**14**	**1**	**9**	
London	**7**	**7**	**2**	**7**	**4**	**9**	**-**	**7**	**-**	**5**	
R. Coll. Science	**3**	**3**	**5**	**8**	**1**	**7**	**-**	**6**	**1**	**7**	
Provincial	**2**	**-**	**-**	**2**	**1**	**2**	**1**	**3**	**-**	**2**	
Scottish	**6**	**6**	**1**	**5**	**2**	**4**	**1**	**6**	**-**	**3**	
Irish	**1**	**1**	**1**	**2**	**-**	**1**	**-**	**1**	**-**	**-**	
Other	**1**	**1**	**3**	**4**	**-**	**4**	**2**	**10**	**2**	**1**	
Private or non-univ.	**-**	**-**	**-**	**7**	**-**	**5**	**-**	**-**	**-**	**-**	
Not known	**13**	**13**	**4**	**5**	**4**	**8**	**2**	**3**	**1**	**11**	
Total	**43**	**43**	**25**	**56**	**19**	**62**	**12**	**60**	**6**	**47**	**105**
Occupation											
Oxford	**-**	**-**	**3**	**2**	**2**	**3**	**1**	**4**	**-**	**3**	
Cambridge	**3**	**3**	**1**	**4**	**1**	**6**	**3**	**6**	**-**	**1**	
London	**11**	**11**	**4**	**10**	**2**	**12**	**2**	**9**	**1**	**7**	
R. Coll. Science	**3**	**3**	**1**	**4**	**1**	**6**	**3**	**6**	**-**	**1**	
Provincial	**4**	**4**	**3**	**8**	**2**	**7**	**-**	**2**	**-**	**2**	
Scottish	**3**	**3**	**1**	**3**	**3**	**4**	**1**	**5**	**-**	**2**	
Irish	**1**	**1**	**-**	**1**	**1**	**1**	**-**	**1**	**-**	**-**	
Govt. research	**5**	**5**	**3**	**6**	**3**	**8**	**1**	**8**	**1**	**10**	
Industry	**-**	**-**	**-**	**-**	**-**	**-**	**1**	**-**	**-**	**-**	
Private or other	**5**	**5**	**5**	**7**	**3**	**7**	**1**	**9**	**3**	**8**	
Unknown	**8**	**8**	**1**	**7**	**-**	**8**	**1**	**5**	**1**	**6**	
Total	**43**	**43**	**25**	**56**	**19**	**62**	**12**	**60**	**6**	**47**	**105**

* **'Chief contributors' are defined as those who have written five or more leaders. This sample, about whom either names or biographical details are known, accounts for 15 per cent of the 685 known leader writers during the period. This 15 per cent accounted for at least 40 per cent of leaders written during the period.**

Throughout these early years, *Nature* offered its 'leadership' support against obscurancy and obstruction in high places. Once Ray Lankester complained that he had received no credit for the fish gallery he had designed at the Natural History Museum. 'I wish you would let someone', he said, 'write an article on what I did there'.[42] Lockyer did. In 1889, Roscoe wrote from the House of Commons asking for an article on the merits of the Technical Education Bill. 'Can you do this for us?'. A leading article in *Nature* obliged. In 1873, the astronomer Charles Pritchard asked Lockyer to insert a note honouring De la Rue's gift to Oxford, '... if you can you will materially strengthen our hands'.[43] *Nature* agreed. In 1879, David Milne Horne thanked Lockyer for *Nature*'s favourable notice of the Ben Nevis Observatory which, he claimed, brought about a grant of £100 a year from the Meteorological Council. In 1881, *Nature* helped S.P. Langley to sponsor an American solar physics expedition. In 1883 and 1884, in the wake of the Krakatoa eruption, *Nature* appealed to its worldwide readership for observations on atmospheric phenomena which could be forwarded to the Royal Society Krakatoa Committee. Sometimes *Nature* even had to be restrained in its defence of science. 'Politically', Donnelly of South Kensington said, 'it is only possible to get things done by opportunism and *Nature* won't help to get things done by letting Science have a jealous scream ...'.[44]

Public declamation appeared side by side with internal controversies. In 1872, Hooker and Owen clashed violently over the Natural History Museum. In 1873, Tyndall and Tait fumed at each other over Forbes' glacier theory and Clausius' theory of thermodynamics. 'I must add', Hooker assured Tyndall, 'that I think it very wrong of *Nature* to print letters containing personal traductions. It is a sad thought that future generations may light on some of its columns and ask what sort of gentlemen were their scientific forefathers considered to be ....'.[45] In 1874, Tait joined issue with Herbert Spencer on the concept of physical research which Spencer had used (or misused) in his *First Principles*. Tait, usually given to poetry, explained his position and enlivened the debate in a short note to Lockyer:

42 *Lockyer Papers*, Lankester to Lockyer, November 8, no yr.

43 *Lockyer Papers*, Pritchard to Lockyer, November 18, 1873.

44 *Lockyer Papers*, Donnelly to Lockyer, September 3, 1891.

45 *Tyndall Papers* (Royal Institution), vol. 7, f. 2718, Hooker to Tyndall, September 14, 1873.

The printers made one terrible blunder
Correct it *in stanta* and then for the thunder
We shall see a moment if this Mr S[pencer]
Has the ghost of a claim to be called a good fencer
In my view his merits
Seemed ever to dwindle
Since I found him allied
With the great Dr T[yndall]
While I, for my part,
Have grown cockier and cockier
Since I found an ally in yourself
Mr L[ockyer]

(I am indebted for these lines to Professor Herbert Dingle who committed them to memory when he was writing a biography of Lockyer in 1927-28).

There were countless other debates, most of which are now forgotten. All, however - to judge from the vitriol dashed on *Nature*'s pages - illustrated the jealousies and rivalries of science. Tempers frequently wore thin. In 1890, Lankester wrote a long letter in answer to Spencer and Romanes on the mechanism of evolution. Much to Lankester's disgust his reply was not printed. 'The sub-editor writes to me that he cannot insert it', Lankester angrily wrote to Macmillan,

> on the grounds that he wished *this* controversy to be closed ... In any case I am not accustomed to write an article worth several gns. - as my time and efforts in that way go - and have it flung back to me ... Unless I am allowed a fair reply to both Spencer and Romanes, I must necessarily cease to write for *Nature*.[46]

In Lockyer's absence, Macmillan interceded for the journal. Eventually Lankester relented, after his comments were sent privately to Romanes. 'You are quite right not to print my letter about Romanes as it is not argumentative but purely denunciatory ...'.[47] By helping to cool outraged passions, *Nature* wielded a strong but subtle power. Moreover, its editors could choose which controversies to pursue and which to reject. In Lockyer's hands this amounted to autocratic power. As H.J.S. Smith, the Oxford mathematician, once

46 *Macmillan Archives*, Lankester Letters, f. 42, Lankester to Macmillan, March 20, 1890.

47 *Lockyer Papers*, Lankester to Lockyer, September 25, 1890.

remarked, Lockyer at times failed to recognise the difference between the Editor of *Nature* and the Author of Nature.[48]

Lockyer found that he had to accept criticisms himself, but he was powerful enough to avert or soften their impact. In April 1895, William Huggins, indignant at Lockyer's speculations on the meteoric hypothesis, claimed that Lockyer was manoeuvring against him. 'It would not so much matter if it were confined to his own paper', he told G.E. Hale, 'but he is constantly at work against me by underground methods, such as dinners at a club to which he invites editors and literary people ... He has, I fear, got an influence lately on the editor of *The Times*. This could be very serious.'[49] This passage may have told more about Huggins than Lockyer, but it was certainly true, as Huggins later admitted, that 'the scientific men here, for the most part, are too much afraid of him to do anything'. *Nature*'s power in the subtle world of scientific politics was difficult to demonstrate. But it was real, and it reflected the manner of *Nature*'s brilliant, ambitious and controversial editor.

Faithful Mirror to a Profession

During the nineteenth century the leading articles in *Nature* tackled such topics as scientific and technical education, industrial competition from Germany, and the role of the Royal Society. Lockyer was concerned about British organic and industrial chemistry.

Of *Nature*'s notes, reviews, letters and leading articles, perhaps the leaders had the greatest social impact. They were of two general kinds - editorial opinions and review leaders. Between 1869 and 1919, about one fifth of the 2,600 leaders were editorials and it was these, together with selected reviews of non-scientific books, which attracted most general notice. Generally speaking, the editorials were written by Lockyer himself (who wrote sixty-six), his staff members (sub-editors) or a selected band of outside experts. In the editorials, about 70 per cent of which were anonymous, much of the journal's flavour, and its attitudes towards society, can be seen.

In its first decade, *Nature* gave prominence to general issues, including the successive scientific voyages that began with the *Porcupine*, the *Challenger* expedition (1870-1875), and continued through the ill-fated polar expedition

---

48 Anecdote variously cited by W.T. Thiselton-Dyer, 'Memorial Tributes to Sir Norman Lockyer', *Nature*, 106 (1920), 22; J.J. Thomson, *Recollections and Reflections* (London: G. Bell, 1936), 244.

49 *Hale Papers* (Californian Institute of Technology), Microfilm Reel 19, f. 113, Huggins to Hale, April 28, 1895.

of 1874; repeated geographical expeditions and the four eclipse expeditions between 1870 and 1878. Lockyer himself also devoted space to the Franco-Prussian war and its significance for Germany, and to government provision for scientific education, the organisation of local scientific effort and national museums, and the fate of animal physiology at the hands of the vivisectionists.

Between 1870 and 1876, the movement for 'science reform' and the 'endowment of science' was heightened by Lockyer's presence on the Devonshire Commission, by the British Government's increased research grant to the Royal Society and the creation of a Loan Collection which the Commission had recommended. The Devonshire Commission helped launch three other commissions and subsequent legislation on the Universities of England and Scotland, and added impetus to reforms in science teaching in schools and the creation and improvement of university college science faculties at Manchester, Newcastle, Leeds and Bristol.

Lockyer's perennial sermon on the importance of research was not put off by the sarcastic reflections of Henry Smith and Sir John Evans that the 'endowment of research' only camouflaged the pursuit of 'research for endowment'.[50] Contrary to popular belief, however, Lockyer did not himself write *Nature*'s leading articles on the 'Endowment of Science' or the Devonshire Commission - these were written by James S. Cotton and John Keltie.

Within science, Lockyer devoted large space to 'borderline questions', particularly physiography, 'where the physicist, the chemist, and the astronomer find each a common interest'. But particular inventions and discoveries, such as the development of the radiometer, the telephone and the diplex telegraph, the liquefaction of gases, the advance of electromagnetic theory and physical geology, experimental biology, aeronautics and the discoveries of argon and terrestrial helium, all featured in the pages of the journal. Only in organic and industrial chemistry was Lockyer apprehensive and pessimistic about British science. As he wrote in 1878, 'The chemists among us are, for the most part, silent, and chemical theory is almost dead in England; indeed it would appear as if the centre of gravity of this science has gone bodily eastward and Berlin and St Petersburg now replace London and Paris'.[51]

Between 1879 and 1889, in years of increasing foreign competition and industrial depression, *Nature*'s pre-occupations ranged from old debates on vivisection and university reform to new subjects such as the polytechnic and

50 'Memorial Tributes to Sir Norman Lockyer', *Nature*, 106 (1920), 22.

51 'Retrospect and Prospect', *Nature*, 18 (1878), 1.

the provision of a central scientific university, or scientific teachers and scientifically trained artisans for industry. *Nature* kept a close watch on every stage of the Technical Education Act (and the development of texts to be used in School Board science classes). The period saw rapid developments in the provincial universities and the first government grants to help meet the rising costs of higher education; and it saw the first steps in the long series of events leading to the creation of a University of London to replace the institution which had been 'nothing but a mere State examination board supported by a Parliamentary grant ...'.[52]

About 1885, *Nature* also began to insist on the application not only of science but also scientific method to the purposes of statecraft: 'There is surely no reason why political action, the conduct of the state, should not be guided by scientific method quite as much as the conduct of a scientific exploring expedition such as that which has so recently sailed over the North-East passage'.[53] In 1887, Lockyer used the Jubilee as an occasion to remind his readers, in three articles, that in scientific education and 'official science', Britain was fifty years behind; to urge again the need for a national Science Museum; to question the role of the Imperial Institute and to despair at the neglect of science in a Philistine community. Since the death of the Prince Consort, Lockyer bitterly recalled, 'there has been an impassable gulf between the highest culture of the nation and Royalty itself. The brain of the nation has been divorced from the head'.[54] A more optimistic review came from Rücker's leading article in July 1887. The review, 'Professor Tyndall and the Scientific Movement', featured Tyndall's retirement dinner, but it was also a pronouncement for the 'great profession' of science, which had come into existence and was now being organised, regardless of daily vocation, for 'the investigation of truth for truth's sake alone'.[55]

In November 1887, Huxley, then Secretary of the Royal Society, made another definitive statement on *Nature*'s behalf in an anonymous article, 'Politics and the Presidency of the Royal Society', deploring the candidacy of Professor G.G. Stokes, President of the Royal Society, as Member of Parliament for Cambridge.[56] Science was above politics; it would be beneath her dignity and better for her interests to be 'neglected than that they should be promoted by back-stairs agencies'. Not only was this an important threat to science; it also endangered the freedom of the Royal Society:

---

52 'A Teaching University for London', *Nature*, 31 (1885), 145; 'The Place of Science in Education', *Nature*, 22 (1880), 529; 'A New Departure for the University of London', *Nature*, 32 (1885), 265.

53 'Public Opinion and State Aid to Science', 32 (1885), 497.

54 'The Jubilee', *Nature*, 21 (1887), 295.

55 'Professor Tyndall and the Scientific Movement', *Nature*, 36 (1884), 145.

56 'Politics and the Presidency of the Royal Society', *Nature*, 37 (1887), 49.

> Either it will continue its beneficent work for untold ages to come, untroubled by the transitory political and social forces raging around it; or headed by politicians pledged to serve their party, it will gradually be dragged down ...[53].

Stokes stood firm, until a timely by-election saw his defeat. But the position Huxley set for the society remained to guide its action towards any party in power.

Nature's third decade (1889-1899) was one of consolidation in educational matters. The chief social issues were, again, technical education, German competition and the state of the Royal Society, while medicine and public health and the new National Physical Laboratory received rapid attention. The Technical Education Act, 1899, came in for increasing criticism. In February 1890, Llewellyn Smith commented on the new Education Codes which took little notice of the Technical Instruction Act of 1889 and which left technical education in its uncomfortable administrative middle ground. In April, Llewellyn Smith returned to the subjects with two more leading articles in two months in 1890, and two more, 'Educational Aspects of Free Education' and 'Organisers of Technical Education', in June and July 1891. .

In June 1892, while the Royal Commission on the University of London was sitting, V.V. Dickens carried the question further by giving *Nature*'s support to an 'extremely novel scheme for a professional institution run largely by a 'professionate' capable of uniting the colleges, the existing universities and the Medical Schools'.[57]

A fortnight later, T.E. Thorpe opposed the so-called 'Gresham Charter' by which the colleges and medical schools were to be federated without being united. The following week Lockyer remarked that 'scarcely any reference has been made to the higher teaching which ought to be at the disposal of the citizens of the most important and largest city in the world'.[58] Instead he submitted the broad list of subjects being taught at the Sorbonne where he himself had enjoyed a 'professional' education over 30 years before.

In December, Rücker attempted to sort out differences between Huxley and Roscoe, representing the Association for Promoting a Professional University and Karl Pearson representing a plan for a state-supported University of Berlin in London, who had resigned from the Association's Committee. In 1893, the Gresham Committee finished its witnesses, and in March 1894, Rücker revealed recommendations for the 'New University of London'. The main

57 'A Professorial University of London', *Nature*, 46 (1892), 121.

58 'The London University of the Future', *Nature*, 46 (1892), 193.

danger which he foresaw the new university facing was jealousy between semi-independent colleges, in a university non-federal in theory but federal in fact. 'The only safeguard against this', he surmised, 'is that everybody concerned will do his best' for the progress of knowledge.[59]

In South Kensington there was other excitement. In December 1890, W.E. Ayrton announced to *Nature*'s readers that the government had begun 'shaking the foundations of science' by announcing that an underground railway (the South Kensington to Paddington suburban railway) was to be built right under the laboratories of Rücker, Lockyer and Boys at the Royal College of Science. The impending loss in damage to the accuracy of the delicate scientific instruments by the railway was beyond belief. 'The English nation', Ayrton announced, 'must abandon its ... claim of being a practical people. Germany will laugh us to scorn. France will hit us with an epigram and Italy will view us with polite amazement'.[60] Happily the train was re-routed and now passes through South Kensington. In April 1891, Lockyer wrote a leading article on the proposed Gallery of British Art to be built at South Kensington. Lockyer had long believed that the land west of Exhibition Road on the Gore estate should be reserved for science and strongly fought the penetration of art. In the end, the scientists won; space was saved for the Science Museum and Mr Tate sought premises on Millbank instead.

In November 1895, *Nature* congratulated *The Times* for having awakened to the necessity for state action for industrial science 'which we have been preaching for some years'.[61] A hopeful sign brought a second leading article from H.E. Armstrong, on December 5, on the 'Organisation of Science'. Armstrong recalled the Council of Science recommended by the Devonshire Commission 20 years before, and pleaded for the application of scientific method in the public services.[62] The phrase 'organisation' came repeatedly to mind; it was a particular favourite of the German chemist, Wilhelm Ostwald, whom Lockyer, and William Ramsay in particular, admired. Ramsay sent one of Ostwald's letters to *The Times*, which devoted a leading article to it. *The Times*, however, concentrated on the duty of industry to endow research, while Lockyer, writing in August, put the blame on the government. 'The real remedy', he added, 'lies in consistently organising both our peace and war forces ...'. Gregory added a paragraph saying that the existing political leaders were unable to reach a solution; '... what we do criticise', he said, 'is the political system which does not consider it necessary that the educational and

---

59 'Romanes on Weisman', *Nature*, 49 (1894), 49.

60 'Shaking the Foundations of Science', *Nature*, 43 (1890), 145.

61 '*The Times* on the Scientific Situation', *Nature*, 53 (1895), 73.

62 'Societies and Academies', *Nature*, 53 (1895), 96.

scientific welfare of the country should be the business of those who are able to appreciate the work done, to see the necessity of reforms and to know the directions in which developments should take place'.[63] This theme of 'expert' government by an 'organised scientific élite' became commonplace in the next decade.

The German scare reached greater intensity in 1896 when E.E. Williams' *Made in Germany* appeared. Henry Armstrong observed that scientific education had made Germany at once a 'cultured' nation and an industrial leader. 'The application of science to industry has brought the whole world into competition and only those who fully understand can apply all the rules and every detail of the game can hope to succeed in it'.[64] Blame was shared out between apathetic industry, inadequate secondary schools and universities unfitted for research. But in June 1897, O. Henrici talked about the hostility between the *technische hochschulen* and the universities of Germany, and stressed the little remembered stand of the universities against the intrusion of technical studies. But in February 1899, Meldola observed that even if Germany's supremacy in technical chemistry were not equalled by her efforts in inorganic chemistry or other branches of science, Britain's limited chemistry brought only dismay.

The nineties also saw fresh criticism of the Royal Society in the press and a defence in *Nature* by Thiselton-Dyer of Kew in favour of electing non-scientists. In December 1892, Huxley took the chance to defend the society from charges of prejudice in its election procedures and over representation of certain men in its permanent offices. In June 1893, Huxley and a group of Fellows determined to deny election to an eminent man of letters who was not a scientific man. Thiselton-Dyer argued that 'if the Royal Society were simply constituted of professional scientific men, its influence in the country would be vastly diminished. ... A purely expert Royal Society would be treated with a kind of ironical respect but otherwise left alone ...' and isolated from the public. Thiselton-Dyer's defence was upheld and the nominee was elected, but the Society remained a calculated distance away from public involvement. Despite its many tasks for government, it refused to be fettered with any trace of political preference in the interest of the scientific community.

During these months Lockyer himself was not inactive. In February 1892, he defended the Chair of Astronomy at Cambridge, left vacant by the death of Adams, against attempts to disconnect it from the Observatory or to regard it as a sinecure. 'The subject, indeed, is one in which we are at present scarcely

---

63 'Professor Ostwald on English and German Science', *Nature*, 54 (1896), 386.

64 'The Need of Organising Scientific Opinion', *Nature*, 55 (1897), 411.

holding our own', while America and Germany were spending vast sums for the equipment required by new methods of physical analysis. Above all, he felt, the observatory should be saved from becoming 'sleepy willows for mathematicians, however distinguished, who have given no hostages to fortune in the shape of noble astronomical work'.[65]

Lockyer: Editor, Civil Servant and Man of Science

Despite financial difficulties which continued for several years, Lockyer's fame and reputation spread.

Lockyer's life was never a 'sleepy willow'. From the outset Lockyer's time was divided between *Nature*, the War Office and astronomy.[66] In early 1870, he gave a series of lectures in Newcastle, Edinburgh, Darlington and later to the Royal Institution. He impressed Americans particularly: 'I took quite a fancy to Lockyer', wrote J.W. Draper to his son, 'he is the *rising* English Astronomer'.[67] In May 1870, Lockyer was appointed secretary to the Devonshire Commission. His appointment reflected the support and favour of his fellow scientists. The commission had come about largely through the efforts of Alexander Strange, a fellow council member on the Royal Astronomical Society. Three of the five scientific members of the commission, including Sir John Lubbock, T.H. Huxley and G.G. Stokes, were by then close friends; from their point of view, the appointment of a scientific journalist who had access to *Nature* was an inspired arrangement.

In the meantime, Lockyer struggled to support his wife and nine children. Lockyer's early poverty gave most of his literary endeavours a strong economic motive. Much of his early literary work for such papers as the *Saturday Review*, the *Daily News*, the *Electrician*, the *London Review* and the *Spectator* was intended to supplement his War Office salary of £360.[68] When he was seconded to be the secretary of the Devonshire Commission, he received £300 a year after much haggling with the Treasury. Nervous exhaustion plagued him and he repeatedly suffered bouts of depression. He had at least

65 'The Vacant Chair of Astronomy at Cambridge', *Nature*, 45 (1892), 337.

66 A.J. Meadows, *Science and Controversy: A Biography of Sir Norman Lockyer* (London: Macmillan, 1972).

67 J.W. Draper to Henry Draper, December 1, 1870, quoted in Nathan Reingold, *Science in Nineteenth Century America* (London: Macmillan, 1966), 257.

68 Public Record Office, T1/8499/11997, Clerk to First Lord of the Treasury, June 16, 1890.

three breakdowns; at the end of 1864, between 1867 and mid-1868 and in March 1877.

Throughout his editorship, Lockyer remained Macmillan's consulting physician on scientific books and schemes.[69] He seems to have kept offices in the Macmillan building in Bedford Street, the complications of his life probably becoming familiar to his friends, who received scientific notes on government stationery and government notes on *Nature* notepaper. This continued connection with Macmillan was vitally important for two reasons. First, Macmillan could offset the expense of advertising his scientific books and win more potential scientific readers by using the pages of *Nature* as a movable marquee. *Nature*'s international circulation, though small, would enhance the sale of English books abroad. Macmillan explained to Lockyer in July 1873 that 'it is only in this sort of way we can make it pay'. 'Whatever profits we make in America', Macmillan even reassured Lockyer, 'you will share with us …'.[70] Reciprocally, *Nature* could count on the financial backing and personal interest of Macmillan's authors.[71] This commercial interest was unconsciously admitted by William Jack, Macmillan's physics adviser between 1876 and 1879, who felt it routine to 'run' a small book or lecture series through *Nature* first, and then publish it in Macmillan's science series afterwards.[72]

In Lockyer's relations with Macmillans, rough moments were rare. In 1870, on the nomination of Joseph Henry, Lockyer received an offer to take up the directorship of the new 25 inch telescope being built in Virginia at the

---

69 Charles Morgan, *The House of Macmillan* (London: Macmillan, 1943), 69.

70 *Macmillan Archives*, vol. 22 f. 121, Macmillan to Lockyer, July 9, 1873.

71 Correspondence between Macmillan and his authors repeatedly bears on the economic sense of reciprocal arrangements. In 1875, for example, Macmillan, writing to Michael Foster in Cambridge about the prospects of the biology syllabus, says, 'Lockyer who is looking over my shoulder, prophesies big about this...' and because '...Lockyer is sanguine about it ... I have ventures on 1,500'. In return, Macmillan used Lockyer: 'Give me a title page', he told Foster, 'I can get inserted in *Nature*...' (*Macmillan Archives*, vol. 24, f. 248, Macmillan to Foster, January 12, 1875, and f. 767, Macmillan to Foster, January 14, 1875). Indeed, *Nature* was such an ideal advertising medium for the house that its use became public knowledge among Macmillan's scientists. To Balfour Stewart of Manchester, Macmillan once raised some technical doubts about Tait's proposed *Physics*. 'I am ... convinced', he wrote, 'that we should make no announcement [not] even in "on the sly" paragraphs in *Nature* or in London Correspondence till we have a clearer idea of the concrete form it will take' (*Macmillan Archives*, vol. 24 f. 574, Macmillan to Balfour Stewart, December 7, 1875).

72 *Macmillan Archives*, vol. 31, f. 451, Jack to Lockyer, August 19, 1877.

expense of John McCormick, the American farm machinery magnate.[73] Lockyer was continuing to support himself by freelance work in science journalism. Between 1869 and 1873, for example, he continued to contribute to the *Saturday Review*.[74] Lockyer's financial position, uncertain at best, would be transformed overnight and the attractions of a permanent post would be his. The offer was tempting.

At about the same time, Macmillan and Lockyer quarrelled violently - the only known record of any disagreement between them - apparently about Lockyer's status in the firm and a series of books he wished to promote. Sir Archibald Geikie acted as a go-between, and Macmillan eventually offered to make Lockyer the editor of a series (possibly the *Nature* series) and to give him a retaining fee of £250 per year as scientific adviser as well as his salary for *Nature*. Geikie hoped this would save Lockyer from emigration and that Macmillan's plan 'may be the nucleus of future comfort for you'.[75]

In December 1870, Macmillan wrote to Lockyer, away on the eclipse expedition to Sicily. Coupling his note with thoughts for a book on the sun god and his worshippers, Macmillan asked Lockyer to stay:

> When the blessed event takes place, I shall have much pleasure in discussing with a view to increasing your pay for scientific advice higher [sic] and make, of course, the advice itself of more value to us. I am sure that it may be put on a satisfactory footing.

'But above all', Macmillan continued,

> I am very anxious about *Nature*. I can't help feeling that a little more of *something* would make it a success and if so, of course, it would be a permanent benefit to you. I have been thinking of many things. At present we are endeavouring to get it more widely taken at schools, and if we succeed in this will go into some other line.[76]

Whether or not the vision of a financially secure position, carrying with it a powerful hand in editorial matters, settled Lockyer's mind in favour of staying, he decided not to emigrate.

---

73 *Lockyer Papers*, President, University of Virginia, to Lockyer, January 17, 1871.

74 *Lockyer Papers*, Philip Howard to Lockyer, August 9, September 13, October 2, 1869; April 7, 1873.

75 *Lockyer Papers*, Geikie to Lockyer, nd. *c.* 1870.

76 *Macmillan Archives*, vol. 19, f. 792, March 10, 1871 (an abridged form of this letter, omitting the mention of salary or any suggestion of Lockyer's leaving, appears in G.A. Macmillan, *Letters*, 264).

In 1874, the Royal Society awarded Lockyer the Rumford Medal and made him Bakerian lecturer. His gratitude was tempered by the economic strain under which he lived. 'I shall', he told Hooker, 'do all I can to remain true to science ... but I feel ... that the time is coming when I must say *moriturus te saluto* and die the death by forsaking research and by ceasing to be true to science become more true to my wife and children'.[77] In 1874, while still at the War Office and working for the Devonshire Commission, Norman MacLeod and John Donnelly of the Science and Art Department arranged to get him three guineas an evening working at the South Kensington Museum. But even if Lockyer's health could tolerate it, the Treasury would not. As Ralph Lingen, Permanent Secretary to the Treasury, minuted, 'Mr Lockyer is Clerk in the War Office, and Secretary to the Science Commission. ... How many other offices is he ready to fill?'.[78] There seemed no easy answer.

In 1875, after an exceptional recommendation by the Devonshire Commission and a personal application to the Prime Minister (and following a large and influential deputation from the British Association urging government action on the Devonshire Report), Lockyer was lent by the War Office to the Science and Art Department. One of the recommendations of the Devonshire Commission had been the creation of a national physical observatory. Lockyer's intervention was to be the first instalment in the long series of events which led to the establishment of the N.P.L. Ostensibly his duties were to help with the loan Exhibition of Scientific Apparatus and the collections of apparatus in the Science Museum; in fact he was given virtually a sinecure with which to carry on his research. Between 1875 and 1881, he drew at the rate of his old War Office salary, plus fees amounting to £700 a year.

Both *Nature* and Lockyer made heavy financial demands on Macmillan. Lockyer's own astronomy texts were a case in point. In 1877, when his sixth book, *Star-Gazing: Past and Present*, was in the press, he received a sharp scolding from William Jack who did not share Alexander's generous sympathies:

> Do for goodness sake keep the expenses of Star-gazing down. We were rather startled by your saying if it hadn't been your own book you might not have been inclined to advise us to publish it and that you were afraid it might not find a large public ... Of course we can't

77 *Hooker Papers* (Kew), English Letters, 1865-1900, vol. 92, f. 175, Lockyer to Hooker, November 7, 1874.

78 Public Record Office, T1/7402/7067, Margin note (*c.* May 1874).

> publish generally without a prospect of paying, and we should soon cease to publish if we did ... We are all the more anxious because we have been looking at the Application of Physics Accóunt - [presumably Lockyer's *The Spectroscope and its Applications*] ... We have a debt of £1,400 against the book and since midsummer we have sold four and had seven returned from agents. We know, of course, that you took a lot of trouble about the book, but a loss like this on one book is enough to make one's hair stand on end ...[79]

'I suspect the public has a bad taste', added Jack, 'and the Applications may be too severe for them'.[80]

In September 1872, George Craik, Alexander's 'Chancellor of the Exchequer' since 1865,[81] told Lockyer that '... when you return we must have a little talk about your work'.[82] The outcome of this 'little talk' is unknown, but if it was a scolding, Lockyer's connections with South Kensington and the scientific scene gave him a position of informal power within the firm. His advice was almost indispensable.

In 1878, for example, Jack wrote to him confidentially about Professor Tanner of South Kensington. 'Is he a good man? What number of pupils from what kinds of schools does he examine? He has a sort of primer on 'the farm' which we are considering'.[83] The *Nature* Science Class Book and Science Primer series (under Huxley, Balfour Stewart and Roscoe) were also vital to Macmillan's overall scientific programme. In 1876, Jack told him that

> Macmillan is all for going on with the *Nature* series, believing that a first rate book or two would loft it tremendously all round. He supposed that Chisholm on balances is quite ready ... Anything else in wind or ready? ... If you could keep up announcements of one or two popular books in the *Nature* series it might help a good deal.[84]

Lockyer's light teaching schedule at South Kensington fortunately left him time for *Nature* and research. But Lockyer's transference to the Science and Art Department involved him in the running battles between that department and the Treasury. Largely through its museum purchases, the Science and Art

---

79 *Macmillan Archives*, vol. 31 f. 552, 31, Jack to Lockyer, August 29, 1877.

80 *Macmillan Archives*, vol. 31, f. 568, Jack to Lockyer, September 1, 1877.

81 Charles Morgan, *The House of Macmillan* (London: Macmillan, 1943), 185.

82 *Macmillan Archives*, vol. 31, f. 649, Craik to Lockyer, September 14, 1877.

83 *Macmillan Archives*, vol. 32, f. 711, Jack to Lockyer, June 8, 1878.

84 *Macmillan Archives*, vol. 27, f. 882, Jack to Lockyer, July 24, 1876.

Department, under Sir John Donnelly, had acquired notoriously expensive artistic and scientific tastes. At a time when the education estimates were the most rapidly increasing sector of public expenditure, the Science and Art Department was expanding even faster than its 'sister', the Education Department. The Treasury fought hard against 'that omnivorous Donnelly who will soon absorb Hyde Park for his marvellous collection of some good things and a marvellous collection of rubbish'.[85] Lockyer's dealings with the Treasury - which even allotted a special file to him - were compounded by his successive applications for government assistance on eight successive eclipse expeditions, through the Solar Physics Committee and through the Royal Society. He had a gift for applying the resources of one establishment to the needs of another - a gift which did not endear him to the Treasury bureaucracy. 'I am sure I do not know why or in what part of the Vote his salary is now smuggled', wrote Ralph Lingen in 1877 after one particularly exasperating episode.[86] Despite continual harassment, his applications (dismissed cynically by one Treasury clerk as merely 'sops to South Kensington') were repeatedly approved, and successive generations of Treasury officials backed down before his persistence.

In the meantime his scientific positions grew more controversial. Throughout the late 1860s, he had debated with Warren De la Rue about the existence of the chromosphere and had had searching arguments with chemists who believed that helium was a version of hydrogen. In 1872, following a disagreement with Richard Proctor, and intense dissatisfaction with the Royal Astronomical Society over the publication of the 1871 eclipse data, Lockyer (and later de la Rue) resigned from the Council of the Royal Astronomical Society.[87] In 1875, when Richard Proctor attacked Lockyer for the term 'chromosphere' in the *Cornhill*, Lockyer's friends rallied round him. P.G. Tait of Edinburgh wrote in his defence and enclosed a letter 'for *Nature* smashing him [Proctor] to powder for another business. Really these scribblers of venom ought to be scrunched'.[88] To some extent, *Nature* did just this.

In early 1877, in the midst of his struggles, Lockyer fell ill and spoke again of giving up science. Leaving home to do research, he could not deal with the cares of his sick wife and son. Lockyer told Hooker that 'he stood to give up science very soon'.[89] But Hooker reassured him, 'You of all persons

---

85 Public Record Office, T1/8267/9653, Welby to Jackson, December 11, 1886.

86 Public Record Office, T1/7647/20401, Lingen minute (1877).

87 *Lockyer Papers*, De la Rue to Lockyer, November 6, 1872.

88 *Lockyer Papers*, Tait to Lockyer, *c*. February 1876.

89 *Hooker Papers* (Kew), vol. 14, f. 124, Lockyer to Hooker, February 28, 1877.

must not talk of giving up science, no-one will listen to it'.[90] The news quickly got round the scientific circle. Two weeks later Macmillan wrote to Lockyer: 'Brunton [Lockyer's physician] tells Craik that he thinks you should have a quiet rest on the continent. If the enclosed cheque will make this easier for you, the firm desires your acceptance of it with the love of all the members'.[91]

In 1878, Lockyer's son Frank died and his wife Winifred followed the next autumn. The twin tragedy and the responsibility of caring for eight children left their mark on him. He seems to have thrown himself into research and writing whenever he could, for the next twenty-four years. There was little time for recreation or society.

Most of Lockyer's research expenses were met by grants from the Government grant to the Royal Society and by using the chemical and physical equipment and the detachment of Royal Engineers at South Kensington. In 1877, Lockyer was appointed chairman of the Solar Physics Observatory, which meant an additional £500 (eventually £1,000) in research expenses. He was also the lecturer in astronomy at the Norman School of Science at a salary of £450 and was allowed to earn up to £300 in examining fees. His fame and reputation spread. He went to the United States on an eclipse expedition in the summer of 1878, and Thomas Cheney asked him, 'as the most eminent literary exponent of science with whom I am acquainted', to write occasional leading articles for *The Times*.[92] In 1884, he was elected vice-president of the Incorporated Society of Authors.[93]

By the 1880s, to all appearances, his position had never been better. His association with Donnelly proved extremely happy. In 1881 he was made Professor of Astronomical Physics at the reorganised College of Science and his official salary was set at £750 a year. Together with fees, this gave him a salary by 1890 of about £1,000. By 1890 his basic salary was raised to £800 to the level of other full professors in the Royal College of Science, which brought him comparative wealth. It was this which enabled him to build a house and observatory at Westgate on Sea where he and Fowler worked at weekends.

Nonetheless, Lockyer was rarely happy or at ease. In the late 1880s and 1890s, controversy played a larger part in his life and he became obsessed with work. His opinions hardened. 'He does not strike me as a gentleman', wrote John Trowbridge, an American visitor, to President Gilman of Johns Hopkins in 1882.

---

90 *Hooker Papers* (Kew), vol. 10, f. 232, Hooker to Lockyer, November 2, 1877.

91 *Lockyer Papers*, Macmillan to Lockyer, March 21, 1877.

92 *Lockyer Papers*, Thomas Cheney to Lockyer, October 10, 1878; November 16, 1878.

93 *Lockyer Papers*, Sir William Besant to Lockyer, September 26, 1884.

Between 1887 and 1888, Lockyer's theory of the luminosity of comets aroused wide interest, as did his far-sighted meteoric hypothesis published in 1890 with the assistance of Fowler, Gregory and other students at the Royal College of Science. Tennyson thanked Lockyer for his 'splendid hypothesis ... in my anthropological spectrum you are coloured like a first rate Star of Science',[94] but both cost him friends among astronomers and may have led to his resignation from the Royal Astronomical Society. His dissociation hypothesis brought him repeated blows from the chemists. Only Abney, Schuster and Crookes were consistently behind him, although Thorpe and Roscoe were sympathetic. Only with the late 1890s and the work of J.J. Thomson on electrons did some of his more speculative views begin to win more general acceptance, though even then on a different conceptual basis.

One of Lockyer's least celebrated but most damaging controversies took place with William Huggins. Huggins is credited with having devised the stellar spectroscope, and with being the first to apply Kirchhoff's spectroscopic method of chemical analysis. Huggins and Lockyer argued repeatedly over the interpretation of solar spectrum evidence and Huggins disagreed strongly with the dissociation theory applied to stellar bodies. Ironically, Huggins sent his views to *Nature*. There was, as Lockyer explained to Sir George Stokes, little redress for him. 'I am practically bound to print all attacks on my work and I printed Huggins, but I don't propose to print my reply'.[95]

In 1889, William Huggins and Lockyer again collided over the interpretation of certain metal spectra. Huggins, normally a shy and retiring man, claimed that *Nature* misinterpreted his conclusions. 'Nothing is so painful to me as controversy, but if the paragraph is allowed to stand, there is nothing for me but to point out that the statement in it should certainly not be made ...'. Lockyer stood firm: 'I too hate controversy; it wastes too much time, but I do not shirk it with an honourable antagonist'. A compromise was reached and Huggins thanked Lockyer for having spared him from debate.[96] 'I am anxious that we should remain friends however we may differ in opinion'. But this had already proved impossible. The last decade of Lockyer's life saw *Nature* moving away from controversy and towards greater harmony among British scientists. In this new order of business, Lockyer played less and less a part.

94 *Lockyer Papers*, Tennyson to Lockyer, November 17, 1881.

95 *Stokes Papers* (Cambridge), Lockyer to Stokes, January 20, *c.* 1883.

96 *Lockyer Papers*, Hooker to Lockyer, May 6, 1889; Lockyer to Hooker, May 7, 1889; Hooker to Lockyer, May 7, 1889.

In January 1894, Lockyer was knighted KCB in the New Year Honours and in November, *Nature* celebrated its twenty-fifth anniversary.[97] Huxley, who had written the journal's first article, wrote its celebratory retrospect, recalling the closer bonds between speculation and scientific method that rational development and Darwinian evolution had helped inaugurate. The working synthesis of mind and method that Huxley had preached from Goethe's prose text twenty-five years before had not, perhaps, progressed far, but the continued existence of *Nature* - not to mention Lockyer's own speculative research - had kept it within sight.

Nature's birthday was celebrated, by publishing the list of nearly 800 men who had contributed in any way to the journal, and by a huge dinner at the Savoy. Fifty people attended, including Frederick, George and Maurice Macmillan.[98] Craik pronounced it 'the most influential journal of science in Europe'. Huxley praised Lockyer's success - 'it was often better to be an architect building a house under the dual control of a husband and wife or a member of the London School Board than an editor' - criticised recent standards of science writing, and toasted the name of Alexander Macmillan who had retired four years before. Publishers, he said, were generally regarded as *ferae naturae*, to be preyed on by authors; but Macmillan had been not so much publisher, as friend. Macmillan's kindness to Lockyer, editorial and personal finance and freedom were repaid by loyalty to the firm. It was this abiding relationship that gave *Nature* its solid hold on life.

Into the Twentieth Century

On November 6, 1919, *Nature*'s Jubilee, Lockyer relinquished his editorship to Richard Gregory. By now, *Nature* had become a 'great international institution'.

In 1902, Lockyer retired from the Royal College of Science. His fame had placed him above the fear of poverty and his release from constant haggling with the Treasury left him free for cultivating his interests in ancient monuments, the astronomical history of Stonehenge and Egyptian temples. In 1903, at the age of 67, he married again and became widely respected as an elder statesman of science. His retirement from official service also presaged a new career of public activity for *Nature*.

97 'The Fiftieth Volume of *Nature*', *British Medical Journal* (1894), ii, 1262.

98 For a complete list see 'Educational Books', *Nature*, 151 (1943), 234.

In January 1901, he wrote two leading articles, on the 'New Century' and the 'Advancement of Natural Knowledge', in which he set out the plan which governed the next fourteen years of his life. British industrial supremacy seemed at an end, and foreign domination was evident everywhere as a direct consequence of the progress of science abroad and the absence of a 'scientific spirit' at home. Britain's government had to be made scientifically awake. The government's preoccupation with naval force had diverted attention from industrial competition. While Britain was committed to increasing her fleets, 'it seemed the duty of no government department to look after the scientific advances which are the only bases of the commerce which is to provide for the constantly increasing expenditure'.[99] It appeared, moreover, that British intellectual resources were not being used to help retain her 'old position by force of brains'.[100]

From 1890 on, Lockyer had drawn attention to national education at all levels, 'based upon a study of things and causes and effects as well as of words'.[101] In securing this, Lockyer, not a university graduate himself, looked to the universities for the remedy. Beginning in January 1903, *Nature* began a series of five leading articles on 'The University in the Modern State'. In the first, eleven British universities were compared with 134 in the United States and 22 in Germany. State grants to one German university were more than the total grant given to all universities and university colleges in the United Kingdom. In March Lockyer dealt with the Prussian universities. Palmerston is said by Matthew Arnold to have defined the Germany of his day as a country of 'damned professors'. To *Nature* it seemed the 'damned professors' had done their job well. In May Lockyer looked to America, and in July Conway of Wales spoke of the academic plight of higher education. In August Lockyer tried to summarise, and applauded the recent efforts of Sir Gilbert Parker to establish co-operative relationships between the universities of the Empire. At the British Association meeting in Cambridge, Lockyer received a DSc - his first university degree.

As Ferdinand Lot, the French educationalist, remarked, Germany had reached 'a supremacy in science comparable to the supremacy of England at sea'.[102] In 1903 Lockyer returned to the twin theme of fleets and brains in his presidential address to the British Association at Southport, 'The Influence of

---

99 'The New Century', *Nature*, 63 (1901), 223.

100 Ibid., 222.

101 Preface quoted in 'Education and National Development', 438-9.

102 Mary T. Lockyer and Winifred L. Lockyer, *Life and Work of Sir Norman Lockyer* (London: Macmillan, 1928).

Brain Power on History'. Modelled, in part, on Captain A.J. Mahan's 'The Influence of Sea Power upon History', the address was a reply to those who saw national resources only in terms of military armaments. His message was simple. British statesmen had neglected to organise and use the nation's intellectual resources and British industry had turned its back on scientific innovation. The country had spent £120 million to strengthen the navy but less than £1 million on its universities. Only £30 million (or 2 per cent of the national income) was being spent on all kinds of education. 'We are suffering', he said, 'because trade no longer follows the flag, as in the old days, but because trade follows the brains, and manufacturers are apt to be too careless in securing them'.[103]

The remedy was equally simple. '... Largely increased endowments of higher education and research and the utilisation of scientific methods in all branches of administration equal to those at the disposal of competing nations, can alone save us'. Lockyer's address had a wide reception. James (later Sir James) Thursfield, leader writer in *The Times*, applauded it as a reply to protectionism; 'The so-called economic crisis is an affairs of brains, not tariffs. The brains are there all right, but we have no adequate machinery for their scientific cultivation'.[104] Henry Armstrong, a frequent contributor to *Nature*, gave Lockyer prominence in the *Quarterly*:

> British coal and iron are fast diminishing and we have few other resources worth mentioning except brains. But the manufacture of brains into a highly finished and efficient product is an industry in which we take no pains to excel and of which there is no organisation.[105]

In response to an appeal from the British Association and the statistics of Sir Robert Giffen, an additional grant of £75,000 was made to the universities in 1904 and promises were made to double it.

The real disappointment, however, Lockyer thought, was the scientific community itself, which lacked a collective voice. The British Association had been created in 1831 with exactly this intention: it could still fill this role. But Lockyer's efforts to win the support of the British Association or the Royal Society did not succeed, and he began to construct his own pressure group - the British Science Guild. In 1905, catching the wave of interest in the concept

---

103 J.N. Lockyer, *Education and National Progress: Essays and Addresses, 1870 - 1905* (London: Macmillan, 1906), 180.

104 *Lockyer Papers*, Thursfield to Lockyer, August 25, 1903.

105 H.E. Armstrong, 'The Reign of the Engineers', *Quarterly Review*, 198, (396) (October, 1903), 462 - 484.

of National Efficiency which swept Britain after the revelation of widespread public ill-health and maladministration during the Boer War, *Nature* published a review of R.B. Haldane's pamphlet 'The Executive Brain of the British Tongue'. Haldane argued for the enlistment of scientists in national administration as a *corps scientifique* with an advisory body at the head, and Richard Gregory picked out the theme for special comment. From the support of pure science in government for its own sake, *Nature* was swinging towards the more political concept of scientific method and procedure in government.

The Guild was inaugurated in 1905 to bring together, without reference to political party, representatives of science, industry and education to act as a lobby for issues of scientific reform, some of which *Nature* had espoused for thirty years - to apply the methods of science to all branches of life, to draw the government's attention to scientific matters and to promote scientific education. The members of this executive council and the committee were all friends of Lockyer, and virtually all the scientists were among *Nature*'s regular contributors. They included Joseph Larmor, William Abney, Lauder Brunton, Alexander Pedler, Francis Elgar, John Perry, William Ramsay, Raphael Meldola, Edmund Brabrook and William White. Sir John Lubbock (then Lord Avebury), although doubtful of the Guild's prospects,[106] became treasurer and R.B. Haldane president.[107]

The British Science Guild drew upon the *Nature* circle and drew its strongest supporters from *Nature*'s friends. But its success was severely limited for some of the same reasons which delayed even the best of *Nature*'s suggested reforms. Institutions had no reason to listen to a non-political, non-professional body representing no visible interest group in society, and political parties no obligation to respond. The Guild was above professional interests, applying the highest ideals of science objectively, and suffered from having no basis in political reality. Apart from the implicit fabianism emphasised by Sidney Webb's presence on the Education Committee, it was socially *bande à part*. 'You don't hit a note which will have much resonance here', H.H. Turner told Lockyer from Oxford, 'it will simply jar, I fear, but I may be wrong'.[108] As it happened, he was right.

---

106 *Lockyer Papers*, Avebury to Lockyer, August 10, 1906.

107 The British Science Guild struggled for nearly thirty years and did many excellent studies on national education and scientific resources in the constant shadow of financial incapacity, official disinterest and public apathy. It was eventually absorbed into the British Association in 1936. (See Roy MacLeod, 'Science for Imperial Efficiency and Social Change: Reflections on Commonwealth Scientific Cooperation, 1940 - 45', *Public Understanding of Science*, 3 (2), (1994), 155 - 194).

108 *Lockyer Papers*, H.H. Turner to Lockyer, January 31, 1903.

Intellectually, the British Science Guild's prospects were unsatisfying for its potential supporters. As A.J. Balfour had observed,

> I, as you well know, am most anxious to promote the cause of science in every way in my power, but, frankly, I am not sure that I form a clear idea of the exact methods which your Guild proposes to adopt to attain that end. The words 'science and scientific method' are ... loosely used, and especially when applied to such subjects as government or taxation, often mean little more than 'reasonable' or 'rational'. Every human being is desirous that the affairs of his country should be 'reasonably' or 'rationally' conducted. Where mankind quarrel is over the means by which that excellent object can be secured. Science gives us no clear utterance on such subjects.[109]

Through *Nature*, some utterances were made. In the decade ending 1909, the space devoted to scientific education and industry nearly doubled. By 1909, pressure on space had increased enormously, in part because of the growing popularity of scientific letters. For the three years between 1909-1912 (volumes 79-89), *Nature* appeared in three volumes a year instead of two, but Lockyer was obliged to cut back on column space given to addresses and reports and notes (Table 5). In general, *Nature*'s leading articles tended more and more to reflect scientific progress, while its secondary articles gave strength to reports of deputations, meetings and lectures of social significance to science. Lockyer's appearances were fewer; after an article on the death of Edward VII, he virtually retired from the leader column.

The social circumstances of Lockyer's youth had begun to change. Most of the 'Young Guard' were elderly; many had died. Between 1905 and 1914, a new generation of university trained scientific men were coming into view and into *Nature*'s pages. In 1900, it is estimated that there were 2,000 men of science in Britain; by 1914, there were 7,000.[110] Significant secondary articles on industry, which had numbered about twenty in the period 1905-1909, increased to thirty between 1910 and 1914. Logic and philosophy of science, the 'new' physical chemistry and stereochemistry, radioactivity and genetic experiments, mendelism, industrial optics, aeronautics, plant physiology, experimental zoology, Rutherford and Thomson's work on energy and matter,

---

109 *Lockyer Papers*, A.J. Balfour to Lockyer, August 10, 1906.

110 R.M. Pike, 'The Growth of Scientific Institutions and Employment of Natural Science Graduates in Britain, 1900-1960' (Unpublished MSc (Econ.) thesis, University of London, 1961), 137.

together with more frequent biographical notices, obituaries and reviews, all characterised *Nature*'s pages in these years.

Table 5 The composition of *Nature*: Sample Issues 1869-1915

	1869	1875	1880	1885	1890	1895	1900	1905	1910	1915
Leader (or leader-reviews)	29	25	57	24	25	19	27	18	18	15
Reviews	43	70	27	36	54	63	63	63	63	90
Letters	25	24	40	45	48	46	126	18	18	23
Scientific and secondary articles	125	128	117	135	135	130	99	126	180	108
Sci. notes	46	-	36	36	-	-	-	-	-	-
Addresses and reports	-	54	-	63	14	36	-	72	-	99
Obituaries	-	-	13	-	-	-	-	9	9	-
Societies, acads., univ., intell.	20	20	88	63	63	45	54	45	45	36
Total column inches	318	353	413	434	393	430	423	410	387	461

For the first issue, and first issue per calendar year thereafter. Figures given in column inches, at 9 inches=1 column.

Nature continued its tradition of acting on behalf of scientists. It was to Lockyer that J.M. Ludlow turned when collecting scientific signatures in support of Dreyfus' petition in 1899.[111] Lockyer even won a feminist accolade from 'all scientific women' when Hershe Ayrton congratulated him on getting, through *Nature*, Marie Curie's claim to a seat in the Académie de Science.[112]

After 1910, aged over 75, Lockyer himself gave some time to these personal claims, but his successes were clouded by personal disappointments. In 1913, when it became necessary to move his observatory to make room for the new science museum, his plans to take a site in Fosterdown, Surrey, were

111 *Lockyer Papers*, J.M. Ludlow to Lockyer, October 9, 1899.

112 *Lockyer Papers*, Hershe Ayrton to Lockyer, January 16, 1911. Many examples of how useful it was to have Lockyer and *Nature* as an ally can be given: 'Can you help me in *Nature* or otherwise?' pleaded the ageing Professor John W. Judd from the old Royal School of Mines. 'Unless something is done to check the anti-scientific tendencies of the present Treasury officials I fear it will go badly with my late colleagues'. Judd was referring to the prospect of government pensions for his fellow scientists. (*Lockyer Papers*, Judd to Lockyer, August 15, 1906.) It was Lockyer who helped organise a delegation to the Prime Minister when support was needed for the Science Museum buildings to extend from the Imperial Institute Road down the Exhibition Road. Lockyer joined forces against the government and strongly supported the building of the new Museum of Practical Geology being built at South Kensington.

rejected by the Board of Education. Instead, the observatory, with the exception of a few instruments, was moved to Cambridge. A protesting letter to *The Times* reprinted in *Nature* was no use. Lockyer was grievously disappointed by this decision, which seemed evidence of the government's impatience to be rid of his worrying claims for funds. Lockyer's son, W.J.S. Lockyer, left the Solar Physics Committee and moved with his father to a new observatory built at Salcombe Regis with the aid of an appeal fund launched by Sir David Gill. The Norman Lockyer Observatory opened in 1913 and occupied Lockyer's attention for the rest of his life.

The war brought many changes to *Nature*. Its industrial and social content multiplied five times between 1915 and 1919. Reports and articles on science abroad, which had increased by 40 per cent in the five years ending 1914, soared another 15 per cent between 1915 and 1919. Articles on scientific education, which had declined slightly between 1910 and 1914, increased sharply again after 1915. Science reform articles, which had maintained a steady increase over the years, rose again sharply as the government at last approached the question of a Department of Scientific and Industrial Research.

Nature took the lead in publicising the shameful neglect of scientific expertise during the first year of the war. In May 1915, before the announcement of the creation of an Advisory Council on Scientific and Industrial Research, Ramsay urged the government to 'trust the experts' and appoint a 'Chemical Council of State' to be instituted by the Royal Society.[113] Yet throughout the months of organisation which preceded the formation of the Advisory Council, *Nature*'s attitude towards this new development reflected its past preoccupations.[114]

While the journal applauded the government's decision to set up a specific council for the application of science to the war effort, Richard Gregory criticised the new council for its failure to appoint scientists to administer it.[115] The advisory council which Lockyer had recommended since the 1870s was never intended to be an official creation administered by civil servants. In the end *Nature* did not get its way and the Department of Scientific and Industrial Research came to be a civil service department.

But *Nature*'s leading articles did have a strong influence abroad. In 1916, George E. Hale, one of the founding members of the National Research Council of the United States, looked to wartime Britain for experience and *Nature* for guidance. 'I have followed with great interest', he wrote, 'your leading articles ... they served me well and I sent ... the full list of them to President Wilson

---

113 'Science and the State', *Nature*, 95 (1915), 309.

114 R.M. MacLeod and E.K. Andrews, 'The Origins of the DSIR: Reflections on Ideas and Men, 1915-16', *Public Administration*, 48 (1), (1970), 23-48.

115 'Science in National Affairs', *Nature*, 96 (1915), 195.

and also to Justice Hughes (the Republican Candidate for the Presidency) to impress them with the importance of science to a nation'.[116]

As strenuous as its work for the war was *Nature*'s abrupt *volte face* towards Germany. In his last leading article, 'The War - and After', Lockyer wrote that those who had admired Germany had now learned that her 'culture' and admirable organisation were not acquired 'for the purpose of advancing knowledge and civilisation but ... in order that a military caste ... shall ride roughshod over Europe'.[117] William Ramsay, once a close friend and admirer of Ostwald, turned sharply against him and all Germany besides, in a long series of leading articles in *Nature* which did much to shape British attitudes. In October 1914, Ramsay reported Rosebery's remarks on 'Germany's Aims and Ambitions'. Suddenly 'faults' in the British political character were seen as virtues, and the 'fair dealing Anglo-Saxon spirit' was the justification for a democracy which waited to grapple with each difficulty as it arose. Suddenly, German cultural accomplishments were no longer as splendid as *Nature* itself had described them. '... The originality of the German race', wrote Ramsay, 'has never, in spite of certain brilliant exceptions, been their characteristic. Their métier has been rather the exploitation of the inventions and discoveries of others and in this they are conspicuous'. 'We in England have always been more intent on welcoming a discovery than in inquiring into the nationality of the discovered', he added, 'indeed it is a common saying that science is international. But we are beginning to revise our verdict'.[118] Allied victory would 'release the world from a deluge of mediocrity'.

'It is not for a scientific journal', Lockyer added, 'to chronicle the progress or to predict in detail the possible consequences of the war'. But it had indeed been *Nature*'s task to awaken the nation to the prospect of war. This it had done perhaps too well. Once started, Lockyer knew from experience the great difficulties British manufacturers would face in making good their delays. *Nature*, no more nor less than any periodical of the time, contributed to the general view that competition between nations existed and almost 'had' to exist for obvious scientific and progressive reasons. *Nature* had never seriously questioned the direction in which this rivalry was leading; its remarks against armaments were not criticisms of the military so much as advice which would with luck give Britain a better chance in the 'game'.

It was difficult to prove that science, which for thirty years had, by *Nature*'s own standards, been contributing to German power, was not

116 *Lockyer Papers*, G.E. Hale to Lockyer, August 14, 1916.

117 'The War - and After', *Nature*, 94 (1914), 29.

118 'Germany's Aims and Ambitions', *Nature*, 94 (1914), 137.

responsible for the evil of war. Scientific truth and technical education were not enough - the 'moral sense of a nation had to be weighed in the assessment of a cultural heritage'.[119] *Nature* followed this by deprecating German industrial management and recommending the British spirit of 'co-operation' as an alternative.[120] Science and technology had helped create an unprecedented total horror. Before the war it had been possible for *Nature* to speak of the conduct of science and the obligation of society to support it; science was, as Huxley and others had argued, above politics. The war had shown that, far from being above politics, science had a grave obligation to society. This realisation and its consequences cast a long shadow over the career of Lockyer's successor and the future of *Nature*.

The war took a heavy toll on publishing and no less on *Nature*. There was a sharp falling off in leading articles by 'chief contributors' as busy men left on war service and the space was shared by either younger or older men. After several years of profit, its subscriptions fell sharply. But Lockyer was optimistic. 'Of course it is unfortunate that we now for the first time show a loss', he wrote Frederick Macmillan, 'but when you state that the circulation has kept up and the results from advertisements are better than last year, we have to congratulate ourselves. And as the paper is, as you say, 600 per cent above pre-war rates, we may be thankful that matters are not very much worse'.[121] From October 1918, Lockyer reluctantly agreed to the increased subscriptions (from £1 8*s* 0*d* to £2 2*s* 0*d*) and advertising rates (from £6 6*s* 0*d* to £8 8*s* 0*d*), the first increases since 1878. Even these proved insufficient and in 1920 subscriptions were raised to £2 12*s* 0*d* or a shilling an issue.[122]

Nature's line by 1919 was perhaps predictable, but the policies it had recommended had become common experience. But its voice was no longer Lockyer's. In October 1918, Lockyer told Frederick Macmillan that he wished his assistant, Richard Gregory, to assume command. 'The way ... he has carried on the duties of Secretary and Deputy Chairman of Committees ... have shown him to be possessed of administrative abilities of a high order'. Macmillan and Lockyer had discussed the desirability of filling his place with a Fellow of the Royal Society. But Gregory's work, particularly the idealism and objectivity of his book *Discovery*, published in 1916, had won a rapturous reception from a public confused and terrified by the evils unleashed by the poison gas, high explosives, and bombs and mines, in scientific warfare. In any

119 'The War', *Nature*, 94 (1915), 527.

120 'Germany's Aims and Ambitions', *Nature*, 94 (1914), 139.

121 *Macmillan Archives*, Lockyer Letters, f. 166, Lockyer to Frederick Macmillan, September 22, 1918.

122 *Macmillan Firm Letters* (Reading), f. 1569, R.A. Gregory to Maurice Macmillan, September 28, 1918.

case, Lockyer reminisced, 'Fellows of the Royal Society are very difficult to lead and very few of them have the slightest idea of administrative work or business capacity'.[123] Moreover in 1919 he was elected to the Athenaeum and knighted; he could have the continuity he needed with the community of science.

The huge issue of November 6, 1919 (106 pages), was dedicated to *Nature*'s Jubilee, but was also a festschrift to Lockyer. Lockyer lived to see his 'valedictory memories' go down besides appreciations by H. Deslandres and retrospects of the fifty years by Archibald Geikie, Ray Lankester, Canon Wilson and close friends and contributors, including Professor T.G. Bonney, Sir Clifford Allbutt, Sir Edward Sharpey-Schafer, Dr C.J. Martin, Dr Smith Woodward, Professor W. Bateson, Professor J. Cossar Ewart, Sir Edward Thorpe, Professor Henry Armstrong, Professor H.B. Dixey, Professor J.C. Philip, Sir J.J. Thomson, Sir Ernest Rutherford, Professor Frederick Soddy, Professor J.J. Townsend, Professor Arthur Fowler, Professor W.H. Bragg, Dr. A.C. Jordan, Professor J.A. Fleming, Dr W.C. Unwin, Professor H.C. Carpenter, L. Bairstow, C.H. Lees, W.H. Dines, Sir John Keltie, Chapman Jones, Emery Walker, Sir William Tilden, Sir Alexander Hill, J.H. Reynolds and Sir Richard Gregory. Congratulatory messages were received from forty British, ten foreign societies, and twelve British universities together with thirty-eight personal testimonials.

The same day Richard Gregory became by title what he had been in fact since the First World War, the new editor of what was already a 'great international institution', a scientific brains trust, a movable feast of tradition and iconoclasm purveyed for the nourishment of ideas and the place of science in society. In August 1920, Lockyer died at Sidmouth, leaving four sons and two daughters of his first marriage. His influence had been profound. One of the greatest testaments to his influence was the way in which *Nature* preserved from 1869 to 1919 the same form and general arrangement. Through him, *Nature*'s influence on science, as H.G. Wells later wrote, 'has been far stronger and far more humane than that of any other journal on the ordinary course of human affairs'.[124]

---

123 Macmillan Archives, Lockyer Letters, f. 167, Lockyer to Frederick Macmillan, October 23, 1918.

124 H.G. Wells, *The Record* (February 1931); *The World of William Clissold* (London: E. Benn, 1933), 396.

I thank Dr E. Kay Andrews and Mrs A. Berridge for their help in the preparation of statistical data. The *Tyndall Papers* are quoted by courtesy of the Managers of the Royal Institution; the *Huxley Papers* by kind permission of the Archives of Imperial College; the *Stokes Papers* and the *Ludlow Papers* by courtesy of the University of Cambridge; the *Hooker Papers* by courtesy of the Librarian of the Royal Botanic Gardens, Kew. I am very grateful to Lady Dorothy Gregory for the use of the *Gregory papers*, and to the Norman Lockyer Observatory, the University of Exeter, and Professor A.J. Meadows, University of Leicester, for the use of the *Lockyer Papers*. I also thank the Archivist of the California Institute of Technology and the Director of Mount Wilson Observatory, Pasadena, California, for use of the *Hale Papers*, and Dr Nathan Reingold, editor of the *Joseph Henry Papers*, for permission to consult the Lockyer letters in the Smithsonian Archives. The *Lubbock Papers* and the *Macmillan Archives* are in the British Museum, while the collection styled *Macmillan Firm Letters* is held at the University of Reading.

XII

Science, Progressivism, and "Practical Idealism:" Reflections on Efficient Imperialism and Federal Science in Australia, 1895-1915

Introduction[1]

> The task of Empire is the ... scientific conquest of its physical, and, shall we not be bold and say, ultimately of its political problems. [We must] endeavour ... to acquire that knowledge in scientific manner, and by scientific methods, which shall enable us to appreciate, in the first place, the vast, the incalculable natural resources which are present in our possession under the Flag—the means of utilising these instruments of material power for the benefit of our race.[2]

Thus Alfred Deakin, Prime Minister of Australia, in the presence of Sir Wilfrid Laurier of Canada, responded to the toast of 'Our Guests' at a combined session of the British Empire League and the British Science Guild in May 1907. The occasion—the Colonial Prime Ministers' meeting in London—presented a particularly appropriate moment for the condensation of typically volatile sentiments about the organic unity of science and empire, that had suffused the atmosphere of imperial relations for a generation. Today, the occasion gives Australians a point of departure, and reflection from which to assess, looking backwards, the picture of colonial science at the turn of the century, and forward to the promise in store for science in the service of a federal nation, loyal to Empire, but increasingly conscious of its own priorities.

Between 1900 and 1915, Australia was *per capita* one of the most prosperous nations on earth. This came about partly by foreign investment, partly by protective tariffs, and partly by the convenient absence of expensive defence commitments. But it also arose from the wealth generated by increasingly productive sectors of primary industry, in agriculture, wool, and, intermittently, minerals. Above all, it was a success story symbolised by 'Federation wheat,' developed by William Farrer and F.B.Guthrie—men who combined the skills of 'practical genetics' and systematic chemistry.[3]

1 This essay derives from studies in the history of Australian science assisted by the Australian Research Council. I am grateful to Richard Jarrell for his comments, and to Stan Veitsman for his assistance in completing a final text.

2 'Science and the Empire,' *Nature*, 76 (9 May 1907), 37.

3 See C.W. Wrigley and A. Rathjen, 'Wheat Breeding in Australia,' in D.J. and S.G.M. Carr (eds.), *Plants and Man in Australia* (New York: Academic Press, 1981); Colin Wrigley, 'F.B.Guthrie: Australia's Pioneer Cereal Chemist,' *Chemistry in Australia* (September

Theirs was the triumph of the 'practical man,' in a setting sensitive to the applications of science.[4] But it was a period also highly successful in the applications of 'scientific method' and rational analysis—in health, the social services, and in secondary education, where reformers like Francis Anderson sought to put Australian institutions on a par not with the least in Britain, but with the best in the world.

Just as Australians, fifty years on, are persuaded to see Gallipoli as a 'coming of age' of a federated nation, its young men fighting as Australians, rather than as Queenslanders or Victorians, welding different colonial traditions in a single sacrifice, so it is tempting to see a 'coming of age' in Australian science in the southern summer of 1915, when Australia created its first national Advisory Council for scientific research and its first Institute for Science and Industry. On 22 December, the day Prime Minister W.H. ('Billy') Hughes announced the miraculous withdrawal of Australian forces from ANZAC Cove, the same man took by storm a luncheon given for him at the University of Melbourne, striding into the room—as Currie and Graham put it—'as a prize fighter the ring'—'determined to challenge the scientific prowess of the Teuton warrior'.[5]

Hughes' acceptance and endorsement of proposals for a national research institute were consistent with the actions of his political contemporaries in Canada and Britain.[6] But they performed an act of constitutional magic, in making 'federal' what had hitherto been merely 'federated'. His words on that day led to the establishment of the Institue for Science and Industry, and eventually to the Council for Scientific and Industrial Research (CSIR)—whose delayed formal beginnings in 1926 signfied a sure, if early, token of Australia's cultural maturity.[7] In 1915, science had played, and not for the first time, a role in Australia's 'march to nationhood'.[8]

The war, which provoked substantial reconstruction in the machinery of government science in Britain, brought similar changes to Australian science, with consequences equally far reaching. From the war, came the Commonwealth Serum Laboratory and the Munitions Supply Laboratory, both set up in 1916. In recruiting large numbers of Australian scientists to the British muni-

1989), 306-311.

4 Roy MacLeod, 'The Practical Man: Myth and Metaphor in Anglo-Australian Science,' *Australian Cultural History*, 8 (1988), 24-49.

5 George Currie and John Graham, *The Origins of CSIRO: Science and the Commonwealth Government, 1901-1926* (Melbourne: Melbourne University Press, 1966).

6 Roy MacLeod and E.K. Andrews, 'The Origins of the DSIR: Reflections on Ideas and Men,' *Public Administration*, 48 (1970), 23-48.

7 Geoffrey Searle, *From Deserts the Prophets Come: The Creative Spirit in Australia* (Melbourne: Heinemann, 1973).

8 Roy MacLeod (ed.), *The Commonwealth of Science: ANZAAS and the Scientific Enterprise in Australasia, 1888-1988* (Melbourne: Oxford University Press, 1988).

tions effort, wartime experience contributed substantially to postwar manufacturing industry, just as wartime medical service benefited Australian research in parasitology and pharmacology.[9] It is undoubtedly correct to see the war as a turning point in the relations of science and government in Australia, as elsewhere in the Empire, in America and in Europe.[10]

But if we cast our eyes back to the decade before the war, we may find a rich soil of incipient traditions, germinating with ideas which were not only to give shape to wartime decisions, but also to foreshadow later Australian attitudes towards the use of science for national purposes. Ten years ago, I outlined, in a schematic way, a series of historical stages, or phases, through which I considered that Australian (and possibly other Dominion) tendencies in science had developed.[11] I intended this to be not so much an explanatory device, as a heuristic framework for testing hypotheses. Since that time, I have been filling in the picture where I can, and falsifying the 'model' where I cannot. I have also been trying to identify nodal points where one can see intersections between the discourses of science as culture, and as cultural agency.

In so doing, I have followed cultural historians in attempting to articulate the associations of science within what Michael Roe has called the 'bourgeoisie hegemony,' the system of socio-economic ideas that dominated the contrary imaginations of colonial Australia since the early nineteenth century.[12] It is debatable whether any single theory of culture can be applied to all Australia, with its urban population of city-states and its pastoral economy, its mythologies embracing, in the words of Francis Adams, coastal plain and inland desert.[13] Australia has been well-described as having not one but three 'founding' cultures—the convict-cum military culture of New South Wales which spread to Tasmania, Victoria, and Queensland, the gentleman-farmer colony of Western Australia, and the squire-farmer colony of South Australia. Indeed, given the intellectual distance between Melbourne and

9 Roy MacLeod, 'The Arsenal in the Strand: Australian Chemists in the British Munitions Effort, 1916-1919,' *Annals of Science,* 46 (1989), 45-67. See also Rod Home (ed.), *Australian Science in the Making* (Sydney: Cambridge University Press, 1988).

10 Roy MacLeod and Kay MacLeod, 'The Social Relations of Science and Technology, 1914-1939,' in Carlo Cipolla (ed.), *The Fontana Economic History of Europe:* Vol. 5: *The Twentieth Century,* Part I (London: Collins/Fontana, 1976), 301-335.

11 Roy MacLeod, 'On Visiting the "Moving Metropolis": Reflections on the Architecture of Imperial Science,' *Historical Records of Australian Science,* 5 (1982), 1-16, reprinted in Nathan Reingold and Marc Rothenberg (eds.), *Scientific Colonialism: A Cross-Cultural Comparison* (Washington, D.C.: Smithsonian Institution Press, 1987). See also Roy MacLeod, 'Passages in Imperial Science: From Empire to Commonwealth,' *Journal of World History,* 4 (1), (1993), 2-29.

12 Michael Roe, *Nine Australian Progressives: Vitalism in Bourgeois Social Thought, 1890-1960* (St. Lucia: University of Queensland Press, 1984), 319.

13 Francis Adams, *The Australians: A Social Sketch* (London: T.Fisher Unwin, 1893).

Sydney, it is doubtful that one culture can be said to embrace all Australian intellecutal life, whether in the 19th century or since.[14] But Roe is certainly right in asking us to think about the affiliations and transmissions of ideas that historically linked Australians in all the colonies to their counterparts in Britain and the United States, especially in the years leading up to Federation, when Australians began to sail as a nation into uncharted constitutional seas.[15]

1. 'Efficient Imperialism' and Progressive Australia

For this purpose, I want to consider a particular formative phrase in the 'passages of imperial science' occurring between Federation and 1915—a period in which we find within science an important interplay between inter-colonial commitments to Federalism—as economic and political motive, linked symbolically and in fact with the interests of national defence—and continuing Empire loyalties, best expressed in terms of 'imperial efficiency.' In Australia, this period saw a new phase of 'federal' science, inspired by a similar concern for 'national efficiency' that emerged in Britain during the 1890—in response partly to the challenges of American, German and Japanese competition, and partly to the tragedy of the Anglo-Boer war, and its devastating indictments of national mishandling.[16] A conceptual strategy for 'Efficient Empire' developed as a natural counterpart of national efficiency, with portable doctrines criss-crossing frontiers of ideology and party. In Britain, R. B. Haldane, A. J. Balfour, Lord Rosebery and the Webbs found resonance with Lord Milner and Joseph Chamberlain in seeking to set Britain right through reforms in administrative, military, commercial and academic life. 'Wake up Britain!,' the Prince of Wales had declaimed. Now that Edward was King, the colonies were to help the Empire by helping themselves, so that the whole would be the sum of more than its parts.[17] In this imperial design, science and its instruments of method were essential scaffolding.

Historians from Bernard Semmel to Geoffrey Searle have written of the 'social imperialism' of the period in Britain, with its lateral endorsement of social Darwinism, eugenics, and the application of 'scientific method' to all walks of life.[18] Whilst absorbing these doctrines, Australians saw similar values emerg-

14 See the workshop entitled 'St Petersburg or Tinsel Town? Science and Melbourne: Their Different Styles and Changing Relations,' in Jim Davidson (ed.), *The Sydney-Melbourne Book* (Sydney: Allen and Unwin, 1986).

15 For the period, see Gavin Souter, *The Lion and the Kangaroo: The Initiation of Australia, 1901-1919* (Sydney: Collins, 1976).

16 Geoffrey Searle, *The Quest for National Efficiency: A Study in British Politics and Political Thought, 1899-1914* (Oxford: Basil Blackwell, 1971; rev.ed., Ashfield Press, 1990).

17 See H.C.G. Matthew, *The Liberal Imperialists: The Ideas and Politics of a Post-Gladstonian Elite (Oxford: Oxford University Press, 1973).*

18 Bernard Semmell, *Imperialism and Social Reform* (New York: Anchor Books, 1968); see

ing in the American 'Progressive' movement, employing the rhetoric of scientific method from the conservation of woodlands and the creation of national parks to the Taylorite perfection of manufacturing industry.[19] Aethestics, utility, and ecology formed the secular trinity of America's 'efficiency cult,' and Australia was not to be spared its missionary influences.

Although differences between the two countries were many, similarities reached deeply. Progressivism in America grew not from the circumstances of separate colonies but within what Robert Wiebe has called a 'distended society,' during three decades in which a quilt of regions emerged from the destruction of the Civil War and a postwar depression to confront a new continental destiny. From a country of small town values and regional dependencies grew a nation shaped by industrialisation, mechanisation and urbanisation, with vast consequences for international trade and political affairs. New institutions at local, State and Federal levels valorised middle-class expectations of 'continuity and regularity, functionality and rationality, administration and management.' At the heart of the progressive impulse lay an impatience with patchwork government, and a commitment to reform through bureaucracy. Whether in urban reform, public education, or the protection of the 'race,' government conducted along scientific principles would bring 'opportunity, progress, order and community' to the nation as a whole. 'Organic cities' of brotherhood and settlement would become functional, efficient places to live. American society, with its complex legacy of colonialsm and immigration, would improve not by accident, but by design.[20]

To an extent guided less by philosophy than by comparison, Australia offered a complementary vision, translated into the language of home and empire. From the 'gas and water socialism' and public health services of Britain's leading municipal boroughs, to the Crown's Development Commission and its strategy for scientific agriculture, came an Australian response to the half-finished work of Victorian reform. Community, order, and progress could be the expectation of Britons, home and overseas, and by implication, that of 3.7 million Australians (few mentioned the 95,000 Aborigines) as well. The tempo of progress would be set by modern science and its methods.

also Michael Worboys, 'The British Association and Empire: Science and Social Imperialism, 1880-1940,' in Roy MacLeod and Peter Collins (eds.), *The Parliament of Science: Essays in Honour of the British Association for the Advancement of Science* (London: Science Reviews, 1981), 170-188.

19 See Samuel P. Hays, *Conservation and the Gospel of Efficiency: The Progressive Conservation Movement, 1890-1920* (New York: Athenaeum, 1972); S. Haber, *Efficiency and Uplift* (Chicago: University of Chicago Press, 1964).

20 Robert H. Wiebe, *The Search for Order, 1877-1920* (London: Macmillan, 1967), viii, 166-170.

Perhaps the most talismanic evidence of science in aid of Empire lies in the archives of the once widely-known but now almost forgotten British Science Guild. Established in 1905, under the chairmanship of Sir Norman Lockyer, editor of *Nature*, and long-standing propagandist for the 'scientific movement', the Guild became a rallying point for the efficiency movement, within Britain and overseas.[21] In his presidential address to the British Association Congress at Southport in 1903, Lockyer had seized upon the title of Captain Alfred Mahan's celebrated book, to awake Britain to the 'Influence of Brain Power on History.' Disappointed by the inertia of the BAAS, Lockyer diverted his energies into establishing a new 'League, a Guild, call it what you may,' a new organisation in any case, which soon became the darling of the progressive wing of British science. Indeed, judging by Lord Haldane's inaugural address at the Mansion House in the City of London, the Guild's polychromatic program of reforms shone with the brilliance of modernism.[22]

Emblematically, the British Science Guild took its message south, and established imperial branches in NSW in 1907 and in South Australia in 1911. In Sydney, Archibald Liversidge (a friend and confederate of Lockyer), Thomas Anderson Stuart, J.T.Wilson, and William Warren—all University of Sydney scientists and engineers—played leading roles in the Guild's work.[23] They were inspired in part by the imperial sentiments issuing from the Royal Society of London, whose Biological Secretary, the distinguished physiologist, Sir Michael Foster, encouraged steps to strengthen the unity of British science, home and overseas.[24] They worked mostly through private lectures, and sometimes through the Royal Society of New South Wales. If their organised numbers were small—the total membership of the Royal Society of NSW in

21 For a history of the Guild, see Roy MacLeod, ' Science for Imperial Efficiency and Social Change: Reflections on the British Science Guild, 1905-1936,' *Public Understanding of Science*, 3 (1994), 155-193 and W. H. G. Armitage, *Sir Richard Gregory, His Life and Work* (London: Macmillan, 1957). For the Guild's Australian branches, see the *Journal of the British Science Guild* (1907-1936),which includes annual reports from NSW and South Australia. Papers from the South Australian branch are held in the Library of the Academy of Science (Canberra), MS 6l (BSG: SA Branch).

22 Roy MacLeod, 'The Social Function of *Nature* in its First Fifty Years,' *Nature*, 224 (November 1969), 441-446; Peter Collins, 'The British Association as Public Apologist for Science, 1919-1946,' in MacLeod and Collins (eds.), *op.cit.* note 14, 211-237; A.J. Meadows, *Science and Controversy: A Biography of Sir Norman Lockyer* (London: Macmillan, 1976). Admiral Mahan's famous book is entitled *The Influence of Sea Power on History* (London edition: Sampson, Low, 1890).

23 For Liversidge and his circle, see Roy MacLeod, *Imperial Science under the Southern Cross: Archibald Liversidge, FRS and the Culture of Anglo-American Science* (forthcoming). The NSW branch of the BSG ended soon after Liversidge returned to England in 1908, but the South Australian Branch continued until at least 1934. The BSG was folded into the British Association in 1936.

24 See *ibid.*, chapter 8.

1905, of which they could command a small proportion, was only 344—their influence on the University community, at least, seems measurable.[25] Their local program was specifically Australian, and they spoke to a context receptive to their message. In America, the zealous pursuit of nature, and Frederick Jackson Turner's frontier interpretation of American history, complemented Australia's search for a manifest destiny of its own, which lay not only in the conquest of external nature, but also in the taming of human nature, and in the provision of social justice. W.G. Spence's *Australia's Awakening* of 1909 brought labour on side with science, a model for Lockyer's attempt to unite labour and capital in the mutual enjoyment of scientific method. Deakin embraced the British message, but he welcomed the American ethos even more, just as he welcomed the American fleet, without Colonial Office approval. Australia's destiny would be in the hands of its scientists. With ultimate confidence, Deakin even let his daughter marry one.[26]

With science would come progress, and with progress, profit. The lesson was a legacy of the Enlightenment and the Industrial Revolution. But the familiar messages of rationality hid deep contradictions. Progressivism shared their moral ambiguity, overtly equivocal in matters of distribution and reward. But progressivism as a mentality suited the empirical temper, and translated to Australia, its assumptions and ambiguities gained a hold at once more compelling and less well defined than in either Britain or the United States.The 'search for order' in Wiebe's America led equally in Australia to the application of 'method' to all departments of state and every discipline of knowledge. The rise of a federal bureaucracy, where there had been none coincided with a new interest in guiding the applications of science to agriculture, public health, weather forecasting, railway transportation, electrification and radio communication. That Australia's future was problematic, was preached by reports of falling fertility rates, and widespread fears of the country's becoming 'unfit' to continue in the race among nations.[27] By common reasoning, notionally unfettered university science, whether in physics or psychology, Arts or anthropology, must be turned to public service; and scientific research must be assessed at least in part by practical outcomes. Such pragmatic progressivism gave a particular connotation to educational reform in the early years of the

25 In the same year, the membership of the Royal Society of Victoria was 163, South Australia 85. To these should be added the (approximate) 100 members of the Royal Society of Queensland, 140 of the Royal Society of Tasmania, and perhaps 100 of the Royal Society of Western Australia. See F. Yuan, 'Education and Science' in Wray Vanplew (ed), *Australians: Historical Statistics* (Sydney: Fairfax, Syme and Weldon, 1988), 345.

26 That is, David Rivett, Melbourne chemist, Rhodes Scholar, FRS, and later Director of the CSIR. See Rohan Rivett, *David Rivett: Fighter for Australian Science* (Melbourne: privately printed, 1972).

27 Neville Hicks, *'This Sin and Scandal': Australia's Population Debate, 1891-1911* (Canberra: Australian National University Press, 1978).

Commonwealth, notably among the four universities in Melbourne, Sydney, Adelaide, and Queensland, which prided themselves on their effusive 'practicality,' but which sometimes failed to convince a sceptical public of their utilitarian value.

2. Federation and the Cult of Science

The period between Federation and the Great War is remembered for many important developments in the relations between science and society in Australia. Between 1901 and 1915, as 'federated' science became Federal, senior public servants and academics added new glosses to the gospel of efficiency. Under the Australian Constitution, the jurisdictions of agriculture and mineral resources, education (including universities) remained with the States. The States ceded power over customs at once, and over other domains—including 'postal, telegraphic, telephonic and other like services'—by proclamation.[28] Over 90% (16,000) of new federal public servants were employed by the Postmaster General's Department, which took on responsibilties for the science of communications. Federation also highlighted the importance of securing national uniformity in statistical measures. Each of the six colonies previously had separate patent laws, but the Patents Act of 1903 established a uniform standard. In 1906, the Commonwealth Bureau of Census and Statistics was formed, and became responsible for the first national census (1911), the first household-based expenditure survey, and the recording of data on wealth, prices, and wages. Within the states, reviews of technical education, water management systems, land tenure and land reservation became hallmarks of the era. Government agriculturists, chemists, and botanists took up where colonial surveyors and explorers—men of 'straight lines,' as David Denholm once put it—had left off.[29]

In some cases, there was a simple change in form: colonial botanists gave way to State botanists; and one must not exaggerate either the extent, or the consequences, of these constitutional changes for funding activities which the States retained. Nor did Federation mean the end of British connections in many scientific agencies. Sydney's Royal Mint, for example, continued until 1927 in the pay of the British Treasury. But with the passage of time, came shifts in orientation. In agriculture, mining and education, eyes had already turned to America, where Australians looked for instruction in domains which would earlier have been defined as British.[30]

28 See Wray Vanplew (ed), *op.cit.*, xv.

29 David Denholm, *The Colonial Australians* (Harmondsworth: Penguin, 1980).

30 Roy MacLeod, 'Of Men and Mining Education: The Establishment and Early History of Sydney's School of Mines, 1894-1914,' presented to the Third International Mining History Conference, Golden, Colorado, June, 1994.

Looking at this conjuncture from the distance of nearly a century, and quickened by the coming centenary of Federation, historians are only now beginning to ask how these changes were associated with an adaptive, pragmatic Progressivism, which both gave shape to Federation Australia, and tailored policies in which science had a special role to play. We may begin, perhaps, with a few observations, by way of suggesting why this association of ideas is both plausible and important.

First, whether in America, Britain or Australia, ideas associated with progressivism, like those associated with social imperialism, or the idealism of T. H.Green, gave an unquestioned role to the state. That role, however, was not arbitrary but accountable. Progressive advocates, including members of the British Science Guild never tired of arguing that parliamentary democracy was necessary, but that it could only be made useful through the exercise of a strong bureaucracy. In Australia, reforms affecting housing, health and the environment could not wait upon the hidden hand of *laissez-faire*, much less upon the undependable factionalism of party politics. Instead, science itself waited upon bureaucracy, which, with expanding State departments of agriculture, public health and education, gradually took form in the first years of Federation.[31]

Second, for progressives everywhere, the key-word was 'efficiency,' a concept suitably class-less and egalitarian, rooted in equality of access and opportunity, yet a concept that was intrinsically opposed to no interest. In Britain, as Lord Rosebery put it, there were Progressives who were not Liberals, but there were no Liberals who were not Progressives.[32] The same might be said of 'One-Nation' Tories. The argument had a pleasing democratic ring. Its application, to be sure, could harbour other messages. Manifestly, the efficient behaviour of one class could be easily translated into the profit of another; the technical preferment of the workman, into the pockets of the manager. In Australia, similar contradictions emerged. Efficiency appealed to the apostles of applied learning and technical training, and to those in the workforce who by the early years of the century far outnumbered those in university education. Efficiency became an agreeable platform for coalition conservatives, and a 'keyword' in parliamentary debate.

Less desirably, perhaps, the rhetoric of 'efficiency' may also have appealed to the deep, persisting current of anti-scholasticism in Australian life—not quite the same as the 'anti-intellectualism' of Clive James' unreliable memoirs, but a current with similar consequences, just the same. Australian universities, as Eric Ashby later recalled, could no longer justify themselves as oases in a cultural desert, bastions of civilisation, whose purpose was to enlighten the heathen sons of merchants and squatters. Above all, they were public institu-

31 See MacLeod, *op.cit.* chapter 8.

32 Quoted (1898) in Raymond Williams, *Keywords: A Vocabulary of Culture and Society* (London: Fontana, 1976), 244.

tions; they gained their credibility from the 'practical' services they performed, in return for the public money they were given.[33] For those who approved of this, progressivism proved to be pragmaticism in a positive light. For a generation taught under Francis Anderson at the University of Sydney, practicality became as conventional a test of philosophy, as efficiency was of a mechanical engine.[34]

Third, progressivism appealed to the particular ethos of Australian science, reflected historically in its community of learned amateurs; men who, like John Smith and W.B.Clarke, Archibald Liversidge and Robert Etheridge, prided themselves on being 'objective' collectors of natural and human artefacts, and who valued their contributions most highly, when they were assimilated within the knowledge structures of Europe and North America. For them, as well as for the practical men, for the inventors on the land as in the cities, as well as for overseas interests which doubled the number of patents taken out in Australia during this period—celebrants of what Ann Moyal has called 'colonial technology'— there was no more useful social philosophy than progressivism.[35] For scientific agriculturists, the extensive grazing and farming regions of the country became, as J.M. Powell has shown, gigantic practical laboratories for *in situ* innovations, many using methods of scientific control, if not new scientific knowledge itself.[36] Not all impulses traced their origins to the applications of science, still less to any systematic philosophy, but many did, and those that did, were advertised. The diffusion of refrigeration into dairy farming proceeded alongside the introduction of superphosphates and improved grasses; with relatively little licence, both could be heralded as scientific and methodological achievements.

In the first decade after Federation, the tradition of Australia's neutral, objective observers, who had made themselves a necessary virtue in colonial times, now gained credibility from the establishment of new federal agencies, notably in telecommunications, meteorology (in 1907), and tropical medicine (1909).[37]

33 Eric Ashby, 'Universities in Australia,' *The Future of Education*, No. 5 (Sydney: Australian Council for Educational Research, 1944).

34 See G.V.Portus, *Happy Highways* (Melbourne: Melbourne University Press, 1953); W.M.O'Neil, 'Sir Francis Anderson (1858-1941),' *Australian Dictionary of Biography, 7* (1979), 53-54.

35 Ann Moyal, 'Invention and Innovation in Australia: The Historian's Lens,' *Prometheus*, 5 (1), (1987), 92-110.

36 J.M. Powell, 'Protracted Reconciliation: Society and the Environment,' in Roy MacLeod (ed.), *The Commonwealth of Science: ANZAAS and the Scientific Enterprise in Australasia, 1888-1988* (Melbourne: Oxford University Press, 1988), 250.

37 See K.T. Livingston, 'Anticipating Federation: The Federalising of Telecommunications in Australia,' *Australian Historical Studies*, No. 102 (April 1994), 97-117; R.W. Home and K.T. Livingston, 'Science and Technology in the Story of Australian Federation: The Case of Meteorology, 1876-1908,' (in press); Roy MacLeod and Milton Lewis (eds),

Guild progressives gained much from their example, and their achievements appealed to politicians of all hues. The weather favoured no class or state; while disease showed no ideology—or rather, spoke with an ideology of its own—in the typhoid-laden streets of Melbourne and Sydney, and in rural areas made uninhabitable by dengue fever. State activity surfaced—NSW to set up a Bureau of Microbiology in 1908, prompting Queensland to do likewise—where federal writ failed to run. But as between states, as in the federal agencies, Australian science crafted 'tools of federalism,' no less impressive and as far reaching as Britain's imperial 'tools of empire.'[38]

Progressivism in Australia appealed to a quiescent Americanism among some Australians, including some of the most assertively 'transplanted Britons.' George (later Sir George) Knibbs, a protégé of Liversidge, was a case in point. Self-educated, then apprenticed to the public service, Knibbs briefly taught at Sydney University in mathematics and surveying before moving, under Liversidge's watchful eye, into the public service. In 1902, he help set up regulations for administering Rhodes Scholarships; later, following progressive criticisms of the New South Wales' educational system, he undertook a roving commission to 'enquire and report' on developments in secondary education in Europe and North America. His review, drawing particularly on American experience, laid a framework for the restructuring of public education that took place in New South Wales after 1912. For Knibbs, that event marked the beginning of a long career in public service. In 1905, he became superintendent of technical education, and in 1906, first Commonwealth Statistician and Director of the Commonwealth Bureau of Census and Statistics. His progressive, 'federalist' impulse inspired demands for uniformity from the States in statistical returns, and helped bring about the first *Commonwealth Yearbook*, which set benchmarks for State and Federal policy-making. Almost self-designed for the role, Knibbs later became the first director of the Commonwealth Institute of Science and Industry in 1921.[39]

While it would be incorrect to speak of Knibbs as 'American' in either manner or taste, his strategies for science, measurement and social improvement were drawn straight from the preoccupations of Washington and New York. His vision of a progressive Australia appealed equally to Americans—from Theodore Roosevelt, who in 1905, proclaimed, 'Next to my own country, I am

Disease, Medicine and Empire: Perspectives on Western Medicine and the Experience of European Expansion (London: Routledge, 1988); and Roy MacLeod and Donald Denoon (eds), *Health and Healing in Tropical Australia and Papua New Guinea* (Townsville: James Cook University Press, 1991).

38 Daniel Headrick, *The Tools of Empire: Technology and European Imperialism in the Nineteenth Century* (New York: Oxford University Press, 1981).

39 Susan Bambrick, 'Sir George Knibbs (1858-1929),' *Australian Dictionary of Biography*, 9 (1983), 620-621.

interested in the progress, success and safety of Australia, that great democratic island-continent,'[40] and sent a Great White Fleet to prove it; to Herbert Hoover, sometime mining engineer in Kalgoorlie, whose conservative sympathies proved ill-suited to the social policies needed by Depression America. Ironically, it was an inability to see the inherent weaknesses in social engineering that led to progressivism's mixed press. Even as W.H. Warren, professor of engineering at Sydney University, pressed his fellow Australian engineers to work for the public benefit, university anthropologists relied upon scientifically-based arguments to forecast the inevitable extinction of the Aborigines—unless they were 'protected' by the government.[41] Indeed, factored out of the Federal census, Aborigines became non-persons for the next sixty years, until a reforming Federal administration began the process of restoring them their full civil rights.

Even at the time, progressivism as a doctrine appealed clearly to the interests of Australian nationalism, even in its less progressive forms. Australian members of the British Empire League and British Science Guild were frequently united in their approval of the methods and rationale of colonial anthropology, of eugenic policies, and of policies of ethnic immigration designed to ensure a continuing White Australia.[42] In this, they were not necessarily in harmony with their British counterparts. But in speaking to certain wider imperial interests, notably imperial defence, they could speak with one voice. Thomas Laby, Liversidge's student at Sydney, and later a distinguished professor of natural philosophy at Melbourne, was archetypical. Keen advocate of the Australian branch of the Round Table, friend and correspondent of J.J. Thomson and Ernest Rutherford, who sent no fewer than thirteen of his students as '1851 Exhibitioners' to England, most of them to the Cavendish, Laby reflected what Barry Butcher has called an 'imperial vision,' which parlayed a progressive Australian patriotism into a higher imperial loyalty.[43]

40 Quoted by Roe, *op.cit.* 1.

41 See Tigger Wise, *The Self-Made Anthropologist: A Life of A.P.Elkin* (Sydney: George Allen and Unwin, 1985).

42 Here, the public positions taken by Dr Richard Arthur, MP for NSW, John Chalmers Eldridge, a member of the Federal Labour Government, and Robert Francis Irvine, professor of economics at the University of Sydney—key figures in the eugenics movement in Australia—would appear to be highly relevant. For these interconnections, see the forthcoming doctoral study 'Striving for National Fitness: Eugenics in Australia, 1880-1950,' by Diana Wyndham, Department of History, University of Sydney.

43 Barry Butcher, 'Science and the Imperial Vision: The Imperial Geophysical Experimental Survey, 1928-1930,' *Historical Records of Australian Science*, 6 (1), (1984), 31-44. For Laby, see Cecily Close, 'Thomas Howell Laby (1880-1846),' *Australian Dictionary of Biography*, 9 (1983), 640-641; H.. S.W. Massey, 'T. H. Laby, FRS - The Laby Memorial Lecture,' *Australian Physicist* (December 1980), 181-187.

Finally, progressivism was successful within a wide intellectual circle in Commonwealth politics, governed for much of its first decade by a succession of coalitions, both radical and conservative, without being demonstrably biased. Federal parliament, operating from Melbourne, conveyed an image both of liberalism and restraint. The admission of women to the vote was a case in point. Federation also suited educated men of business and commerce who had little confidence in politicians, and even less in an under-educated public; but who believed in the power of knowledge to shape a 'conservative liberalism' for Australian society. Their 'conservative liberalism' neatly mirrored the continuing tradition of 'liberal conservatism' which the late John Manning Ward finds deeply structured in the history of Australian society, and of which he was himself a formidable exponent.[44] As such, it was a natural progressivism, requiring no revolutionary change. Social justice could be compatible with 'bourgeois hegemony.' And with this progressivism *sans doctrine*, science formed a willing partner, in the elaboration of what Powell has called the 'complex texture of legislation, refinements of bureaucratic management procedures, and a battery of improvements in transport and communications, tenure arrangements, credit provision and bank facilities' that typified the progressive impulse.[45]

If I am correct, the net effect of these affinities was to foster in Australian science a vision of what I have elsewhere called 'practical idealism,' a vision of progressivism adapted to Australian traditions and circumstances. For many Australians, the biennial Congresses of the AAAS provided the easiest forum for its advocates, including Liversidge, the Association's 'founding father.' In his inaugural address in 1888, Henry Russell, the Government Astronomer of NSW, the first Australian-born FRS, and Fellow of the University's Senate, foreshadowed the necessity to conserve the resource endowments of his 'new country.'[46] In 1890, Baron von Mueller of Melbourne delivered what some historians consider Australia's first public appeal for the reservation of selected land for the benefit of future generations.[47] The AAAS made room for debate on such issues, complementing the progressive position, while it highlighted the findings of social science, the new biology, and the applications of forestry, agronomy, public health, and chemistry to Australia's economic future.

44 John M. Ward, *James Macarthur, Colonial Conservative, 1798-1867* (Sydney: Sydney University Press, 1981).

45 Powell, *op.cit.* 252.

46 H.C. Russell, President's Address, *Report of the Australasian Association for the Advancement of Science*, I, (Sydney, 1888), 12-13.

47 Powell, *op.cit.* note 25, 251; also J. M. Powell, *Environmental Management in Australia, 1788-1914: Guardians, Improvers and Profit: An Introductory Survey* (Melbourne: Oxford University Press, 1976).

Indeed, the AAAS proved at least as progressive as its British counterpart. It was by the Medical Section of the AAAS, for example, that Australians were taught to prevent tuberculosis, to consider the sources of mental deficiency, and to improve provision for child health. It was, perhaps, an elitist philosophy—yet, given a fair press, these messages touched many whom the colonial learned societies had failed to reach. Guided first by Liversidge, then by J.H. Maiden of Sydney and David Orme Masson of Melbourne, the AAAS successfully transformed itself from being the principal agent of 'federated', intercolonial science,' into becoming the principal publicist for 'federal' science. Years later, it also became the nursery for the nation's first Academy of Science. Its practical idealism *sans doctrine* contrived to bring progressive reforms onto the platform of all political parties—at least, before the apocalypse of Gallipoli, Flanders and the Somme, and the depression of the interwar years, put paid to so many of Australia's pre-war visions of progress.

3. The Contradictions of Progress

It would be wrong to suggest that this half-movement to America, let alone the acceptance of any broad, progressive mandate, was the product of any consistent or systematic reasoning on the part of Australian government, industry, or academia. Progressivism in its Australian form was a blank cheque of limitless possibilities, on which statesmen and scientists could draw airily upon ideals, without worrying too greatly about realities; it was a platform of convenience, on which scientists, social scientists and politicians could for certain purposes join forces, while on others, go their separate ways. Nor can it be said that progressive policies, even when enacted by consensus, always had the effects their advocates anticipated. The political economy of Australia did not fundamentally change. Farming areas that were marginal in colonial times, remained so under Federation. Official advice on cropping and grazing was not always taken, or well received. Mining interests, in times of commodity crisis, failed to keep pace with the lessons of scientific management. Bubonic plague may have been banished from Sydney's Rocks in 1901, but the health administrators of NSW retained a legacy of problems in sanitation, water supply and sewerage.

Indeed, progressivism had yet to witness its greatest victories when, in 1914, Australia played host to its first (and only) visit by the British Association for the Advancement of Science. Nothing was more indicative of the country's embrace of Empire and nationalism, than the reception accorded the Association. The visit, in planning since the BA's first successful overseas Congress, at Montreal in 1884, was taken as an imperial gesture, Australia's scientific 'coming of age.'[48] (It may be asked whether it signalled a similar meaning for

48 Peter Robertson, 'Coming of Age: The British Association in Australia, 1914,' *Australian Physicist*, 17 (2), (1980), 24.

Canadian science, a generation earlier).[49] Arriving on the eve of war, with some of its European guests destined for Australian internment, the BA travelled from one capital city to another, much as it has done in South Africa, when, following the Anglo-Boer War, it had consolidated the community of scientists in a symbolic act of Union.[50] In Australia, its presence prompted a review of science as an function of imperial obligation and national sovereignty.

Historians have yet to assess the impact of the BA's visit on the discussions of 1915, which led, past Gallipoli, to the creation of the Institute for Science and Industry. But the coverage given its meetings, coupled with the revelations of the war itself, conveyed a sense of national urgency to the business of science, and projected its scientists, now made known to the public, into the political arena. The war, which threw greater entitlements upon the Federal Government, also required the Federal Government to assume responsibility for meeting imperial mandates, issued by Britain, which only a coherent policy for science and industry could achieve. The prospect was one of opportunity and uncertainty, and the practice, one of experiments, that became the substance of postwar debate. But the events of 1915-1916, while uniting the nation behind its scientists, also revealed tensions with important ramifications for 'Commonwealth science'. Both then and afterwards, these were to undermine and frustrate the political agenda of progressivism. Among these frustrations, I will mention four.

The first—and in a contemporary Canadian context, most compelling—tension that threatened progressivism derived from the relatively weak nature of Australian Federalism. Scarcely debated before they arose, Federal-State jurisdictions in science divided national efforts and, when complicated by private enterprise, produced strange and frequently 'inefficient' alignments. For example, the Federal government assumed control of postal and telegraph services, but interstate telephone services (the first, between South Australia in 1902, and between Sydney and Melbourne in 1907) remained in private hands, under State regulation. In 1905, the Commonwealth took responsibility for regulating experiments in radio communication; but food and drug regulation remained the province of the States.

Agriculture, similarly, remained the province of the States, but rabbit plagues, which devastated South Australia in 1904, and Paterson's Curse, in 1906, were not confined by nature to that unhappy jurisdiction. Prickly Pear demanded a national, coordinated response. In 1916, when Orme Masson and his newly-

49 See Vittorio de Vecchi, 'Science and Government in Nineteenth Century Canada' (Unpublished Ph.D. dissertation, University of Toronto, 1978); 'The Dawning of a National Scientific Community in Canada, 1878-1896', *Scientia Canadensis*, VIII (1), (1984), 32-58, and 'Scientists and Scientists in Government, 1878-1896 - Part I,' *Scientia Canadensis*, VIII (2), (1984), 112-142; Part II, *idem*, IX (2), (1985), 97-113.

50 MacLeod and Collins (eds.), *op.cit.* Introduction.

established Advisory Council for Science and Industry visited Brisbane and Sydney to coordinate the war against pests, they faced an uphill struggle.[51] On the claws of the miserable tick, hung a vital issue of Commonwealth jurisdiction, with tangled issues of divided responsibility for quarantine and laboratories confronting State and Federal Governments. Mineral exploration presented a similar problem: mining regulation was a State jurisdiction, and continues to be so. Not until 1946, when a Commonwealth Bureau of Mineral Resources was established, did Australia have a central agency to coordinate scientific resources for exploration and mapping for the good of the country as a whole.

There were other important problems bearing upon Commonwealth policy , on which scientific advice might have had a beneficial influence. As Bruce Davidson points out, at none of the thirteen AAAS Congresses before Federation was there any discussion of the division of powers affecting agriculture, although the States' retention of commodity price-fixing powers, the control of Crown land, and the transport system on which agriculture depended, were of obvious economic and scientific importance. Indeed, the AAAS had other failures on its account books; its Agricultural Section failed to confront the devastation of Prickly Pear until 1913, by which time the weed covered 8 million hectares in Queensland and NSW.[52] Perhaps its greatest failure, although the one for which it was least responsible, was its act of self-suspension during the war, when decisions were taken that put thousands of inexperienced ex-servicemen on the land, on small farms in marginal areas. That romantic vision was destined to run its course, and thoroughly disillusion the prophets of progress, well before more rational counsels prevailed.

Questions of federalism affected even scientific organisations. Pressure to make the AAAS (later ANZAAS) and the Australian Association of Chemists (later the RACI) fully representative of scientists from the different states at times threatened efficient self-government. Before the Federal government's translation to Canberra, 'national' organisations of scientists (as of other groups) had to choose their capital allegiances, and by their choice, much was decided. Locating the headquarters of the AAAS in Sydney was initially convenient in 1888, but it was not necessarily the best permanent home for Australia's 'federated scientists'. The city-state identification of Australian scientists remained powerful, long after Federation. The residues are with us today. There remains something faintly anachronistic about a 'National' Museum being in Melbourne, but there is similarly no logic, other than that of history, in an 'Australian' Museum being in Sydney.

51 See Len Weickhardt, *Masson of Melbourne: The Life and Times of David Orme Masson* (Melbourne: RACI, 1989), 87.

52 Bruce Davidson, 'Developing Nature's Treasures: Agriculture and Mining in Australasia,' in MacLeod (ed.), *The Commonwealth of Science, op.cit.* 276-277.

A second frustration grew from the unresolved position of the universities. The years between 1900 and 1914 were ones of dramatic expansion in student numbers, and in the prospects for generating a Huxleyan 'army of science.' The six universities, including the two established in this decade, were left to maintain themselves, and their increasingly expensive science departments, by a mixture of fees and block grants from the State governments. At first, the universities were slightly better off, as certain State charges were taken up by Federal revenues, and the benefits were returned in the form of larger university endowments. But this soon wore off, and the increased pressure on buildings, arising from increasing student numbers in all faculties, taxed the unit of resource to the limit. Neither the States nor the Federal Government had a solution to this; and a weary repetition of reformist failures was left to haunt the country for nearly two decades after the war.

A similarly unsatisfactory prospect greeted progressive advocates of large scientific projects, including expeditions to the Antarctic. Scientific societies, languishing in membership and in appeal, had to raise funds for such enterprises from State governments in much the same way as their predecessors had done from colonial governments, and with the same limited, *ad hoc* results. The AAAS sought remedies; as did the British Science Guild, but without success. Hopes would brighten in the 1920s; but as late as 1914, solutions eluded Guild members, and their worthy research committees.

It can also be argued that the progressive impulse, which drew much from Australia's obligation to an 'efficient' Empire, left unresolved the character of Australia's obligation to Britain. The tradition of sending the country's best and brightest students to England, accelerated by the Rhodes Scholarships and the 1851 Exhibition awards, remained a feature of the landscape for a generation to come. Rhetorically, such links strengthened the crimson thread of kinship. More realistically, as Liversidge recognised, they tended to place Australians in permanent jobs overseas. This was perhaps an inevitable function of size and distance. And it would be wrong to see the ' brain drain' gradient towards Britain in terms of an inevitable and continuing intellectual deference. As Allan Maccoll has shown, a powerful Melbourne-Bloomsbury axis in inorganic chemistry worked to the extreme advantage of Britain, and more than repaid British investment in the education of the young Australians concerned.[53] Similar accounts could be presented at the tally of research in agriculture and anthropology. But whatever the Guild progressives might have wished, Britain retained its monopsonistic hold on Australian science. British government advice was sought and followed; British standards were studied and applied; and Britain remained Australia's official source of recognition, and font of reward. British models

53 Allan Maccoll, 'Australian Chemists at University College London, 1899-1988,' *Ambix*, 36, Part 2 (1989), 82-90.

continued to have an overwhelming influence on the culture of Australian government, science, and industry.

Finally, the rational program of progressivism could not completely ovetake the inertia of public sentiment. Romantic myths of heroic, pioneering individuals exploring the outback and the 'dead heart,' retained their hold on the Australian imagination and in some ways inhibited the acceptance of scientific research as a public responsibility. A belief in boundless riches around the corner, awaiting discovery—an endless frontier of opportunity—proved difficult to displace, as the experience of T. Griffith Taylor, Sydney's professor of geography, bitterly revealed. In his hands, 'environmental determinism' forced attention to the real limitations of white settlement in Australia. It was, ironically, the famous Canadian 'possibilist,' Vilhjamur Stefansson, who became the darling of the environmental imperialists, while Griffith Taylor, true to his progressive views, was forced to move to the less provincial environment of the University of Chicago.[54] The 'lucky country' remains a powerful myth, one with which advocates of science have effected an uneasy alliance.

Conclusion

These observations suggest ways of looking afresh at the years 1895 to 1914 as a critical period in the history of Australian science. Throughout the industrial world, this was a period of experimentation in the organisation of scientific services, as in the application of physics, chemistry, and the biological and earth sciences. In Australia, between the 'march to nationhood' of the 1890s and the massive changes prompted, intensified, and accelerated by the Great War, emerged a set of ideas and practices—some imported, others promoted from within—that set Australia on a course that would ultimately absorb 'efficient imperialism' into a perspective distinctly Australian. Following the war, Australia's science, defence, and political economy entered a period dominated by British policies aimed at encouraging imperial self-sufficiency. In these arrangements, what may be called a new phase of 'Australian Commonwealth science' emerged, in which Australian scientists found a continuing role within the Empire self-serving in several respects.[55] An apparent willingness to remain a sub-set of British science, identified with particular forms of British progressivism, would remain a characteristic and debated feature of Australian science for the next fifty years. Only with the 1960s would Australian science take on an increasingly international frame of reference, with implications we are only now beginning to understand.

54 For this account, see Powell, *op.cit.* 254.

55 Boris Schedvin, *Shaping Science and Industry: A History of Australia's Council for Scientific and Industrial Research, 1926-49* (Sydney: Allen and Unwin, 1987), 18.

Here, however, the history of science meets science policy, and social history meets the predicament of Australia in the world of 'big science' and multinational competition. The progressive impulse of the early Commonwealth years raised expectations that two world wars, a Depression, and many years of low investment in science have failed to dispel. As we begin to recover the origins and bearing of 'Commonwealth science,' especially between now and the celebration of Federation in 2001, it will be useful to reconsider the contributions of such concepts as practical idealism, progressivism and imperial efficiency in creating a 'state of mind' within which an independent—and interdependent—national science came into being.[56] Arguably, such an intellectual framework remains relevant, and its understanding awaits a generation of historians who are ready to turn the axis of received wisdom, until it catches the light.

56 R.W. Home and Sally Gregory Kohlstedt (eds), *International Science and National Scientific Identity: Australia between Britain and America* (Dordrecht: Kluwer Academic Publishers, 1991), reviewed in *Historical Records of Australian Science*, 9 (1), (1992), 67-70.

INDEX